T0326382

Envisioning a future without food waste and food poverty

Societal challenges

Envisioning a future without food waste and food poverty

Societal challenges

edited by:
Leire Escajedo San-Epifanio
Mertxe De Renobales Scheifler

Wageningen Academic
P u b l i s h e r s

EAN: 9789086862757
e-EAN: 9789086868209
ISBN: 978-90-8686-275-7
e-ISBN: 978-90-8686-820-9
DOI: 10.3921/978-90-8686-820-9

Cover photo:
Freshly Baked Traditional Bread
© Paulgrecaud | Dreamstime.com

First published, 2015

© Wageningen Academic Publishers
The Netherlands, 2015

International Scientific Committee

López Basaguren, Alberto. Constitutional Law, Faculty of Social Sciences and Communication, University of the Basque Country, Spain

Merino Maestre, Maria. Applied Mathematics, Faculty of Science and Technology. University of the Basque Country, Spain

Padrones Eguzkiagirre, Natalia. Sortarazi Association, Spain

Poveda, Alaitz. Physical Anthropology, University of the Basque Country, Spain. Department of Clinical Sciences, Genetic and Molecular Epidemiology Unit, Lund University, Sweden

Rebato Ochoa, Esther. Physical Anthropology, University of the Basque Country, Spain

Simón Alfonso, Lola. Social Psychology, School of Social Work, University of the Basque Country, Spain

Suñén Pardo, Ester. Immunology, Microbiology and Parasitology, Pharmacy Faculty, University of the Basque Country, Spain

Urcelay González, Itxaso. Sortarazi Association, Spain

This book was produced with the assistance of:

Sortarazi
Claretian Association for Human
Development (Spain)

University of the Basque Country
(Spain)

Urban Elika
Studies for Food and Society

Euskampus Foundation
Rocandio de Pablo, Ana. Bromatology and
Human Nutrition, University of the Basque
Country, Spain

Table of contents

Part 4. Proposals and strategies which empower consumers

Part 5. Social Responsibility of universities: contributing to reduce food waste and food poverty

Part 6. Sociology and anthropology of food waste and food poverty

Part 7. Food banking, food redistribution and social supermarkets

Part 8. Food sovereignty and food security

Foreword

Food management, as a fundamental resource for human survival, presents both old and new dilemmas. During the first hundred thousands of years of human history survival was essentially the goal. Among other aspects, survival depended, and still depends, on procuring our necessary daily food, and we progressively developed different skills and techniques to obtain more and better food and drink. With the first tools *Homo habilis* began a journey towards what – many thousands of years later, and thanks to other milestones such as the ability to control fire – was to become the food industry. In addition to such relevant transformations as livestock rearing and agriculture, evidence of incipient food technologies have been found in archaeological ruins from thousands of years ago: knowledge and practical applications of the scientific principles of food preservation, processing, preparation, packaging, storage and transportation. Products of the earth, now improved by agriculture and animal husbandry, were transformed into more easily consumed and tastier products. For example, 8,000 years ago harvests were stored in holes in the ground and then covered with clay (Al-Fayyum, Egypt), and 7,000 years ago cheese was being made in Europe.

As pointed out in the literature, throughout time, human intervention has gradually overcome several biological barriers – albeit not all of them – to produce most of the foods humans need where and when it was convenient. This fact has historically conditioned food consumption and its possibilities for trading which, in turn, depend on geographical and climatic locations and on the existence of adequate systems for its conservation and transport. In our age, the chain from those who grow the raw materials to the consumers hardly knows time, climatic or distance limitations. Nowadays research and development of agricultural foodstuffs has exponentially improved the quantity, safety and quality of the food available all over the world.

Food waste and food poverty are jointly discussed in this volume, hoping for a world free from both: a world with an optimal management of food resources and the necessary initiatives and strategies to achieve it.

But those complex management networks do not provide an answer to the near 800 million people currently enduring food poverty. They also raise difficulties for people and nations who wish to produce healthy and culturally appropriate food, through ecologically sound and sustainable methods. The truth is that feeding a population is much more complex than the mere ingestion of essential nutrients. In addition to a multitude of processes and infrastructures, our food systems also have inherent values and regulations, as well as ways that allow certain individuals to exercise different forms of power over the food that others need. And we are progressively aware of the absolute need to introduce changes in our food systems if we want to guarantee adequate access to healthy food to all human beings.

In recent decades our awareness of food systems has been continually growing, and we are progressively incorporating parameters such as their environmental sustainability or their equity from the socio-economic point of view. We also pay attention to the food relationships not only with health in the widest sense, but also with other values such as cultural diversity or the development of personal identity. With that background, two big issues are examined here which paradoxically attain scandalous proportions today: food waste and food poverty. The high levels of food loss and food waste do not allow us to continue maintaining that hunger is due (in general terms) to lack of food. According to the Food and Agriculture Organization of the United Nations (FAO) report 'Global initiative on food loss and waste reduction' (2014): *just half of what is lost or wasted could feed the world alone.*

Back in 2000, 189 countries of the United Nations agreed to achieve 8 Millennium Objectives by 2015. The first of them, before others related to education, equality or sustainability, is the objective to reduce extreme poverty and hunger by half. According to 'The millennium development goals report 2014', extreme poverty has indeed been reduced by half, thus fulfilling that goal. But the FAO report 'The state of food insecurity in the world, 2015' clearly states that 795 million people were estimated to be chronically undernourished in 2012-2014. Undernourishment is the lack of access to sufficient nutrients to carry out an active life. Since the period 1990-1992, the number of undernourished people has decreased by 15% but it is still a very high figure. Moreover, and despite overall progress, marked differences across regions persist, disease rates associated with inadequate nutrition are still very important and, consequently, additional efforts are needed. Food insecurity and hunger are very important problems in the world. The combined consequences of the financial crisis, market instability and increasing food prices severely aggravated the situation during the last several years, also in the wealthy developed world. According to Caritas-Europe, 79 million people are food insecure in the EU (Caritas Report, 2014: http://tinyurl.com/puz3r55).

Is it possible to have a world where nobody is hungry? That is the challenge that Ban Ki-Moon announced as the next 'Zero Hunger Challenge'. A world in which all people have access to the necessary food in order to enjoy a happy and healthy life and in which all food systems are sustainable. This objective to improve food management, of enormous importance for the survival and health of people, but also for the sustainability and sovereignty of food systems, is a highly complex challenge, with socio-political, economic, environmental, ethical and social justice implications. Eliminating hunger involves investments in development and social protection as well as contributions to peace and political stability. Furthermore, the sustainability of food systems and food chains is considered a crucial issue, because food poverty is present in an area that wastes up to half of the world food production (see, among others: the FAO report 'Global food losses and food waste: extent, causes and prevention', 2011). Thus, the paradox occurs in our planet because hunger and malnutrition coexist with the production of more than enough food.

According to the EU Parliament, currently 179 kg food per capita are wasted in EU27 (89 million ton per annum) and if nothing changes it will get worse, with major impacts in the environment and the economy as well as in ethics. Food waste could rise to over 120 million ton by 2020. Members of the European Parliament called on the Commission to implement a coordinated strategy combining EU-wide and national measures to improve the efficiency of food consumption chains sector by sector and to tackle food waste *as a matter of urgency* (EU Parliament, 2011).

Without compromising food safety standards in relation to both human and animal health, Member states and stakeholders are looking for every opportunity to prevent food waste and strengthen the sustainability of the food system. The aim is to reduce food waste by at least 30% by 2025 ('Towards a circular economy: a zero waste program for Europe', and related legislative proposals).

What strategies can we devise to confront these challenges? How can we increase the awareness of the states, stakeholders and citizens and what good practices can be promoted? With the objective of contributing to solve these questions from within the university and the scientific-technological environment, the multidisciplinary group URBAN ELIKA-Elikagaiak denontzat (Urban Nutrition-Food for All) of the University of the Basque Country (Spain) opened up an international, multidisciplinary deliberative forum in 2014. Researchers were invited to share their knowledge, stimulate collaboration with colleagues from across disciplines and expand their networks. Some of the fields of inquiry included: rural development, agronomy, engineering, environmental sciences, human nutrition, law and governance, sociology and philosophy. A large part of the results obtained in the aforementioned deliberations are presented in this book.

We wish to express our gratitude to all contributors, as well as to those who have participated as reviewers, members of the scientific panel of the book, and its editorial management board. We also acknowledge the financial support of the University of the Basque Country and the EUSKAMPUS Fundazioa.

Leire Escajedo San-Epifanio and Mertxe de Renobales Scheifler

Part 1. Challenging the governance of food systems

1. Food loss and waste: some short- and medium-term proposals for the European Union

L. González Vaqué
China-European Union Food Law Working Party. He served at the FAO and the EU Commission

Abstract

The worldwide double crisis – financial and food – has considerably worsened the existing situation of poverty and hunger, which is now even affecting certain more vulnerable groups in developed countries. This paper analyses one of the factors that undoubtedly contributes to aggravating the situation, affecting availability of and access to food: namely, the losses produced throughout the food chain. Basically concentrated in the European Union, the author seeks to identify those legislative measures that might be adopted in the short- or medium-term with a view to reducing this wastage.

Keywords: FAO, EU Commission, EU Parliament, legislative measures

> Obviously, the solution is not for rich countries to send old tomatoes or stale bread over to poor countries after saving them from the rubbish bin (Tristram Stuart, 2009).

1. Introduction

It is mid-2015 and the available data with regard to hunger and food insecurity in the world is still alarming. The worldwide double crisis – financial and food – has considerably worsened the existing situation of poverty and hunger, which is now even affecting certain more vulnerable groups in developed countries. As far as food loss and waste is concerned, despite the fact that the alarm bells started ringing several years ago and although, as I shall explain later, the population is becoming aware of the magnitude of the challenge, and the implementation of numerous initiatives is beginning to alleviate the problem in some sectors, it is difficult to produce a positive assessment of the scenario.

Every year, the Food and Agriculture Organization of the United Nations (FAO) warns us that the number of undernourished people in the world continues to grow: the global food equation has undergone changes: increasing food demand needs increasing food supply (availability) in order to meet the food needs of the whole population. We know today that the problem of hunger and food insecurity is largely a result of the scourge of war and the massive displacement of refugees; meanwhile, even in those regions where peace *reigns*, the lack of access to resources and the low income of poor families are the major cause of famine and malnutrition.

This is neither the time nor the place for me to suggest means of eliminating, nor even to outline, the consequences of the adversities to which I have just referred. I shall confine myself to analysing one of the factors that undoubtedly contributes to aggravating the situation, affecting availability of and access to food, mainly (but not exclusively!) at a family level: namely, the losses produced throughout the food chain, bearing in mind that the amount of food lost and wasted on a global scale is considerable.

As I shall point out, there are significant losses during the harvesting, handling, processing, transport and storage of food prior to its reaching the consumer's table. Further loss and wastage occurs at a personal or family level, particularly as a result of food preparation. Equally important is the quantity of products thrown away after cooking, both in the home and in restaurants, bars, canteens and other collective food consumption areas.

Such a complex reality and so thorny a problem prompt me to limit the scope of this chapter: territorially, as I will basically concentrate upon food loss in the European Union (EU); and thematically, given that I will seek to identify those legislative measures that might be adopted in the short- or medium-term with a view to reducing this wastage. For I am convinced that 'quien mucho abarca poco aprieta', in other words, *do not bite off more than you can chew ...*

2. Basic data

Before focusing on the EU's activities vis-à-vis this question, in order to offer an overall picture of the situation and the importance of the problem it poses, I think it appropriate to refer to 'key findings' (FAO, 2014; González Vaqué, 2015) upon which is based the FAO's SAVE FOOD programme, a worldwide initiative aimed at the reduction of food loss and waste, in which I participate in a modest way:

- Roughly one third of the food produced in the world for human consumption every year – approximately 1.3 billion tonnes – gets lost or wasted.
- Food losses and waste amounts to roughly US$ 680 billion in industrialized countries and US$ 310 billion in developing countries.
- Industrialized and developing countries dissipate roughly the same quantities of food – respectively 670 and 630 million tonnes.
- Fruits and vegetables, plus roots and tubers have the highest wastage rates of any food.
- Global quantitative food losses and waste per year are roughly 30% for cereals, 40-50% for root crops, fruits and vegetables, 20% for oil seeds, meat and dairy plus 30% for fish.
- Every year, consumers in rich countries waste almost as much food (222 million tonnes) as the entire net food production of sub-Saharan Africa (230 million tonnes).
- The amount of food lost or wasted every year is equivalent to more than half of the world's annual cereals crop (2.3 billion tonnes in 2009/2010).
- *Per capita waste by consumers is between 95-115 kg a year in Europe and North America* (emphasis added by the author), while consumers in sub-Saharan Africa, south and south-eastern Asia, each throw away only 6-11 kg a year.
- Total per capita food production for human consumption is about 900 kg a year in rich countries, almost twice the 460 kg a year produced in the poorest regions.
- In developing countries 40% of losses occur at post-harvest and processing levels while in industrialized countries more than 40% of losses happen at retail and consumer levels.
- *At retail level, large quantities of food are wasted due to quality standards that over-emphasize appearance* (emphasis added by the author).
- Food loss and waste also amount to a major squandering of resources, including water, land, energy, labour and capital and needlessly produce greenhouse gas emissions, contributing to global warming and climate change.
- The food currently lost or wasted in Latin America could feed 300 million people.

- *The food currently wasted in Europe could feed 200 million people* (emphasis added by the author).
- The food currently lost in Africa could feed 300 million people.
- Even if just one-fourth of the food currently lost or wasted globally could be saved, it would be enough to feed 870 million hungry people in the world.
- Food losses during harvest and in storage translate into lost income for small farmers and into higher prices for poor consumers.
- In developing countries food waste and losses occur mainly at early stages of the food value chain and can be traced back to financial, managerial and technical constraints in harvesting techniques as well as storage and cooling facilities. Strengthening the supply chain through the direct support of farmers and investments in infrastructure, transportation, as well as in an expansion of the food and packaging industry could help to reduce the amount of food loss and waste.
- In medium- and high-income countries food is wasted and lost mainly at later stages in the supply chain. Differing from the situation in developing countries, the behaviour of consumers plays a huge part in industrialized countries. The study identified a lack of coordination between actors in the supply chain as a contributing factor. Farmer-buyer agreements can be helpful to increase the level of coordination. Additionally, *raising awareness among industries, retailers and consumers as well as finding beneficial use for food that is presently thrown away are useful measures to decrease the amount of losses and waste* (emphasis added by the author).

3. The policy to prevent and avoid food waste in the EU

General considerations

Of course, not only the European Commission (European Commission, 2015) but all the Community institutions are today aware of the magnitude of the problem posed by food waste.

With regard to the European Commission, criticised by some authors for having concerned itself to date primarily with waste per se (Vidreras Pérez, 2014), the following documents can be cited:
- The Communication from the Commission to the European Parliament, the Council, the European Economic and Social Committee and the Committee of the Regions entitled 'Roadmap to a resource efficient Europe' (COM(2011) 571 final of 20.9.2011)
  - In connection with the subject of this chapter, the aforementioned roadmap stated that 'by 2020 [...] disposal of edible food waste should have been halved in the EU' and the Commission pledged to continue evaluating 'how best to limit waste throughout the food supply chain, and consider ways to lower the environmental impact of food production and consumption patterns'.
- The Communication from the Commission to the European Parliament, the Council, the European Economic and Social Committee and the Committee of the Regions 'Setting up a European retail action plan' (COM(2013) 36 final of 31.1.2013)
  - In this Communication the Commission announced that, in the context of the existing platforms in the EU, it would support retailers in the adoption of measures to reduce food wastage without compromising food security (awareness campaigns, communication, enabling food banks to redistribute food, etc.), for example via the 'Retail agreement on waste' (http://tinyurl.com/ndtrlyv); it also committed itself to the development of '... a long

term policy on food waste ...' (see action number 6 in section 3.4.2 of the Communication, entitled 'Reducing food waste').

Given that neither do I wish to devote too much time to analysing this Communication, I shall confine myself to partially transcribing what is set out in the aforementioned paragraph 3.4.2:

> 'Building awareness and helping consumers fight food waste are vital to achieving sustainable patterns of consumer choices. Such waste stems mainly from: (i) supply-chain inefficiencies; (ii) stock-management inefficiencies; (iii) marketing strategies; and (iv) marketing standards (product rejection due to packaging issues, where neither food quality nor safety is affected).
>
> [...]
>
> The role of retailers as intermediaries is crucial for food-waste reduction. The Retail Forum for Sustainability is a multi-stakeholder platform set up in order to exchange best practices on sustainability in the EU retail sector and to identify opportunities and barriers that may further or hinder the achievement of sustainable consumption and production. It was set up following the Sustainable Consumption and Production and Sustainable Industrial Policy Action Plan and it is already tackling key environmental issues, including food waste through common voluntary action.'

The said paragraph is completed with the provision of two actions:

- no. 6, to which I have referred; and
- no. 7:

> 'Through dialogue with stakeholders, the Commission will define best practices to make supply chains more environmentally-friendly and sustainable and minimise the energy consumption of retail outlets. The Commission will encourage retailers in the context of existing fora to apply these best practices.'

- The Communication from the Commission to the European Parliament, the Council, the European Economic and Social Committee and the Committee of the Regions 'Towards a circular economy: a zero waste programme for Europe' (COM(2014) 398 final of 2.7.2014)

    I shall not dwell on the analysis of this Commission, strongly criticised by the literature, which has gone so far as to describe as misguided the idea of applying in the EU a circular economy (Mayoral, 2014; Vidreras Pérez, 2014) and because the corresponding legislative proposal has been withdrawn in 2015 (OJ C 80, 7.3.2015, p. 17).

    In this context, the Commission has announced that it will put forward a new legislative proposal on waste targets: the precise scope of the new initiative on the circular economy (?) is currently being analysed by the Commission (European Commission, 2015).

Brevitatis causae, I shall not refer to other Community documents that directly or indirectly address food waste: The 7th General Union Environment Action Programme 'Living well, within the limits of our planet', which runs from 17 January 2014 until 2020 (González Vaqué, 2015); or the Commission's more recent Communication 'Tackling unfair trading practices in the business-to-business food supply chain' (COM(2014) 472 final de 15.7.2014), in which it is stated that the direct effects of the aforementioned unfair commercial practices, especially when applied unexpectedly, and unforeseen modifications of the contractual terms, may give rise to excess production and therefore lead to *unnecessary food wastage* (González Vaqué, 2015).

On the other hand, I consider it wholly appropriate to devote the following section to an important Resolution of the European Parliament (19 January 2012) which is a source of valuable information of particular use in the formulation of the necessary legislative proposals in order to reduce food loss and waste in the EU.

The European Parliament resolution of 19 January 2012 on how to avoid food wastage: strategies for a more efficient food chain in the EU (2011/2175(INI))

Scope and importance

The above mentioned European Parliament resolution of 19 January 2012 on how to avoid food wastage: strategies for a more efficient food chain in the EU (2011/2175(INI)) is a document the importance of which has not escaped the literature (Mayoral, 2014; Vidreras Pérez, 2014).

- Amongst the various reasons the European Parliament (EP) listed in its recitals, the following are of particular note:
- Every year in Europe a growing amount of healthy, edible food – some estimates say up to 50% – is lost along the entire food supply chain, in some cases all the way up to the consumer, and becomes waste.
- A study published by the Commission estimates annual food waste generation in the Member States at approximately 89 million tonnes, or 179 kg per capita, varying considerably between individual countries and the various sectors, without even considering agricultural food waste or fish catches returned to the sea.
- The figures disclosed by the FAO, according to which 925 million people around the world are currently at risk of undernourishment are alarming.
- That less food waste would mean more efficient land use, better water resource management, and positive repercussions for the whole agricultural sector worldwide, as well as boosting the fight against undernourishment in the developing world.

From the contents of all these considerations, the EP concluded that:

- 'the fact that a considerable amount of food is being discarded on a daily basis, despite being *perfectly edible* is unacceptable, because that food waste gives rise to both environmental and ethical problems and economic and social costs, which pose internal market challenges for both business and consumers' (emphasis added by the author);
- 'food waste has a number of causes: overproduction, faulty product targeting (unadapted size or shape), deterioration of the product or its packaging, *marketing rules* (problems of appearance or defective packaging), and inadequate stock management or marketing strategies' (*idem*);
- 'in order to reduce food waste as much as possible, it is necessary to involve *all participants* in the food supply chain and to target the various causes of waste sector by sector' (*idem*);
- there is a degree of confusion 'around the definition of the expressions *food waste* and *bio-waste*; believes that 'food waste' [although it] is generally understood to mean all the foodstuffs discarded from the food supply chain for economic or aesthetic reasons or owing to the nearness of the *use by* date, but which are still perfectly edible and fit for human consumption and, in the absence of any alternative use, are ultimately eliminated and disposed of, generating negative externalities from an environmental point of view, economic costs and a loss of revenue for businesses';

- 'quality requirements regarding appearance, whether imposed by European or national legislation or by internal company rules, which stipulate the size and shape of fresh fruit and vegetables in particular, are at the basis of many unnecessary discards, which increase the amount of food wasted';
- 'the initiatives already taken in various Member States aimed at recovering, locally, unsold and discarded products throughout the food supply chain in order to redistribute them to groups of citizens below the minimum income threshold who lack purchasing power' are to be commended, as is 'the valuable contribution made, on the one hand, by volunteers in sorting and distributing such products [locally] and, on the other, by professional companies that are developing anti-waste systems and measures'; and
- 'investing in methods leading to a reduction in food waste could result in a reduction in the losses incurred by agri-food businesses and, consequently, in a lowering of food prices, thus potentially also improving the access to food by poorer segments of the population' (see points 3, 4, 6, 14, 18, 22 and 25 of the EP resolution; see also point 24 concerning 'the work of companies and professional partnerships in the public, private, academic and community sectors in devising and implementing, at European level, coordinated action programmes to combat food waste'.

Legislative proposals

Upon the basis of what is proposed and suggested in the said European Parliament resolution of 19 January 2012 it is possible to make a list of steps that need to be developed and which, in principle, involve the adoption of legislative measures. Allow me to add before developing the list in question that, due to the ambiguous drafting of some EP proposals and to the very broad concept of legal measures of EU Law, this is only a provisional proposal, and is not, of course, exhaustive ...

Issues that demand the adoption of legislative measures:
- the approval of 'practical measures towards halving food waste by 2025 and at the same time preventing the generation of bio-waste' (point 3 of the EP resolution);
- the adoption of 'a policy of enforcement with regard to food waste; hopes that a waste treatment enforcement policy right along the food chain will be adopted by applying the *polluter pays* principle' (*ibidem*, point 5);
- the establishment of 'specific food waste prevention targets for Member States, as part of the waste prevention targets to be reached by Member States by 2014, as recommended by the [Directive 2008/98/EC of the European Parliament and of the Council of 19 November 2008 on waste and repealing certain Directives (OJ L 312, 22.11.2008, p. 3)]' (*ibidem*, point 9);
- 'the retargeting of support measures at EU level regarding the distribution of food products to the Union's least-favoured citizens, Community aid for the supply of milk and dairy products to schoolchildren, and the programme for encouraging the consumption of fruit in schools, with a view to preventing food waste' (*ibidem*, point 13);
- the introduction of a harmonised definition of food waste via the Commission's presentation of (González Vaqué, 2015) 'a legislative proposal defining the typology of *food waste* and in this context also to establish a separate definition of food residuals for biofuels or biowaste, which are separate from ordinary food waste since they are reutilised for energy purposes' (point 15 of the EP resolution);

- the inclusion of 'ambitious measures in the next legislative proposals on agriculture, trade and distribution of foodstuffs aimed at making efficient use of resources in order to fight food waste' (*ibidem*, point 17);
- the drafting of '... guidelines on the implementation of Article 5 of the [aforementioned] Waste Framework Directive (2008/98/EC), which defines by-products, given that the lack of legal clarity under EU legislation regarding the distinction between waste and non-waste could hinder the efficient use of by-products' (*ibidem*, point 19);
- implementing 'possible amendments to the public procurement rules on catering and hospitality services so that, all other conditions being equal, when contracts are awarded, priority is given to undertakings that guarantee that they will redistribute free of charge any unallocated (unsold) items to groups of citizens who lack purchasing power, and that promote specific activities to reduce waste upstream, such as giving preference to agricultural and food products produced as near as possible to the place of consumption' (*ibidem*, point 28);
- the evaluation and promotion of 'measures to reduce food waste upstream, such as dual-date labelling (*sell by* and *use by*), and the discounted sale of foods close to their expiry date and of damaged goods' (see point 30 of the EP, which also mentions '... that the optimisation and efficient use of food packaging can play an important role in preventing food waste by reducing a product's overall environmental impact, not least by means of industrial eco-design, which includes measures such as varying pack sizes to help consumers buy the right amount and discourage excessive consumption of resources, providing advice on how to store and use products, and designing packaging in such a way as to increase the longevity of goods and maintain their freshness, always ensuring that appropriate materials which are not prejudicial to health or to the durability of products are used for food packaging and preservation'); and
- the assignment, 'in the future, the necessary funding to promote stability in the primary sector, for example by means of direct sales, local markets and all measures to promote low or zero food miles' (point 33 of the EP resolution).

4. Conclusions

In writing this chapter I am aware there is still a considerable amount of data and numerous reports to be assembled and analysed regarding food loss and waste, including the sphere of the EU. It is true that the statistics at our disposal are often plentiful, but always heterogeneous and even incongruent (as I have said, this is due *inter alia* to the fact that we lack a common definition of *food wastage*). Moreover, in my opinion it is necessary to put science and the latest technology to work as soon as possible in order to calculate the levels and volumes of waste at each stage of the food supply chain and research into the causes and possible means of preventing loss and waste. This affirmation, based on my own personal opinion, may appear to run counter to a certain neo-romantic tendency to look to the past in the search for bucolic solutions (self-sufficient markets, etc.) to serious present-day problems. I will not deny that some of these solutions may be effective and make a valuable contribution to the necessary increase in public awareness ... In any case this is not the right place to get bogged down in pointless discussions on methods and systems. I will quote several authors who have underlined that, whether we like it or not, the food supply chain is becoming increasingly globalised: some food products are produced, transformed and consumed in very different parts of the world, so *the impact of growing international trade upon food loss is still in need of better evaluation* (Cederberg *et al.*, 2012).

What I have just said with regard to the need to use science and technology is equally valid for legislation: it is necessary for the Community legislator to adopt with the utmost urgency the regulations with regard to the issues that I have listed. The prevention of the continued increase of food loss requires more than awareness campaigns and *Soft Law* provisions.

It is true that such campaigns may be useful *complementary* measures in addition to those I propose (provided their diffusion is not limited and the leaflets published are not left to gather dust in the Ministry's storerooms, along with the publications from previous campaigns ...). In fact, in the so often cited European Parliament resolution of 19 January 2012 on how to avoid food wastage: strategies for a more efficient food chain in the EU, there are some interesting suggestions to be found:

- With regard to the Council, the Commission, the Member States and the agents of the food chain, the EP proposed the actions listed below which, in principle, do not require the adoption of legislative measures:
  - 'to address as a matter of urgency the problem of food waste along the entire supply and consumption chain and to devise *guidelines* for and support ways of improving the efficiency of the food supply chain sector by sector, and urge them to prioritise this within the European policy agenda' (emphasis added by the author);
  - to adopt a 'a coordinated strategy followed by practical action, including an exchange of best practice, at European and national level, in order to improve coordination between Member States with a view to avoiding food waste and in order to improve the efficiency of the food supply chain [which] could be achieved by promoting *direct relations between producers and consumers* and *shortening the food supply chain*, as well as by calling on all stakeholders to take greater shared responsibility and encouraging them to step up coordination in order to further improve logistics, transport, stock management and packaging' (*idem*);
  - a 'joint action by way of investment in research, science, technology, education, advice and innovation in agriculture with a view to reducing food waste and educating and encouraging consumers to behave more responsibly and deliberately to prevent food waste' (Segrè, 2014);
  - to formulate 'guidelines to address *avoidable food waste* and to implement greater resource efficiency in their section of the food supply chain, to continuously work to improve processing, packaging and transporting so as to cut down on *unnecessary food waste*' (emphasis added by the author); and
  - information campaigns in which the stakeholders 'recognise and explain the nutritional value of agricultural products of imperfect size/shape in order to reduce discards' (see points 2, 10, 17, 18 and 20 of the EP resolution; see also point 21 in which the EP, with the same objective 'urges the Commission and the Member States to encourage the exchange of best practice and promote awareness-raising campaigns to inform the public of the value of food and agricultural produce, the causes and effects of food waste and ways of reducing it, thereby fostering a scientific and civic culture guided by the principles of sustainability and solidarity').
- It also requested of the Commission:
  - the study of 'the causes and effects of the disposal, wastage and landfilling annually in Europe of approximately 50% of the food produced and to ensure that this includes a detailed analysis of the waste as well as an assessment of the economic, environmental, nutritional and social impacts';
  - the assessment of 'the impact of a policy of enforcement with regard to food waste';

- 'an analysis of the whole food chain in order to identify in which food sectors food waste is occurring most, and which solutions can be used to prevent food waste';
- cooperation 'with the FAO in setting *common targets* to reduce global food waste' (emphasis added by the author);
- consideration, 'when drawing up development policies, [of whether it is appropriate to support] measures aimed at reducing waste along the entire food supply chain in developing countries where production methods, post-harvest management, processing and packaging infrastructure and processes are problematic and inadequate';
- support for 'the modernisation of [the] agricultural equipment and infrastructure in order to reduce post-harvest losses and extend the shelf-life of food';
- more concretely defined 'ways and means of better involving agri-food businesses, wholesale markets, shops, distribution chains, public and private caterers, restaurants, public administrations and NGOs in anti-waste practices';
- working with Member States to 'issue recommendations regarding refrigeration temperatures, based on evidence that non-optimal and improper temperature leads to food becoming prematurely inedible and causes *unnecessary waste*' (emphasis added by the author);
- clarification, again with Member State support, of the 'meaning of the date labels (*best before*, *expiry date* and *use by*) in order to reduce consumers' uncertainty regarding food edibility and to disseminate accurate information to the public, notably the understanding that the minimum durability *best before* date is related to quality, while the *use by* date is related to safety, in order to help consumers make informed choices'; and
- the publication of a 'a user-friendly manual on the use of food close to expiry dates, while ensuring food safety in donation and animal feed, and building on best practices by stakeholders in the food supply chain, in order, for instance, to match supply and demand more quickly and effectively' (see points 3, 5, 6, 7, 12, 25, 31 and 32 of the EP resolution).

This is a quick list of more or less urgent proposals. In any case, we cannot confine ourselves to complaining and reporting the problem. It is necessary to act, in many cases in the legislative field, at every level: with respect to production, trade and catering. To restrict our actions to consumption, which, of course, must also strive to avoid loss, would constitute an error with unforeseen consequences.

References

Cederberg, C., Gustavsson, J. and Sonesson, U. (2012). Global food losses and food waste. FAO, 72 pp.

European Commission. (2015). Food waste. Available at: http://tinyurl.com/oyn7gag.

FAO. (2014). Key facts on food loss and waste you should know! Available at: http://tinyurl.com/krrrnge.

González Vaqué, L. (2015). Food loss and waste in the european union: a new challenge for the food law? European Food and Feed Law Review 1: 20-33.

Mayoral, A. (2014). FAO: El Foro Global sobre Seguridad Alimentaria y Nutrición. BoDiAlCo 9: 23-32.

Segrè, A. (2014). Gaspillage et éducation alimentaire. Dipartimento di Scienze e Tecnologie Agro-alimentari Alma Mater Studiorum, Università di Bologna, Bologna, Italy, 5 pp.

Stuart, T. (2009). Waste: uncovering the global food scandal. Penguin, 451 pp.

Vidreras Pérez, C. (2014). Estudio sobre las pérdidas de alimentos desde una perspectiva española. BoDiAlCo 8: 24-33.

2. Edible but unmarketable food: some legal problems to be solved on food waste prevention

A. Arroyo Aparicio

Commercial Law, Universidad Nacional de Educación a Distancia, Madrid, Spain; aarroyo@der.uned.es

Abstract

In recent years concern about food waste has grown in the EU. As food losses occur in the chain from producer to the final consumer for several reasons, several proposals are needed. These proposals would involve changing the EU Members legislation and practices on food donation. For instance, an Italian Law of 2003 applies only to the purposes of charity donation and free distribution of food to the most deprived persons. The non-profit organizations are treated, within the limits of the service provided, as *final consumers* and as such they have reduced responsibilities regarding the state of preservation, transportation, storage and use of foods. It must be stated that the approach to non-profit organizations as final consumers is a very ground-breaking perspective and legislation that could serve as a model for other national legislations. Some legal thoughts with a view to determining the liability of those involved and to find the best solutions to the edible but unsellable food are to be analysed.

Keywords: edible but unmarketable food, food donation, Samaritan Food Donation Laws, consumer, waste prevention

1. Introduction

Food loss and waste can occur at each stage of the supply chain: during the production, handling, processing, distribution or consumption (Lipinski *et al.*, 2013). This waste represents an expense of environmental and economic resources which is a serious cost that needs to be urgently addressed. Despite this compelling need, the development of a policy framework is undermined by the complexity of monitoring food waste.

First, as far as definition is concerned, the expression 'edible but unmarketable food' corresponds to food donation with respect to waste prevention. In fact, this topic links to the edible food that was designated to be wasted due to various reasons such as wrong size or shape, overproduction, low prices on the market, etc. (Schneider, 2013). All these products can be redistributed by organizations (generally nongovernmental organizations) and these activities are also called 'donation of edible food'. Just a part of the legal aspects related to food donation as waste prevention are the target of this work, with special regard being given to the 'edible but unmarketable food'.

Secondly, in view of historical aspects, the origins of modern donation of food to people in need can be traced in America. The name of John Von Hengel, a retired businessman that worked as volunteer at a 'soup kitchen', is linked to the first 'food bank' since the late 1960s. From this time to the present, the American and European initiatives are particularly well-developed. The most relevant organizations are 'Feeding America' and the European Federation of Food Banks (FEBA). The FEBA was established in 1984 and is operating in 21 European countries.

This broad development in practice has no legislative comprehensive framework at European level. In contrast, an example of American legislative facilitation of food donation is represented by Bill Emerson Good Samaritan Food Donation Act (US Public Law 104-210) signed in 1996. The Act provided, at a federal level (Morenoff, 2002), a limitation to donor's liability (Haley, 2003).

As mentioned, there is no common specific legislation at European level; however, the European Parliament adopted a resolution on 19 January 2012 on how to avoid food wastage (2011/2175; INI) which recommended that the European Commission take practical measures towards halving food waste by 2025. At this stage, the Commission also promised a Communication on Sustainable Food. Some points of the future Communication have been outlined in a legislative proposal published in July 2014. This proposal includes several amendments to the Waste Framework Directive 2008/98/EC. It can be seen the EU's body of waste policy is under review by the Commission and it should be said that some links between food waste and other waste policies are clear, as noted by the House of Lords Report 2014 related to legal perspective in United Kingdom (see p. 8 and Chapter 5).

Thus far, there are in the food waste debate context some legal requirements to be exposed. On the one hand, the EU General Food Law (Regulation 178/2002) applies to all food and organizations placing food on the market, including food banks and non-profit organizations. According to this, food operators have to respect the safety of foods at all stages of the food chain and they must ensure the legal requirement within their area of responsibility (Comparative Study, 2014). Such legal requirements could hinder food donation and Italy is the only EU Member which has put in place a better regulation.

Another important problem to be solved is to determine the parties in the questions related to donation of 'edible but unmarketable food'. For instance, the Italian Law cited above applies only for the purposes of charity and the free distribution of food to the most deprived persons where the non-profit organizations are treated, within the limits of the service provided, as *final consumers* and as such they enjoy reduced liabilities regarding the correct state of preservation, transportation, storage, and use of foods. The question arises whether a similar interpretation can be concluded from the actual Spanish legal framework (*infra II*).

Several others proposals should be considered (*infra III*), apart from fiscal incentives or codes of practice on food donation, which are not analysed in this work. Some brief conclusions are presented at the end (*infra IV*).

2. Good Samaritan Laws and Italian Law compared to the Spanish Legislation

As has been mentioned previously, food losses in the chain to the final consumer occur for several reasons. It is not only because this food could have visual defects, but also because of consumer behaviour. Furthermore food losses may also be minimised through the use for other purposes outside the market, such as for the support of the most deprived persons. However, the food recovery by non-profit organizations creates new risks related to *company's reputation*, for example in the case of products that are closer to the expiration date (Aiello *et al.*, 2014). It has rightly been pointed out that food donors are exposed to detriment of brand image or to face fines in the case of food poisoning (Comparative Study, 2014).

Related to this, donors, food banks and non-profit organizations are interested in obtaining a legal framework on the basis of 'Good Samaritan Laws' (Laws which were devised as a way of protecting those who come to the aid of others). This legal framework of American origin, as noted before[1], can also be linked to some prior German and French judicial doctrine. But Italy is the only European Member State where the Italian Federation of Food Banks succeeded to promote a legal perspective. In fact, Law No. 155 was enacted on 16 July 2003. This Italian Law applies only to the purposes of charity, free distribution of food to the most deprived persons, and the non-profit organizations are considered, within the limits of the service provided, as *final consumers* in regard to the correct state of preservation, transportation, storage and use of foods.

The L. 155/2003 is revolutionary in its simplicity. The Law includes a single article with wide reaching consequences. According to L. 155/2003, non-profit organizations that retrieve food are not held to the rules on food safety as an entrepreneur. This framework facilitates the donation of food to charities. Italian Food Banks counted 55.888 tons distributed in 2014. In addition, the Fondazione Banco Alimentare O.N.L.U.S published a Position Paper highlighting the compatibility of the L. 155/2003 with the EU hygiene legislation (Position statement della Fondazione Banco Alimentare O.N.L.U.S.). Moreover, the Italian Law of 27 December 2013 no. 147/127 has set essential safety responsibilities for food donation.

It must be stated that the approach to the non-profit organizations as final consumers is a very ground-breaking perspective and legislation. Indeed, the notion of a consumer as it is known in a legal sense differs from the concept of consumer as used in other matters (e.g. sociology or marketing). In the legal sense, in a contractual way for instance, a precise definition of the 'consumer' is essential in order to delimit the persons entitled to extended legal protection in relations with traders. The wider the circle covered by the definition of consumer, the more extensive the scope of consumer law provisions is and the less reason there is to speak about consumer law as a special regulation (Arroyo, 2013). On this basis, the Italian use of the notion of consumer differs from this last scope. In all concrete notions the functional elements distinguish the consumer from the professional. Conversely, in the Italian Law, the equivalency between non-profit organizations and consumers implies the application of the same treatment to both parties.

Regarding this notion of 'consumer', could a similar interpretation be concluded from the Spanish legal framework? According to the European Consumer Protection Directives, 'consumer' means any *natural* person who, in contracts covered by the Directive (applicable in the specific case), is

---

[1] The Bill Samaritan Law on Food donation is an American and federal law, which limits the liability exposure of operators that make good-faith donations of products that they know to be fit for consumption at the time of the donation. In this sense, a food operator *'shall not be subject to civil or criminal liability arising from the nature, age, packaging, or condition of apparently wholesome food or an apparently fit grocery product that the person or gleaner donates in good faith to a non-profit organization for ultimate distribution to needy individuals'.*

acting for purposes which are outside his trade, business, craft or profession. The expressed reference to natural persons must be highlighted as discussed below.

Under Spanish Law, there is a General Law for the protection of consumers together with other specific Laws. The purpose of the General Law is to establish the regime for the protection of consumers and users, and the General Scope of application are 'the relations between consumers or users and entrepreneurs' (Article 2). For the purposes of this General Law, consumers and users are natural or *legal* persons acting in a sphere that falls outside entrepreneurial or professional activity (Article 3). In the opposite corner, for the purposes of this law, 'entrepreneurs are considered to be those natural or legal persons acting in the context of their entrepreneurial or professional activity, whether this be public or private' (Article 4) and producers are considered to be 'goods manufacturers, service providers, intermediaries thereof, or importers of goods or services into the territory of the European Union, as are persons presenting themselves as such by providing their name, trademark or other identifying feature along with the goods or services, whether on the container, wrapping or any other protective or presentational component' (Article 5). Lastly, Article 7 gives the concept of supplier as 'entrepreneur that supplies or distributes products on the market, irrespective of the entitlement or contract by virtue of which such distribution is carried out'.

Article 3 of General Law for the protection of consumers provides that legal persons and entities without legal personality acting *non-profit in a foreign field to a trade or business* are also consumers (*for the purpose of the Law*). For instance, from EU Law perspective, 'Member States may decide to extend the application of the rules of the Directive (in this case, Dir. 2011/83) to legal persons or to natural persons who are not consumers within the meaning of this Directive, such as non-governmental organizations, start-ups or small and medium-sized enterprises' (Dir. 2011/83, Whereas 13). It can therefore be stated that Spanish General Law concerning consumer notion has an extended notion including not only natural but also legal persons.

In view of the above, it is relevant to bring up the Italian Law now. The purpose of the Italian Law is directly related to the duty of care to be expected of the non-profit organizations concerned with food donations. By contrast to that specific legal framework, the Spanish Law is a general one. This different scope of application undermines a totally identical interpretation. There is at least in general the possibility under Spanish Law to consider non-profit organizations as consumers. This consideration presents, however, two obstacles. The first concerns the scope of the term 'legal persons' covered by the Spanish Law. The second refers to the application of this Law only to the relations between consumers and professionals.

Without a clear specific legal framework, under Spanish legislation, the liability related to food donations could be considered under the general provisions (contractual or non-contractual liability) or rules on civil liability on defective products. But in relation to defective products legislation it must be stated that the transfer of food to a charity by an operator could not considered as totally equivalent to an operation (sale e.g.) in the sense of the Directive 85/374/EEC (Planchenstainer, 2014).

Lastly, without a specific legal framework, it is clear that the parties implied in food donations are subjected to food safety and hygiene rules (178/2002 and 852/2004 European Regulations).

3. Other initiatives

Food durability and labelling

The Comparative Study 2014 demonstrates that fiscal incentives encourage food donations. And it is also shown, in terms of food durability and labelling, that although the donation of products past their 'best before' date is allowed under European Law, some national legislations have introduced barriers to donating food when this date is exceed.

This is why in some countries 'good practices' guidelines for safe handling of food are published. In this sense it can be said that Belgium has introduced a circular with a non-limiting list of foods. This list gives the food that can be used by food banks and charities and it serves to guide the conservation of food after they reach or exceed the date of minimum durability.

Another policy initiative was adopted in 2012, when the German government launched a 'too good for the trash' campaign and the country has also pioneered 'food-sharing', using the Internet to distribute produce recovered from store rubbish while still in good condition ('Zu gut für die tonne').

Recent French legislation

As part of a broader Law on energy and the environment, the French government agreed in May 15 to include a provision requiring supermarkets of over 400 square meters to sign contracts by July 2016 to donate unsold but edible food to charities or for use as animal feed or farming compost.

4. Conclusions

On the assumption that the reduction of food waste across the EU is essential, some legislative measures seem to be appropriate. They include the Italian Law which is a very effective way to promote food donation. Moreover, and as seen from above, it appears evident that the Spanish legal framework does not clearly contain a similar regulation.

The new French regulation seems to be a breakthrough approach for a major change in the way edible but not marketable food is redistributed.

Finally, the two most relevant studies dealing directly with EU Food Waste Prevention (Comparative Study 2014 and The House of Lord Study 2014) aimed at identifying the most appropriate measures. They noted, *inter alia*, the relevance of best practices codes.

As previously stated, fiscal incentives are not within the scope of this article at this stage, even though such incentives have been taken in Belgium and France (Comparative Study 2014). In any case it can be concluded that the adoption of legal measures is a relevant step to prevent food waste.

References

Aiello, G., Enea, M., Muriana, C. (2014). Economic benefits from food recovery at the retail stage: an application to Italian food chains. Waste Management 34: 1306-1316.

Arroyo, A. (2013). Who is a 'consumer' on food law: some reflections on the notion of consumer and the EU food law. In: Röcklinsberg, H., Sandin, P. (eds.) The ethics of consumption. The citizen, the market and the law. Wageningen Academic Publishers, Wageningen, the Netherlands, pp. 465-470.

Deloitte (2014). Comparative study on EU Member States' legislation and practices on food donation, Final Report, June 2014 (European Economic and Social Committee). Available at: http://tinyurl.com/p4v92jn.

Haley, J. (2013). The legal guide to the Bill Emerson Good Samaritan Food Donation Act, August 8, 2013, Ark. L. Notes, 1448.

House of Lords – UK Parliament, 2014. European Union Committee, 10th Report of Session 2013-14, Counting the Cost of Food Waste: EU Food Waste Prevention, April 2014.

Lipinski, B. Hanson, C., Lomax, J., Kitinoja, L., Waite, R., Searchinger, T. (2013). Reducing food loss and waste. Working Paper. Available at: http://tinyurl.com/notoewv.

Morenoff, D.L. (2002). Lost food and liability: the good samaritan food donation law story. Food & Drug Law Journal 57: 107-132.

Planchenstainer, F. (2013). They collected what was left of the scrap: food surplus as an opportunity and its legal incentives. Available at: http://tinyurl.com/pgncbvr.

Schneider, F. (2013). The evolution of food donation with respect to waste prevention. Waste Management 33: 755-763.

3. Reducing obesity, food poverty and future health costs in Ireland – a proposal for health-related taxation

C. Loughnane[1] and M. Murphy[2*]

[1]Irish Heart Foundation, 50 Ringsend Road, Dublin 4, Ireland; [2]Social Justice Ireland, Arena House, Arena Road, Sandyford, Dublin 18, Ireland; michelle.murphy@socialjustice.ie

Abstract

Obesity and food poverty impact on people's diets and there is a clear relationship between poor diet and disease. Ireland is experiencing high levels of both – 7% of children rising to 36% of older people are obese and food poverty affects almost one in eight citizens. Unless obesity and food poverty rates are reduced it is predicted that there will be a significant impact on quality of life, life expectancy and healthcare costs in Ireland. In order to reduce obesity and food poverty and their associated social and economic costs, health related taxation must be considered as a policy instrument. The Department of the Taoiseach's *National Risk Assessment Reports 2014 and 2015* identify the increase in chronic disease as one of the five social risks facing Ireland. Obesity is a major modifiable risk factor for chronic diseases, including cardiovascular disease. These reports note that the rise in childhood obesity and other trends can be seen as an indicator of future rises in chronic diseases. In order to plan ahead and initiate change to reduce obesity and food poverty through preventative measures, consideration must be given to introducing health-related taxes. Health related taxation has the potential to be a key policy instrument in addressing the problems posed by food poverty and obesity. The revenue generated by a sugar sweetened drinks tax for example could be used to develop effective obesity prevention programmes and initiatives to eradicate food poverty.

Keywords: obesity, health-related taxation, food poverty

1. Obesity in Ireland

Obesity is an excess of fatness above the ideal for health (Waters *et al.*, 2011). Body mass index (BMI; BMI calculates weight in kilogrammes divided by height in metres, squared) cut-off points are generally used to identify obesity in individuals and across populations. Prevalence of obesity in Ireland has increased significantly in the last two decades. Between 1990 and 2011, obesity rose from 8 to 26% in men, and from 13 to 21% in women (Irish Universities Nutrition Alliance, 2011). Currently, 24% of adults (18-64 years) are obese. The percentage classified as obese increases with age – 13% of both women and men aged 18-35 years were classified as obese, rising to 31% of women and 42% of men aged 51-64 years. Amongst Irish children, rates of obesity accelerated from the 1970s and continued to increase into the 'Celtic Tiger' period (Perry *et al.*, 2009). Depending on the definition of obesity and the cut-off points used, there has been a dramatic two-to-fourfold increase in obesity in Irish children aged 8-12 years since 1990 (Table 1; O'Neill *et al.*, 2007).

Leire Escajedo San-Epifanio and Mertxe De Renobales Scheifler (eds.) *Envisioning a future without food waste and food poverty*

39

DOI 10.3920/978-90-8686-820-9_3, © Wageningen Academic Publishers 2015

Table 1. Obesity rates in Ireland by age.[1]

	Age	Obesity rate
Pre-schoolers	2-4 years	3%
	3 years	6%
Children	7-11 years	7.2%[2]
Teenagers	12-17 years	7.5%
Adults	18-64 years	24%
Older people	Over 50's	36%

[1] There is no single source of data on obesity rates in Ireland. Data is available from a number of sources, including the national pre-school nutrition survey, children's food survey, the teen's food survey and the adult's food survey, the growing up in Ireland study, the WHO childhood obesity surveillance initiative, SLÁN and the Irish longitudinal study on ageing.

[2] Average over 3 studies of 7, 9 and 11 year olds.

Obesity trends

Ireland's experience of rising obesity is mirrored across the world. Data from the World Health Organisation (WHO) shows consistent increases in obesity from 1980 to 2008 in almost all countries (European Heart Network and European Society of Cardiology, 2012). As in many areas of public health, there is a pronounced social gradient in obesity, apparent in Irish children as young as three years of age (Williams *et al.*, 2013). People experiencing economic disadvantage are more likely to be an unhealthy weight than more advantaged peers(Williams *et al.*, 2013).

Childhood obesity

Obesity rates increase with age (Irish Universities Nutrition Alliance, 2011) and obesity in childhood tends to continue into adulthood. There is a significant likelihood that some obese children will have multiple risk factors for cardiovascular diseases, type 2 diabetes and other co-morbidities before or during early adulthood (National Heart Forum, 2008). The Growing Up in Ireland study shows that social inequalities increase the risk of overweight and obesity from an early age. At 9 years of age, children from disadvantaged areas are much more likely to be obese (Layte and McCrory, 2011). The incidence of obesity in Ireland has led the Government's Special Rapporteur on Children to term childhood obesity 'a vital child protection issue and a challenge to implementation of the right of children to the highest attainable standard of health in Ireland' (Shannon, 2014).

Health inequity, obesity and cardiovascular disease

Health inequity means that Irish people on lower incomes are sicker and die earlier than those with higher incomes. Mortality from cardiovascular disease is a major element of health inequity in Ireland. As the social gradient in health means that obesity is more prevalent in disadvantaged communities, we are likely to see an increase in socio-economic inequalities in deaths from cardiovascular disease

(CVD). (For a discussion of obesity rates and inequality in coronary heart disease mortality in England see Bell *et al.*, 2012).

In society, there is a tendency to blame poor lifestyle behaviours for ill-health, sometimes leading to a blame culture directed at those who are obese. Yet, individuals cannot be solely responsible for lifestyle factors which lead to ill-health. These lifestyle factors – including poor diet – are influenced by social, economic, cultural and political factors. Unequal experience of the social determinants of health across the population impacts on the distribution of poor health, including obesity (Bell *et al.*, 2012). To tackle the health inequity in obesity and CVD, government should address the accumulation of disadvantage across people's lives in the conditions in which they live, work and age.

2. Food poverty in Ireland

In addition to high levels of obesity among all ages, a considerable proportion of the Irish population is living in food poverty. Food poverty is the inability to have a nutritious diet due to affordability or accessibility. It is primarily the result of low incomes. The immediate impact of food poverty is poor diet and lack of nutrients. In the longer term, food poverty may lead to diet-related diseases, including obesity. The connection between diet and health may explain why people with a chronic illness in 2010 were more likely to be experiencing food poverty (14%) than those without a chronic illness (9%) (Department of Social Protection, 2014).

In 2013, 13.2% (one in eight) of the population were experiencing food poverty (Department of Social Protection, 2014). This was an increase from one in ten people in 2010 (Table 2; Carney and Maître, 2012). Food poverty is defined as: having missed a meal in the previous fortnight due to lack of money; inability to afford a meal with meat or a vegetarian equivalent every second day; or inability to afford a roast, or vegetarian equivalent once a week (Carney and Maître, 2012).

Social stress, such as unemployment, and cost of food are barriers to healthy eating for Irish people with lower socio-economic status (*safe*food, 2008). Healthy food is more expensive, with healthier diets, such as Mediterranean diets high in fruits, vegetables, fish and nuts tending to be more expensive than less healthy diets, high in processed foods and refined grains (Morris *et al.*, 2014; Rao *et al.*, 2013). As a result, people living in food poverty tend to spend a higher proportion of their income on food, yet due to the high cost of healthy food find it difficult to meet healthy eating guidance (Friel and Conlon, 2004).

Research in Ireland has found that low-income households in Ireland are most at risk of inadequate diet and its negative effects on health and well-being (Friel and Conlon, 2004; Harrington *et al.*, 2011).

Table 2. Rising food poverty in Ireland (Department of Social Protection, 2014; Social Monitor, 2013).

2010	2011	2012	2013
10.0%	11.4%	11.8%	13.2%

People in low-income households know the foods which are healthy, but are restricted by financial and physical constraints in following such a diet (Friel and Conlon, 2004). Research shows that calories from healthy foods (fruit, vegetables, lean meats, etc.) are up to ten times more expensive than from foods high in fat, sugar and salt (Food Safety Authority of Ireland, 2011).

Families with children in Ireland are three times as likely as those without children to be affected by food poverty(Carney and Maître, 2012). In addition to lacking the money to purchase food and facilities to prepare it, people experiencing food poverty may also have difficulty accessing affordable healthy foods in shops in their neighbourhood. Research examining the impact of food environments and the availability of healthy foods for 10,000 people living in Ireland found that lack of available healthy food in disadvantaged areas may further impede disadvantaged groups from following a healthy diet (Layte *et al.*, 2010). Madden outlines the link between dietary energy density and dietary energy cost for people living in Ireland (Madden, 2013). Refined grains, added sugars and added fats are amongst the cheapest sources of dietary energy, whereas the more nutrient dense foods such as lean meats, fish, vegetables and fruit are more expensive. The impact of low incomes on dietary choice is that low income consumers with limited resources are more likely to select diets with high contents of refined grains, added sugars and added fats as the most cost-effective way to meet daily calorific requirements. In his paper examining nutrition-related taxation, Madden finds that a 'fat-tax' accompanied by a subsidy on healthy food such as fruit and vegetables would have a neutral or negligible poverty impact and could address the problem of food poverty and obesity simultaneously by making a healthy diet more affordable to low income groups.

3. Obesity projections

Without any policy intervention it is projected that there will be an additional 717,950 overweight or obese adults in 2030 when compared to 2010 (Figure 1). If this projected trend becomes reality, it will result in a significant increase in the number of people with chronic diseases and a knock-on impact in a significant increase in direct healthcare costs. These include GP visits, drug costs and in-patient costs among others. The burden of chronic disease that is primarily associated with and driven by obesity is CVD, type 2 diabetes and some forms of cancer. The direct cost of overweight and obesity to the Irish health service is already almost € 400m (*safe*food, 2012) every year. This cost, which is driven mainly by obesity related issues, is projected to increase significantly as the percentage of the population who are obese continues to rise. Webber *et al.* (2014) also note that while there has been a reduction in the burden of chronic disease across countries over the past decade, the increase in obesity is now threatening these gains.

Projected direct cost of obesity in Ireland in 2020 and 2030

As previously noted, the direct cost of overweight and obesity in Ireland is almost € 400 million per annum. The indirect cost of obesity is € 728 million per annum. Direct costs of obesity are the costs to the healthcare system in terms of chronic diseases as a result of the levels of obesity across the population. The indirect costs of obesity are associated with reduced productivity, due to premature mortality, work absenteeism and a reduced effective labour force. This indirect cost does not fully capture the social and economic cost to society and to the individual of obesity related illnesses, low self-esteem and mental health issues.

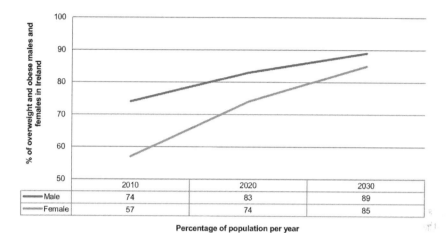

Figure 1. Projected prevalence of overweight and obesity in the Irish population 2010-2030 (Perry et al., 2013).

As Figure 1 shows, obesity levels in Ireland are projected to increase dramatically, and so will the direct obesity related costs to the healthcare system. The main drivers of increased healthcare costs as a result of obesity related diseases are the increased number of CVD cases. Figure 2 shows the projected cost in 2020 and 2030 if the present trends continue.

As Figure 2 shows, if present trends continue and no policy interventions are made the cost of obesity will rise to over €4.3 billion in 2020 and to € 5.4 billion in 2030 (Perry *et al.*, 2013) The increased costs of coronary heart disease and stroke are projected to account for 92% of obesity related costs in 2030 (Perry *et al.*, 2013). When the indirect costs (reduced productivity, due to premature mortality,

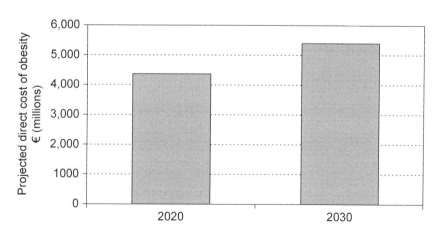

Figure 2. Projected cost of obesity in 2020 and 2030 if present trends continue (Perry et al., 2013).

work absenteeism and a reduced effective labour force) are included (projected to be € 9.88 billion in 2030), obesity will cost the state more than €15 billion by 2030. These projected costs represent a huge social and economic burden to the state. The projected impact of reducing BMI on obesity costs

Reducing BMI of a population can substantially reduce the health burden of obesity, and in turn reduce the cost of CVD and chronic diseases on the health system. A micro simulation model showing the impact of a 1 and 5% reduction in BMI shows that this would substantially reduce the health burden in terms of the reduction of cases of CVD, type 2 diabetes and chronic diseases (Webber et al., 2014). Some immediate and simple gains that would result from a reduction in BMI would be a reduction in medical care costs; a reduction in drug costs for chronic illnesses; and a reduction in prescription costs. In the Irish case, achieving a 5% reduction in BMI by 2020 would achieve significant savings of € 394m and would have an immediate and direct impact on the health of the population and the cost to the health service. It would also have long term social and economic benefits. The level of savings and the health benefits to the population are substantial. These savings should be directed into improving the health service and other public services, which in turn will improve economic performance and quality of life.

Policy proposals

The implications of obesity and projected future levels of obesity are significant and daunting. Policies to address this challenge must be developed now. The authors propose that government adopt a target to reduce the BMI of the population by 5% by 2020 in order to significantly improve the health of the population and to guarantee significant long-term savings to the exchequer. Such a target would require a change in strategy, but it would also yield significant results in terms of exchequer savings in the short to medium term and have immediate social and economic benefits.

To reach this target by 2020 will require a concerted whole-of-government approach to improve the health of the population requiring significant commitment and cross departmental responsibility.

Health related taxation

As noted earlier chronic diseases, including cardiovascular disease, are the biggest cause of death in Ireland. Obesity is one of the major modifiable risk factors for chronic diseases, and as such governments should plan ahead and initiate change to reduce obesity through preventative measures such as price policies(Webber et al., 2014). Primary prevention is crucial if increasing obesity trends are to be averted (Webber et al., 2014).

In order to make significant progress towards reducing BMI by 5% by 2020 the authors propose that government introduce a sugar-sweetened drinks (SSD) tax and use the revenue generated to address food poverty and promote health policies to prevent and reduce obesity. Such forms of health related taxation are entirely justifiable in order to recover the costs of obesity to the state (Yale Rudd Center, 2012). The 2011 report of the Secretary general of the United Nations General Assembly on the prevention and control of non-communicable diseases (NCDs) states that there are many cost effective and low cost population wide interventions that can reduce risk factors for NCDs, including the introduction of food taxes and subsidies to promote a healthy diet.

4. Conclusions

Without policy intervention to address obesity and food poverty there will be a significant increase in chronic diseases in Ireland. The increased costs of coronary heart disease and stroke are projected to account for 92% of obesity related costs in 2030 (Perry *et al.*, 2013). Food poverty and obesity are related issues and account for significant social and economic costs. Health related taxation (designed appropriately) can be used as a policy instrument to address both problems. Reducing BMI of a population can substantially reduce the health burden of obesity. In order to make significant progress towards reducing BMI by 5% by 2020, the authors propose that government introduce a SSD tax and use the revenue generated to address food poverty and promote health policies to prevent and reduce obesity. Such forms of health related taxation are entirely justifiable in order to recover the costs of food poverty and obesity to the state (Yale Rudd Center, 2012).

Acknowledgements

An extended version of this paper was published by the authors in June 2015.

References

Bell, R., Allen, J., Geddes, I., Goldblatt, P., Marmot, M. (2012). A social determinants approach to CVD prevention in England. University College London Institute of Health Equity. Available at: http://tinyurl.com/npgkvq4.

Carney, C., Maître, B. (2012). Constructing a food poverty indicator for ireland department of social protection. Department of Social Protection, Dublin, Ireland. Available at: http://www.welfare.ie/en/downloads/dspfoodpovertypaper.pdf

Department of An Taoiseach (2014). National Risk Assessment 2014. Stationery Office, Dublin, Ireland.

Department of An Taoiseach (2015). Draft National Risk Assessment 2015. Stationery Office, Dublin, Ireland.

Department of Social Protection (2014). Social Inclusion Monitor 2013.

European Heart Network and European Society of Cardiology (2012). European cardiovascular disease statistics. 2012 edition.

Food Safety Authority of Ireland (2011). Scientific recommendations for healthy eating in Ireland. FSAI, Dublin, Ireland.

Friel, S., Conlon, C. (2004). Food poverty and policy. Combat Poverty Agency, Cross Care and St. Vincent de Paul, Dublin, Ireland.

Harrington, J., Fitzgerald, P., Layte, R., Lutomski, J., Molcho, M., Perry, I. (2011). Sociodemographic, health and lifestyle predictors of poor diets. Public Health Nutrition. DOI: http://dx.doi.org/10.1017/S136898001100098X.

Irish Universities Nutrition Alliance (IUNA) (2011). National adult nutrition survey.

Layte, R., McCrory, C. (2011). Growing up in Ireland – Overweight and obesity among 9-year olds. The Stationary Office, Dublin, Ireland.

Madden, D. (2013). The poverty effects of a 'fat-tax' in Ireland. UCD, Dublin, Ireland.

Morris, M., Hulme, C., Clarke, G., Edwards, K., Cade, J. (2014). What is the cost of a healthy diet? Using diet data from the UK Women's Cohort Study. Journal Epidemiology Community Health. DOI: http://dx.doi.org/10.1136/jech-2014-204039.

National Heart Forum (2008). Call for evidence – assessing the impact of the commercial world on Children's Wellbeing.

O'Neill, J.L., McCarthy, S.N., Burke, S.J., Hannon, E.M., Kiely, M., Flynn, A., Flynn, M.A., Gibney, M.J. (2007). Prevalence of overweight and obesity in Irish schoolchildren, using four different definitions. European Journal of Clinical Nutrition 61(6): 743-751.

Perry, I.J., Keaver, L., Dee, A., Shiely, F., Marsh, T.,Balanda, K. (2013). Application of the UK Foresight Obesity Model in Ireland: the health and economic consequences of projected obesity trends in Ireland. DOI: http://www.ncbi.nlm.nih.gov/pmc/articles/PMC3827424/.

Perry, I., Whelton, H., Harrington, J., Cousins, B. (2009). The heights and weights of Irish children from the post-war era to the Celtic tiger. Journal of Epidemiology and Community Health 63(3): 262-264.

Rao, M., Afshin, A., Singh, G., Mozaffarian, D. (2013). Do healthier foods and diet patterns cost more than less healthy options? A systematic review and meta-analysis. BMJ Open 3: DOI: http://dx.doi.org/10.1136/bmjopen-2013-004277.

*Safe*food (2008). 'Food and nutrient intake and attitudes among disadvantaged groups on the island of Ireland – summary report'. Available at: http://tinyurl.com/az9b74x.

*Safe*food (2012). The cost of overweight and obesity on the island of Ireland. Safefood, Dublin, Ireland.

Secretary General of the United Nations General Assembly (2011). Report on the prevention and control of non-communicable diseases (NCDs). Available at: http://tinyurl.com/o3a62y3.

Shannon, G. (2014). Seventh Report of the Special Rapporteur on Child Protection. Available at: http://tinyurl.com/osmgq3q.

Waters, E., de Silva-Sanigorski, A., Hall, B.J., Brown, T., Campbell, K.J., Gao, Y., Armstrong, R., Prosser, L., Summerbell, C.D. (2011). Interventions for preventing obesity in children (review). Cochrane Database of Systematic Reviews 2011, 12.

Webber, L., Divajeva, D., Marsh, T., McPherson, K., Brown, M., Galea, G., Breda, J (2014). The future burden of obesity-related diseases in the 53 WHO European-Region countries and the impact of effective interventions: a modelling study. Available at: http://bmjopen.bmj.com/content/4/7/e004787.full.

Williams, J., Murray, A., McCrory, C., McNally, S. (2013). Growing up in Ireland – development from birth to three years. The Stationery Office, Dublin, Ireland.

Yale Rudd Center for Food Policy & Obesity (2012). Sugar-sweetened beverage taxes: an updated policy brief. Available at: http://tinyurl.com/nfcvs22.

4. Food laws and labelling as a contributor to food waste

E. Lissel

Faculty of Law, Lund University (and Laws and Standards Specialist at IKEA Food Services AB), Sweden; e_lissel@hotmail.com

Abstract

Food law, and food labelling, establishes the rights of consumers to safe food and to accurate and honest information. Although food safety is at the core of food information, the broad principle of the food law is about telling the truth about products so that the consumer can make a choice about what they choose to eat. However, food law and labelling can also become a trade barrier and thus contribute to food waste. Countries over the world have different rules regarding how food labels should look like and what information should be included. A product fit for one market cannot easily be sold at another market. An issue like 'use by date' is interpreted differently in different countries as well as the sugar content in jam. Also, the size of the text on the label is different in different markets. Combined with a short shelf-life the products that are perfectly fit for eating, has to be scrapped when stopped at the border. Consequently, different kinds of trade barriers can thus contribute to food waste even though the content is perfectly fit for consumption. The relevant questions are thus, should we strive for a more homogenous food law and labelling or strive for more compliance at local markets? This paper argues that we ought to strive for a more homogenous food law and labelling and thus eliminate some trade barriers when it comes to food exports. Compliance cannot easily be achieved if using a label fit for one market and using it for another market. The aim of this paper is to compare labelling in different countries in the world in order to describe the differences and thus to illustrate the difficulties when it comes to labelling. This paper will not discuss issues such as food safety, but rather show how safe and edible food cannot be sold due to label issues. Also, this paper cannot cover all issues on labelling but only touches upon some of them. This means that the focus of this paper will be on some examples which often have variations in different markets. These examples are allergens, date marking and country of origin. First, an introduction to labelling will be presented.

Keywords: food regulations, date marking, allergens, country of origin

1. Labelling in the world

All food needs some kind of labelling in order to demonstrate the content. However, labelling can be portrayed in numerous ways. There are though a number of similarities concerning the content of labelling in the world. These can be summarized in for example:
- name of the food (product identity);
- ingredients list;
- net weight/quantity/content;
- allergens;
- date marking;
- place of business.

Leire Escajedo San-Epifanio and Mertxe De Renobales Scheifler (eds.) *Envisioning a future without food waste and food poverty*

DOI 10.3920/978-90-8686-820-9_4, © Wageningen Academic Publishers 2015

47

These requirements ought to in one form or another be present in the label. However, there are different interpretations or definitions on what this means and how this should be portrayed. Some examples which constitute differences is requirements regarding the name of the food, ingredients that are allowed or not allowed to be included as well as additives, nutrition declaration, health claims, allergens and date marking. In the following we will focus on allergens and date marking.

When it comes to labelling of allergens, there are some differences. Korea has for example identified 13 allergens, the EU has 14 and USA has identified 8 allergens to mention some examples. Japan has a number of mandatory allergens and a list of 20 recommended ingredients. India has no provisions yet, but a draft regulation which is similar to the US legislation. China's list of allergens is also similar to the one used in USA.

Table 1 indicates that there are a number of different allergens in the world. How the allergens shall be portrayed is also diverse. In China, it is recommended that the allergen is declared in the ingredients list by a name easily recognisable. In the US, there are two options. Either include the name of the food source in parenthesis following the common or usual name of the major food allergen, or place the word 'Contains' (with a capital 'C') followed by the allergen (FDA, 2004). In the EU, the allergen shall be emphasised through a typeset that clearly distinguishes it from the rest of the list of ingredients, for example by font, style or background colour (Regulation (EU) No. 1169/2011). There do seem to be dissimilar views on what this means though as some countries in the EU seem to interpret this as using 'bold'. In the Gulf Cooperation Council (GCC) countries, the allergens must be declared as 'contain' or 'may contain' and highlighted preferably in bold. Thus, the products labelled as in the EU does not fulfil the GCC labels nor the US requirements either. A standard which could be used in these countries ought to be by highlighting the allergens in the ingredient list as well as using a 'Contain' statement below the ingredients list with the ingredients

Table 1. Examples of allergen labelling in South Korea, EU, USA and Japan.

South Korea	EU	USA	Japan
egg	egg	egg	egg
milk	milk	milk	milk
wheat	cereals	wheat	wheat
peanut	peanuts and others	peanuts	peanuts
soybean	soybeans	soybeans	buckwheat
shrimp	crustaceans	crustaceans	shrimp/prawn
crab	fish	fish	crab
mackerel	nuts	tree nuts	
pork	mustard		
buckwheat	lupine		
peach	molluscs		
tomato	celery and others		
SO_2 >10 mg/kg	sulphur dioxide		
	sesame seeds		

in bold. Thus, the ingredients are highlighted according to the regulations in this comparison. It is however not clear if this practice complies with all markets.

There are some varieties as regards date marking in different markets. Some countries (such as China, India, Korea, and Malaysia) require that the date of production shall be printed as well as the best before date. Some countries also separate between 'use by' and 'best before'. In Singapore, the date marking shall be specified in one of the following ways: 'use by', 'sell by', 'expiry date' or 'best before' in day, month and year according to the Sale of Food Act. In USA product dating is not generally required by Federal regulations. However, if a calendar date is used, it must express both the month and day of the month (and the year, in the case of shelf-stable and frozen products) and immediately next to the date must be a phrase explaining the meaning of that date such as 'sell-by' or 'use before'. Canada however, has rather strict requirements concerning the date. If the year is shown, then it must be shown first and the month following in words after. Showing the year is only mandatory when its declaration is needed for clarification such as when the durable goes into the next year. Since Canada is bilingual, there are symbols for the months in two letters which represent the different months in accordance with the Canada Food and Drug Regulation B.01.007 (http://tinyurl.com/q46ensc). These symbols or acronyms are however not universal. In Japan, the year, month and day must be printed if the product has a shelf-life less than 3 months and only year and month if more than 3 months (Hayashi, 2011). In the GCC countries however, the day, month and year must be printed if less than 3 months durability and month and year if more than 3 months.

Thus, since the date marking can be printed with the year first or last as seen above in Table 2, this can sometimes be confusing about the durability.

So far, this paper has illustrated some of the differences and similarities in some markets when it comes to allergens and date marking, but why do products have to be scrapped?

2. What differences can encompass scrapping of the products?

There can be several reasons for scrapping food. Missing ingredients such as additives or processing aid seem to be the most common reason for scrapping, but this issue is not included here. Due to the focus of this paper, only scrapping due to misbranding will be presented. As mentioned earlier, allergens and date marking can be portrayed in several ways. What this means in terms of misbranding will be presented below. First however, country of origin will be examined as well since this subject can be regarded as misbranding if not accurately defined.

Table 2. Examples of date marking in USA, Canada, Japan and the GCC countries.

USA	Canada	Japan	GCC
Use before	Best before/meilleur avant	Best before	Best before
12 JUN 28	12 JN 28	2012-06(-28)	(28)-06-2012

The requirements for place of origin, or country of origin which is a more common expression, can be different in different markets. False indication of the place of origin is considered as a punishable act in Japan and the place of origin has to be displayed so that the consumer can easily see and read (Hayashi, 2014). When dealing with food originating from one country and which is processed in another, the approach for labelling is based on the place of last substantial change. Generally, this means that the last country in which a food is substantially changed is the country of origin. However, there are differences in defining 'substantially changed'. In Australia, a 'made in' statement requires at least 50 per cent of the cost to produce the product was incurred in that country (Australian Competition and Consumer Commission: https://www.accc.gov.au/consumers/groceries/country-of-origin). In the United Arab Emirates, the imported product would be said to originate from a particular country if it was wholly produced or contain at least 40% in value added transformation from that country (The Canadian Trade Commissioner Service, UAE Import Regulations; http://www.tradecommissioner.gc.ca/). In the EU, 'goods whose production involved more than one country shall be deemed to originate in the country where they underwent their last, substantial, economically justified processing or working in an undertaking equipped for that purpose and resulting in the manufacture of a new product or representing an important stage of manufacture', in accordance with Article 24 in Council Regulation (EEC) No. 2193/92 (EC, 1992). Also, the Food Information Regulation (No. 1169/2011; EC, 2011) also has provisions on the country of origin. Thus, the country of origin is not an easy task to evaluate and also such a large subject that this paper cannot cover it all here. Important to notice though is that country of origin in one state can be regarded differently in another which might be seen as misbranding even though unintentionally.

Another kind of misbranding is allergens. One reason for this is off course since it is combined with a risk. A person which is allergic to something can react if consuming a product that is not labelled correctly and includes allergens. The focus here is however not on food safety but rather showing how these differences can contribute to food waste. Major food allergens in USA, regardless of whether they are present in the food in trace amounts, must be declared (FDA, 2004). FDA does not define 'trace amounts'; however, there are some exemptions for declaring ingredients present in 'incidental' amounts in a finished food. If an ingredient is present at an incidental level and has no functional or technical effect in the finished product, then it need not be declared on the label. An incidental additive is usually present because it is an ingredient of another ingredient. This ought to mean that if a product contains milk in trace amount due to cross-contamination since the production site cannot avoid it without taking precaution – this would not be considered incidental. Thus, milk as an allergen must be declared even in trace amount. If the product contains milk in trace amount and it is not declared as an ingredient and the producer cannot show satisfactory production processes, then the product needs to be scrapped due to safety precautions. In USA, food shall be considered misbranded if the label is false or misleading, required information is not prominent, if the product is falsely represented or if the product is lacking information (CFR 21, Chapter 9, 1 April 2013; FDA, 2013). This means for example that labels without the prominent proper size of the legal name of the food would be considered misbranded. Also, a label without the correct way of emphasising the allergens would be considered misbranded. Thus, these defaults could require the scrapping of these products, especially the emphasising of allergens.

EU-VITAL has developed a harmonized action level for the labelling of allergens such as milk with the following levels: below 20 mg/kg – no need to declare, 20-500 mg/kg – a 'May contain' statement is allowed, above 500 mg/kg – 'Contain'. Voluntary Incidental Trace Allergen Labelling (VITAL; http://www.eu-vital.org/en/actionlevels.html), originates from Australia/New Zealand which both have same levels, Japan also has similar levels. This means that there are different levels that must be declared in different countries. Some countries allow for a level of trace amount but if a product in the US does not comply with the allergen rules, the product ought to be scrapped. Of course these levels can be discussed whether or not they are safe to eat for consumers with an allergy, but the point here is only to show that these different levels can constitute problems when it comes to labelling.

Lastly, date marking can be considered as misbranding. In Singapore it is prohibited to change the date marking. Removing, erasing, altering, obscuring, superimpose or in any way tamper with any date mark on any pre packed food is prohibited. Infringement constitutes a fine (Sale of Food Act, 2005). Wrong dates on the products are also seen as misbranding in India. If labels contain false or misleading statements it will be considered as misbranded and goods which appear to be adulterated or misbranded can be refused to enter India (FSSA, 2011).

If the best before date or production date is labelled incorrectly on the package in China, it would be considered false labelling to put a sticker with a correct date (Meador and Bugang, 2011). No supplement or amendment is allowed to the declared date. Thus, the product needs to be scrapped. If a label is found to be noncompliant to the Chinese regulations, the body responsible for the noncompliant labels may be subject to one or more of the following consequences: (1) an order to make corrections within a specified period; (2) confiscation, recall or destruction of applicable products; (3) confiscation of illegal income; (4) imposition of a fine; and (5) in severe cases, an order to suspend business or even the revocation of the noncompliant company's applicable permit (Fu-Tomlinson and Wang, 2015).

Food labelling in China should be easily noticeable, durable and legible; be true and accurate, with no false, deceptive, exaggerated or misleading information; and contain no express or implied statement regarding functions of prevention or treatment of diseases (Fu-Tomlinson and Wang, 2015).

These are some examples when products ought to be scrapped due to labelling issues. There are different interpretations and levels of misbranding, and it concerns if the label is false or misleading or if the product is lacking information. Due to different requirements on the label across the world, a label can easily be regarded as misbranded if some facts are missing or labelled differently. This does not necessarily mean that the product is not safe to eat. Also, in many cases it is possible to re-label the food rather than scrapping the food – but this paper portrayed some cases where the food has to be scrapped and in that sense contributes to food waste.

3. Conclusions

As mentioned, this paper only portrays some of the examples of misbranded products. What conclusions can be drawn so far? Some differences concerning labelling in the markets can perhaps be corrected through using a template which covers several requirements and thus amending all these requirements by finding a template that works in several markets. However, some requirements require larger and separate amendments in order to fit all requirements. If for example the date marking and country of origin is interpreted and thus portrayed differently even though the date and country of origin is the same – then this cannot be mended through an overall template. These differences, amongst others, require a more unified interpretation and portrayal – otherwise the products cannot be sold.

Thus, some kind of harmonization is required in order to reduce food waste. Due to the differences in the markets, there is no alternative to strive for compliance in all local markets. One product that is perfectly fit for one market will not be treated the same in another market, thus we ought to strive for a more homogenous food law. This can be achieved by harmonization on a regional level if not on a multilateral. Some regional markets have started the process on harmonization of labelling such as the EU and the GCC but more ought to be done in the world in order to reduce food waste.

References

European Comission (1992). Council Regulation (EEC) No. 2913/92 of 12 October 1992 establishing the Community Customs Code. Official Journal L 302: 1-50.

European Comission (2011). Regulation (EU) No 1169/2011 of the European Parliament and of the Council of 25 October 2011 on the provision of food information to consumers. Official Journal L 304, 22.11.2011: 18-63.

Food Safety and Standards Authority of India FSSA (2011). Section 23 and 24 of Food Safety and Standards Regulations 2011, (FSSA). Available at: http://www.fssai.gov.in/GazettedNotifications.aspx.

Fu-Tomlinson, Y. and Wang, V. (2015). Food Labeling for companies manufacturing or distributing food in China. Food Safety Magazine. Available at: http://tinyurl.com/nqcya9w.

Hayashi, Y. (2011). United states Department of Agriculture Report, Japan Consumer Affairs Agency defines 'Use by date' and 'best before date', JA1048, 12/28/2011.

Hayashi, Y. (2014). United States Department of Agriculture Report, Japan, An overview of the new Food Labeling Standard, JA4043, 12/17/2014.

Meador, M. and Bugang, W. (2011). United States Department of Agriculture Report, China – Peoples Republic of, General Rules for the Labeling of Prepackaged Foods, CH110030, 5/25/2011.

Sale of Food Act (2005). Food Regulations, Rg 1, G.N. No. S 264/2005. Available at: http://tinyurl.com/qjc3sg9.

US Food and Drug Administration (FDA) (2004). The Food Allergen Labeling and Consumer Protection Act of 2004 (FALCPA). Available at: http://tinyurl.com/nswrzek.

US Food and Drug Administration (FDA) (2013). USA, Code of Federal Regulations CFR 21, as updated 1 April 2013. Available at: http://tinyurl.com/abb3.

Part 2. Critical reflections on food waste

5. Values in the trash: ethical aspects of food waste[1]

M. Gjerris[1] and *S. Gaiani[2]*
[1]*Department of Food and Resource Economics, University of Copenhagen, Denmark;* [2]*Department of Agro-Food Sciences, University of Bologna, Italy; mgj@ifro.ku.dk*

Abstract

Food waste is one of the most discussed subjects within food production in recent years. Throwing food away at a time when almost 900 million humans live in hunger and it is becoming more and more clear that food production draws on limited resources of eg. agricultural land and fresh water and is one of the important contributors to climate change, simply seems wrong. Here we discuss three questions in relation of this almost self-evident fact: (1) different definitions of food waste and the difficulties in reaching a global definition, how desirable it might be; (2) different ways of preventing food waste from the individual to the international level and the importance of examining the values behind different strategies; and (3) ethical challenges in relation to food waste and the opportunity to utilize the indignation that many feel when confronted with food waste to re-think the deeper relationship between humans and nature.

Keywords: definitions of food waste, ethical challenges, values

1. Introduction

Food waste is one of the most discussed subjects in recent years within the area of food production (Evans *et al.*, 2013, Segré and Gaiani, 2010, Stuart, 2009). This is hardly a surprise when one looks at the numbers. According to a Food and Agriculture Organization of the United Nations (FAO) report from 2011 approximately 1/3 of all food produced for human consumption is wasted each year amounting to a staggering 1.3 billion tonnes of food (FAO, 2011) and in a DG environment study from the European Commission, the quantity of food discarded in the 27 Member Countries amounts to 89 million tons, or 180 kg per capita per year (European Commission, 2010).

This waste of resources happens amidst a growing recognition that human impact on biodiversity, water resources, environment, climate change etc. is reaching a critical point illustrated by the number of planetary boundaries that have been or are about to be transgressed (Steffen *et al.* 2015), causing irreversible damage to both eco-system services necessary for human thriving and threatening thousands of species with extinction (Ceballos *et al.,* 2015).

Another issue that bring the scale of food waste into perspective is the almost 800 million people living in hunger (FAO, 2015). The daily images of malnourished children from different parts of the world are a very vivid argument against food waste. Finally the interest in food waste has grown considerably in Europe since the economic crisis hit in 2008 and especially the Southern European countries saw the discrepancy between the amounts of food wasted and the amount of people going

---

[1] Please note that this article is a re-written and shortened version of: Gjerris M and Gaiani S (2013). Household food waste in Nordic countries. Estimations and ethical implications. Nordic Journal of Applied Ethics 7: 6-23.

hungry in their own societies, turning into a tangible problem, rather than a somewhat distant issue mainly related to the developing world.

Food waste thus seems to be a very simple problem in the sense that it is easy to agree that something has to be done. Very few people actively support food waste. But is it really so simple? And if it is, is there more to food waste than initially meets the eye? And what is food waste to be exact? There are many competing estimates of the amount of food waste – and to begin finding solutions, it is probably a good idea to know exactly what the problem is.

In this article we will show that the discussions about food waste are more complex and could have wider consequences than one should initially think. First we discuss the problems and importance of defining food waste. Next we briefly describe how the phenomena of food waste demands action from the individual level to international agreements and challenges current agricultural and food production systems. Finally we discuss central ethical issues related to food waste highlighting both issues of social justice and environmental ethics to show the complex values embedded in food waste.

2. Definitions of food waste

Professional bodies, including international organizations, state governments and secretariats have developed a number of characterizations of food waste. There is thus no clear internationally accepted standard definition. This carries with it the obvious disadvantage that it can be difficult to compare figures and numbers from different surveys and different countries.

Designations of food waste vary, among other things, in what food waste consists of, how it is produced and where or what it is discarded from or generated by. Definitions also differ due to cultural variations because what is considered to be waste in some countries may not be considered as waste in others (i.e. intestines and internal organs of some animals) and certain groups do not consider food waste to be a waste material, due to its applications (bio-fertilizers, bio-energy, etc.).

The FAO uses food 'loss' to refer to reductions in edible food mass during production, post-harvest and processing. FAO uses food 'waste' to refer to reductions at the retail and consumer levels. Food waste and food losses together are included in the more general term 'food wastage' (FAO, 2011).

USDA's Economic Research Service (http://www.ers.usda.gov/publications/eib-economic-information-bulletin/eib121.aspx), on the other hand, defines food loss as the edible amount of food, post-harvest, that is available for human consumption but is not consumed for some reasons. This includes cooking loss and natural shrinkage (for example, moisture loss); loss from mould, pests, or inadequate climate control; and food waste, where food waste is defined as the component of food loss that occurs when an edible item goes unconsumed, as in food discarded by retailers due to colour or appearance and plate waste by consumers.

In the European Union, until 2000, food waste was defined by the Directive 75/442/EEC and considered as 'any food substance, raw or cooked, which is discarded, or intended or required to be discarded'. This directive was replaced by Directive 2008/98/EC where there is no specific definition of food waste, but just a broad description of 'categories of waste'.

In the UK the Waste & Resources Action Programme (WRAP) has divided food waste in three categories:
1. Avoidable. Food and drink thrown away that was, at some point prior to disposal, edible (e.g. slices of bread, apples, meat).
2. Possibly avoidable. Food and drink that some people eat and others do not (e.g. bread crusts) or that can be eaten when a food is prepared in one way but not in another (e.g. potato skins).
3. Unavoidable. Waste arising from food or drink preparation that is not, and has not been, edible under normal circumstances (e.g. meat bones, egg shells, tea bags) (WRAP, 2009).

Even though these definitions overlap to some extent, there is little doubt that the initiatives against food waste deemed to be the most efficient and necessary ones depend on the definition at work. Further: what is considered food waste is not just a technical issue, but also depends on the ethical values of those defining it. According to Smil (2004) overeating should be considered a form of food waste – in this case referring to the difference between the quantity of food that each person consumes and what he/she really needs (energetic value) – Smil emphasizes the responsibility of each human being in the fight to reduce food waste, thus placing moral guilt on people who over-eat (with all the difficulties of defining what that is).

Another suggestion, mentioned in a report from the Barilla Center for Food and Nutrition (Barilla, 2012) is to include feed for animals as food waste. This is done as the plant proteins converted to animals proteins could have been eaten by humans and ensured valuable nutrients to many more people. In this scenario not being a vegan means basically wasting food – a proposition that makes sense from a nutritional point of view, but will hardly be acceptable for the food culture associated with meat eating that characterizes many citizens in the world.

Discussing what the definition of food waste should be is thus not just interesting from the point of view of academic clarity. A more uniform global definition would enable more accurate comparative work, but would probably also be controversial because of the unavoidable implicit values that result in the differing definitions in existence today.

3. Possible avenues of action

The costs of decreasing food waste are relatively low, and the potential benefits are substantial. Less food waste would lead to more efficiency, more economic productivity. Whether one opts for one or the other definition of food waste, there is widespread agreement the issue needs to be addressed. Which actions and at what level it should be taken is, however, dependent on the values that one brings into the discussion. Is food waste primarily an individual problem that the consumer has to react to? Is it a problem for the food sector? Or is it a problem that requires political action at both national and international level? Finally, reducing food waste is thus also a question of ideology, e.g. how much can the state meddle with the habits of the consumer?

Here we outline three currently running initiatives whose aim is to address food waste at different levels of the food supply chain – namely households, the grocery sector and the entire national food supply chain – and find possible strategies to reduce it.

To help households tackle food waste and raise awareness on its impacts, the Love Food Hate Waste (LFHW) campaign was launched in 2007 in the UK (LFHW, 2007) by WRAP. Household food waste makes up almost half of all food waste in the UK and it costs £ 12 billion to households. LFHW works with retailers and brand owners, local authorities, businesses, community and campaign groups, and provides tips, recipes, messages, graphics and consumer insights to prevent food waste. The program is designed to reduce the 7 million tonnes of food thrown away annually (60% which could have been eaten). Certainly it is not the only campaign on food waste taking place in a EU country, but it is definitely one of the most successful ones in terms of the targets accomplished (WRAP, 2009).

WRAP is also responsible for the Courtauld Commitment, a voluntary agreement aimed at improving resource efficiency and reducing the carbon and wider environmental impact of the grocery sector. WRAP works in partnership with leading retailers, brand owners, manufacturers and suppliers who sign up and support the delivery of the commitment's targets. The Courtauld is predicted to contribute to a cumulative reduction of 1.1 million tonnes of waste and 2.9 million tonnes of CO_2. Specifically it aims to reduce the carbon impact of the grocery packaging by increasing the recycling rates of all grocery packaging and to reduce household food and drink waste. It supports the UK governments' policy goal of a 'zero waste economy' and the objectives of the Climate Change Act to reduce greenhouse gas emissions by 34% by 2020 and 80% by 2050 (WRAP, 2013).

The Waste Framework Directive from the European Commission 2008, requires all Member States to adopt Waste Prevention Programmes by December 2013, but only few of them have been able to actually set them up. One of the best examples of such a programme is the National Pact against Food Waste, promoted by the French Ministry for Agriculture, Food Industry and Forestry in 2013 (Frenchfoodintheus, 2015). The pact commits to a wide basket of measures, including training programmes on food waste in agricultural and hospitality colleges, food waste criteria in public purchasing programmes, a clarification of the legal framework and liabilities for food donation, the full integration of food waste in public waste prevention plans, the evaluation of food waste in corporate social responsibility performance, a clarification of best before date terminology, a public communication campaign on food waste and the testing of a pilot programme allowing individuals to donate excess food using an online tool.

The three cited initiatives demonstrate that a comprehensive approach – one that brings together the stakeholders of the food supply chain, combines private and public commitment to food waste reduction, meets best performances in terms of sustainability and raises awareness on all the issues and impacts connected to the waste of food and resources – is strongly needed in order to achieve the 50% reduction envisaged by the EU Commission by 2025 (DG EBV, 2010).

4. The ethical challenges

There are many interesting discussions hidden underneath people's immediate negative reaction to food. What if food waste is a necessity to the culture of affluence that many see as part of the good human life? Food waste is then more an inherent problem to consumerism than a mere accidental side-effect that can be easily corrected. And what if food waste is more a symptom of deeper problems connected to the relation between humans and nature than just an issue in itself?

Food waste has obviously negative consequences for humans as it is a waste of nutritional resources that some less fortunate people may sorely need. A reduction of food waste at consumer level, however, can only seldom directly be utilized for alleviating hunger of disadvantaged people. A reduction of food waste could, anyway, benefit them if the money saved by not throwing food away was directed towards their support. This is of course not the case if the money saved from reducing food waste is spent on other kinds of consumer goods (Macmillan, 2009).

Such dilemma raises the ethical question whether those who live in affluence are obligated not to waste food and instead use the saved resources to help others. If the answer to this is yes, the question then becomes why this is so in the area of food waste, but not in other areas of consumption? Why is food so special that food waste seems self-evidently wrong whereas the waste of resources connected for example to fashion is not debated to the same degree? Our answer to this is that long ago – even in affluent Western societies – food was sometimes scarce and hunger a sensation that most had felt.

Throwing away food feels morally wrong, as it is such a tangible sign of richness bordering on decadence. When dealing with food waste as cultural phenomenon and hunger as a global issue, it is hard not to react at such visible inequality in the world. Food waste thus becomes a symbol of human injustice that hits us right in the face (Edwards and Mercer, 2007).

Food waste also contributes to increasing pressure on the ecosystems: desertification, eutrophication, pollution of air, land and water, depletion of scare assets such as fresh water and phosphorous, climate change are just some of the negative consequences that again have effects on humans (FAO, 2011). But besides this rather obvious environmental concern, we tentatively put forth the suggestion that food waste is also seen as inherently problematic as it can be seen as disrespectful to the organisms, ecosystems and biosphere that sustain human life.

Food is one of the most basic ways humans interact with nature. Food can be interpreted as participation in the life of the community constantly situating humans in a very intimate relation with the rest of the world. The often strong reactions toward food waste could thus point to deep seated experiences of human existence as being closely knitted into the beings and rhythms of nature. A nature that is worthy of respect and love not just because it sustains our lives but because it is part of the 'Ecological Self' – as the Norwegian philosopher Arne Næss (1990) named it – and opens up to experiences of gratitude and community. In this light wasting food can be seen as a rejection of nature as a relational subject and a reduction of it to an external object with only instrumental value.

This understanding of community with the planet that sustains our lives has been given a voice by different thinkers within eco-centric religion and philosophy. Few, though, have captured it so distinctly as the poet Gary Snyder who stated: *Every meal is a sacrament* (Snyder, 1990). Food waste can thus be interpreted as a kind of 'sacrilege' (Gjerris and Gaiani, 2014).

5. Closing remarks

In rural cultures, leaving a fruit attached to the tree was seen a sign of empathy with nature. In many cultures thanking for harvest and meals through prayer and celebration is common practice (Counihan and Van Esterik, 1997). In Western consumer societies it seems the practice of eating has gone from establishing social, ecological and religious bonds into simply aesthetic consumerism that deepens the distance between those who have (and can waste) and those who do not and contribute to the destruction of the planetary eco-systems. 1.3 billion tons of food are wasted every year. In industrialized countries, consumers waste roughly 30% of food that they buy. The cost of this waste has economic, social and environmental impacts. As this is becoming more and more evident, it forms the background of the widespread indignation at food waste.

Food waste is a phenomenon of impressive scope and a vicious cycle that needs to be interrupted in order to meet the food demands of a growing population and protect the eco-systems and the myriads of life forms depending on them. Policy-makers, manufacturers, and individual citizens need to take definitive steps to prevent waste in fields, stores, and homes. Continuing to loose food along the food chain means losing the opportunity to help people who suffer from hunger to benefit from this excess production. There is actually enough food for every human on the planet, if we realize how food is wasted and develop the methods to avoid the waste and distribute the wealth. Looking at the increasing world population this becomes only more necessary. Better economic and productive policies, an improved food supply chain and changes in food habits are urgently needed to meet the challenge of food waste and hunger.

Combining these measures with renewed appreciation for the nature that brings forth food and an understanding of its physical limits and ethical importance will help in this transformation process. Nature is an important resource that needs to be treated with care to be able to provide the food we need. But it is so much more than that and losing sight of it, is part of problem we face today.

References

Barilla Center for Food & Nutrition (2012). Food waste: causes, impacts and proposals. Barilla Center for Food & Nutrition

BioIntelligence Service (2012). Preparing a waste prevention programme. Available at: http://tinyurl.com/ay4jx98.

Buzby, J.C., Wells, H.F., Hyman, J. (2014). The estimated amount, value, and calories of postharvest food losses at the retail and consumer levels in the United States. Economic Information Bulletin Number 121. United States Department of Agriculture.

Ceballos, G., Ehrlich, P.R., Barnosky, A.D., García, A., Pringle, R.M., Palmer, T.M. (2015). Accelerated modern human-induced species losses: Entering the sixth mass extinction. Science Advances, 1, 5, e1400253.

Counihan, C., van Esterik, P. (1997). Food and culture – a reader. New York: Routledge.

Edwards, F., Mercer, D. (2007). Gleaning from Gluttony: an Australian youth subculture confronts the ethics of waste. Australian Geographer, 38 (3): 279-96

European Commission (1975). 75/442/EEC Waste Framework Directive. Available at: http://tinyurl.com/ohu8z3d.

European Commission (2008). Directive 2008/98/EC of the European Parliament and of the Council of 19 November 2008 on waste and repealing certain Directives. Brussels: European Commission. Available at: http://tinyurl.com/p83vrul.

European Commission (DG ENV) (2010). Preparatory Study on Food Waste. European Commission (DG ENV). Available at: http://tinyurl.com/4424obc.

Evans, D., Campbell, H., Murcott, A. (2013). Waste matters: new perspectives on food and society. Wiley-Blackwell, Oxford, UK.

FAO (2011). The state of the world´s land and water resources for food and agriculture. Managing systems at risk. Food and Agriculture Organization of the United Nations.

FAO (2015). The state of food insecurity in the world 2015. Food and Agriculture Organization of the United Nations.

Gjerris, M., Gaiani, S. (2014). Food waste and consumer ethics. Encyclopedia of Food and Agricultural Ethics. Springer, Berlin, Germany.

Gustavsson, J., Cederberg, C., Sonesson, U., van Otterdijk, R., Meybeck, A. (2011). Global food losses and food waste. Food and Agriculture Organization of the United Nations.

MacMillan, T. (2009). What is wrong with waste? Food Ethics, 4 (3), 4.

Frenchfoodintheus (2015). National Pact Against Food Waste. Frenchfoodintheus. Org Available at: http://frenchfoodintheus.org/886.

Naess, A. (1990). Ecology, society and lifestyle. Outline of an ecosophy. Cambridge University Press.

Segrè, A., Gaiani, S. (2011). Transforming Food Waste into a Resource, Royal Society of Chemistry, UK.

Smil, V. (2004). Improving efficiency and reducing waste in our food system. Environmental Sciences, vol. 1, pp. 17-26.

Snyder, G. (1990). Survival and sacrament. In: Snyder G: The practice of the wild. Counterpoint, pp. 187-198.

Stuart, T. (2009). Waste: Uncovering the Global Food Scandal. Penguin

Steffen, W., Richardson, K., Rockström, J., Cornell, S.E., Fetzer, I., Bennett, E.M., Biggs, R., Carpenter, S.R., de Vries, W., de Wit, C.A., Folke, C., Gerten, D., Heinke, J., Mace, G.M., Persson, L.M., Ramanathan, V., Reyers, B., Sörlin, S. (2015). Planetary boundaries: Guiding human development on a changing planet. Science DOI: http://dx.doi.org/10.1126/science.1259855.

WRAP (2009). Household food and drink waste in the UK. WRAP. Available at: http://tinyurl.com/qd3mb25.

WRAP (2013). Courtauld commitment. Available at: http://tinyurl.com/ndcvh2f.

6. Agri-food industries and the challenge of reducing food wastage: an analysis of legal opportunities

M. Alabrese*, *M. Brunori, S. Rolandi and A. Saba*
Scuola Superiore Sant'Anna, Pisa, Italy; m.alabrese@sssup.it

Abstract

Food wastage poses environmental and economic as well as social and ethical questions. The significance of these questions places the food wastage issue to the attention of the institutions, civil society and academia. International and EU documents widely discuss the role of some legal disciplines in the generation of wastage, both as a direct cause and as a mean for its reduction. So far, the majority of public and private interventions in Europe focused on the distribution and consumption segment of the food supply chain, while a lack of reduction measures is registered in the previous segments. Nevertheless, the European Commission estimates that the food waste attributable to industrial processing is around 39%. The paper is aimed at analysing the EU legal framework which influences the generation of wastage during the food processing chain. It addresses the emerging legal opportunities in reducing food wastage, recognising two main areas: labelling strategies and by-products valorisation. After a synthetic exposition of the data related to food wastage (Section 1), the paper shows how in the European legislation food safety requirements and food waste reduction strategies interrelate (Section 2). The paper moves on to assess the EU legal framework on waste, in particular the definition and the potential valorisation of food by-products, (Section 3) the food labelling regulation and the contractual practices (Section 4), underlying the opportunities for food wastage reduction. In conclusion, the paper highlights the emerging legal opportunities and the related shortcomings of the potential strategies for reducing the wastage generation during the food processing chain.

Keywords: food wastage, food processing, labelling, by-products, agri-food industry

1. Introduction: few data on food wastage generated during the processing chain

Food wastage poses environmental and economic as well as social and ethical questions. At international level, an ongoing debate discusses causes, impacts and mitigation measures of food wastage, by involving a cross-section of stakeholders, from grassroots organizations to governments and academia (FAO, 2013a,b, 2014).

According to FAO, approximately one-third of all food produced for human consumption is lost or wasted. The monetary value of the food wasted amounts to USD 936 billion, even if it does not account for environmental and social costs of the wastage that are borne by society at large (FAO, 2014).

Food wastage is generated all along the supply chain, from the agricultural production stage to final consumption. International literature provides numerous definitions of food wastage, by relating to particular points of arising and framed in relation to specific environmental controls (Schneider, 2013). For the purpose of the paper, it is worth mentioning the difference between food losses and waste (Parfitt *et al.*, 2010). Food loss is generated at production, post-harvest and processing stages in the food supply chain. Conversely, food waste occurs at the end of the food chain and relates to retailers and consumers' behaviour. Food wastage conventionally refers to both food loss and waste, thus including any food lost by deterioration or discard (FAO, 2014).

European Union is devoting extensive attention to the establishment of prevention and reduction measures of food wastage in order to promote an efficient use of resources. The majority of public and private interventions focuses on the distribution and consumption stages of the food supply chain (Barilla CNF, 2012; European Commission, 2011), while a lack of measures is registered in the previous stages. Nevertheless, the European Commission estimates that the food wastage attributable to industrial processing is around 39% (European Commission, 2010). This data shows the importance of tackling food losses in food processing. Legislation and regulations are usually regarded as obstacles for preventing wastage (Lattanzi, 2014; Waarts, 2011) even though under the current regulatory framework there may be room to implement good practices in food wastage prevention.

Considering the food wastage issue it is worth focusing on the opportunities of food waste prevention in the food processing chain from a legal point of view. This analysis deals with the regulation affecting the behaviours of those who operate in the intermediate phase of the industrial food production, pointedly all food producers and processors, excepted operators of the primary sector and the distribution system. In this segment of the food chain, two relevant points should be analysed for the scope of this investigation: the first concerns the phase in which foodstuffs are produced or processed; the second relates to the moment in which foodstuff is traded. In each of the two moments food wastage can occur, and the instruments that should be looked at for drawing a food wastage reduction strategy are different. In the manufacturing phase the European norms on food safety and the production residues discipline are relevant. As far as the trading moment is concerned food labelling regulation and the private agreements between food business operators acquire a significant role.

2. European food safety regulation and food wastage

The European food regulation provides both norms that can allow preventing food wastage, and norms that can contribute to the reuse of foodstuff that has, for several reasons, exited the food chain. In both cases, the principal act to hold in consideration is the Regulation (EC) 178/2002. This Regulation lays down the general principles and requirements of food law, establishes the European Food Safety Authority and lays down procedures in matters of food safety. According to article 3, 'food business' means *any undertaking, whether for profit or not and whether public or private, carrying out any of the activities related to any stage of production, processing and distribution of food*. Article 14 clarifies that the responsibility of the food business operator pertains to food safety requirements, which are the following: (1) food shall not be placed on the market if it is unsafe; (2) food shall be deemed to be unsafe if it is considered to be: (a) injurious to health; (b) unfit for human consumption.

In line with the purposes of the act, which aims at guaranteeing high standards of consumer's health protection, the Regulation gives an extensive definition of what should be referred to as 'placing on the market' that means *the holding of food or feed for the purpose of sale, including offering for sale or any other form of transfer, whether free of charge or not, and the sale, distribution, and other forms of transfer themselves.*

In the wake of these provisions, the so called Hygiene Package was issued. It provides a first set of rules that can play a role in preventing food loss. Regulation (EC) 852/2004 on the hygiene of foodstuff, which is perhaps the main part of the Hygiene Package, states that food business operators *shall ensure that all stages of production, processing and distribution of food under their control satisfy the relevant hygiene requirements laid down in the Regulation*. Under this framework business operators have to fulfil several obligations that are essential to guarantee the safety of foodstuff and with regard to waste generation two opposite considerations rise. If on the one hand the respect of the hygiene standards and risk control procedures facilitates the prevention of food wastage due to the contamination of the food, on the other hand too rigid requirements may lead to an excessively severe selection of foodstuffs. In this connection, it is important to underline that the Regulation allows a certain degree of flexibility in the application of its principles giving the operators the task of establishing and operating their own food safety programmes and procedures (based on the HACCP principles) in order to comply with the Regulation. In this sense it may be seen as an opportunity to balance the need to guarantee high food security standards and the goal of reducing food waste. A sustainable approach may guide operators towards good practices aimed at assuring food safety without compromising food loss prevention.

For what concerns the use of foodstuff that has already exited the food chain, the causes of food wastage should be previously assessed. There are two main causes for foodstuffs wastage. Firstly, it is due to contamination and the food lack of safety. In this case the human consumption is to be excluded. Secondly, the other reason why a given foodstuff is eliminated from the food chain is that it doesn't fit the industrial standard. In this case, despite the non-compliance with industrial needs, the food is nevertheless edible. With regard to this scenario, it can now be appreciated how the definition of food business is relevant. In fact, even the subjects willing to donate what it cannot be sold merely for aesthetic standards would be bound by the responsibility stated in article 14 of the Regulation (EC) 178/2002. Consequently, the risk of being held responsible for defective products may discourage the operator to engage with charitable associations and donate those products.

3. EU legal framework on waste: a potential opportunity for food business operators

EU framework legislation on waste is aimed at minimising the adverse effects of the generation and management of waste on human health and the environment, by reducing the use of resources and by favouring the practical application of the waste hierarchy (European Commission, 2012). Directive 2008/98/EC sets the basic concepts and definitions related to waste management, such as definitions of waste, by-products, waste hierarchy. According to the waste hierarchy, prevention requires taking measures before a substance, material or product has become waste. According to article 4 of Directive 2008/98/EC, these measures are aimed at reducing the quantity of waste through, *ex pluribus*, the re-use of products or the extension of the life span of products, the adverse impacts on the environment and human health or their content of harmful substances. Under article 3(1), Directive 2008/98/EC defines waste as any substance or object that the holder discards or intends, or is required to, discard. Furthermore, Directive 2008/98/EC defines the subcategory of bio-waste, which means *biodegradable garden and park waste, food and kitchen waste from households, restaurants, caterers and retail premises and comparable waste from food processing plants* (Article 3.4).

At food processing stage, waste prevention needs to be based on comprehensive valorisation of all the processed components. In this context, the classification of a product as waste has relevant practical implication for its potential valorisation, by being subject to the strict obligation for waste management under EU law. A significant opportunity is offered by the notion of by-product. A production residue is defined as a material not deliberately produced in a production process, which may be regarded as being not a waste but a by-product only if it fulfils specific requirements. Such requirements are listed under article 5 of Directive 2008/98/EC. The test is cumulative and all the criteria must be met.

The first condition requires that by-products must result from a manufacturing process, thus excluding a consumption process. The second requires that further use of the substance or object is certain. A residue continues to be considered as waste when there is a possibility that the material is in fact not useable in consideration, inter alia, of the technical specifications or there is no market for the material. However, long term contracts between the holder and its subsequent users can be considered an indication that the certainty of use is present. The third provides that materials must be used directly without any further processing other than normal industrial practice, which may include, *inter alia*, washing, drying and preservation processes. The last requires that the further use of the by-products is lawful, thus fulfilling all relevant product, environmental and health protection requirements for the specific use.

In adopting a case-by-case approach, it will be essential to analyse the further use of the specific production residues, such as pharmaceutical and cosmetic uses, manufacture of animal feed, food additives, biomaterials and energy production.

Nevertheless, economic operators in some cases classify potential by-products erroneously as waste, due to the number of grey zones that surrounds its legal definition (Waarts *et al.*, 2011) especially at national level. This causes, on the one hand, unnecessary costs for the business operator and, on the one hand, a reduction of the economic attractiveness of the production residue.

Against this background, a broad debate was developed in legal literature and in European Court of Justice case-law (Edwards, 2013). European Commission pointed out that the evolving jurisprudence and relative absence of legal clarity has made the application of the definition of waste difficult for competent authorities and economic operators (European Commission, 2012). Therefore, the EU Court of Justice recognises a need for flexibility in adopting a case-by-case approach, by considering all the specific factual circumstances involved. Additionally, the Court provides that the concept of waste cannot be interpreted restrictively in compliance with the aims pursued by Directive 2008/98/EC.

Research developments need to provide encouragement and guidance on developing by-product or feed uses for recurring residues generated during the food processing stage. Amongst them, it is worth mentioning residues generated during the slaughter of animals for human consumption, the processing of products of animal origin, such as dairy products, and the disposal of dead animals.

There is a clear interest for a wide range of applications of animal by-products, which are commonly used in important productive sectors, including pharmaceutical, feed and leather industries (Waarts *et al.*, 2011). Regulation (EC) 1069/2009 indicates how animal by-products need to be processed or definitely eliminated from the food chain. It requires animal by-products to be directed towards safe means of disposal or allows their use for different purposes, provided that strict conditions are applied which minimise the health risks involved. In particular, use as animal feed is permitted under some conditions, following a risk assessment.

4. Opportunities deriving from food labelling legislation

As far as the legislation on food labelling is concerned, it may offer several opportunities under the purposes of this paper. Labelling regulation is acquiring increasing importance in food law. As seen above, labelling requirements could concur to cause food wastage. Nevertheless the subject may also be a tool for food waste prevention. The new Regulation (EC) 1169/2011 on the provision of food information to consumers, in fact, enlarges the relevance of the information of pre-packed food in stages previous to the final consumer. It states in article 8 that *food business operators that supply to other food business operators food not intended for the final consumer or to mass caterers shall ensure that those other food business operators are provided with sufficient information to enable them, where appropriate, to meet their obligations […].* This paragraph can be interpreted in the sense that, despite those subjects not dealing with the consumer are not obliged to provide information *per se*, they are nevertheless bound to communicate to other business operators the relevant information that will allow the latter to comply with their obligation to provide information to consumers.

As the new regulation is putting more emphasis on the food information throughout all the food chain, there could be opportunities for the food business operators to use this tool as a levy for facilitating their social responsibility implementation. For instance, food business operators willing to minimize the food waste production could indicate their efforts through voluntary labelling information, by indicating their commitment to donating or placing on the market also the non-conform but perfectly edible products. Mentioning in the label, or in the accompanying documents, that a product is made following efforts to reduce food wastage could facilitate its market placement. Again, the same strategy could be adopted in case a business operator engages in the opportunities derived from the by-product regulation.

If, on the one hand, the labelling legislation is finalized to realize the right to information of consumers, on the other, taking advantages of the possibilities deriving from it in the intermediate phases of the food chain may create a new space for communication between business operators and create virtuous effects in the campaign for food wastage reduction.

Still another strategy for preventing food wastage may be given by a wise application of the labelling regulation. Indeed, Regulation (EC) 1169/2011 limits the market placement only for products whose 'use by' date is expired, stating in article 24 that *[a]fter the 'use by' date a food shall be deemed to be unsafe in accordance with Article 14(2) to (5) of Regulation (EC) 178/2002.* On the contrary, nothing similar is said for the 'best before' date, therefore there are margins for maintaining in the food chain products for which the operator esteems that food safety and quality requirements are met notwithstanding the passed 'best before' indication. This very rarely occurs because of the excessive cost of the re-packaging and for the necessity to avoid the risk of the food perishing that would lead to responsibility for defective products. Furthermore, if the European law establishes the difference between expiry date and minimum conservation term, the sanctions for the violation of the rules are left to the Member States. The autonomy left by the European legislator to the Member States constitutes an important chance for the design of strategies aimed at reducing food waste while guaranteeing food safety and the consumers' right to information. To this end, as food labelling legislation is becoming increasingly complicated and generates the risk, especially for small food producers, to commit formal mistakes regarding food information, National authorities should bear in mind these difficulties and make efforts to guarantee that the compliance with formal requirements does not result in food waste generation.

5. Contractual practices

In the intermediate section of the food chain there are different needs that have to be complied with. One of them is the selection of the products. In this phase there could be wastage generated by the necessity to comply with strict quality standards, which are private standards, regarding the dimensions, colour, weight, etc. All these characteristics are usually previously agreed on the furniture contracts between the producer and the transformer and the transformer and the buyer. The content of the agreements could reduce the wastage of products imposing less stringent requirements. It is also to be considered that, if the selection is properly done, certain products can be addressed to a different use. If the selection phase does not happen before the sale from the producer to the transformer this generates a higher level of wastage. For this reason is desirable, on the light of the food waste reduction, that the object of contracts in the different phases of the chain are agreed on focusing on a preventive selection of the products with the association of different destination of the them, considering also a different destination in case of the non-compliance with the terms of the contract (Waarts, 2011).

6. Conclusions

The brief overview provided so far touches on some of the strategies that, given the current legal framework, may be implemented in order to tackle food wastage in the food processing sector. The opportunities related to the food safety and the food labelling legislation have been described as well as the chances offered by the waste legal framework and the private negotiation. However all the proposals discussed are not binding, they are behaviour-driven instruments that can be lawfully acted within the relevant legal context. It means that they need proactive behaviours from the food business operators who may be encouraged by economical aspects, that is saving costs or increasing income (such as in the hypothesis of voluntary labelling information). But overall, solutions to be truly effective must be supported by a real operators' commitment towards a more sustainable food system. Such commitment, in turn, can be strengthened by the raising consumers' awareness on the importance of reducing food waste, as they are the final target of each business operator's efforts.

References

Alabrese, M. (2013). Alla ricerca di una distinzione tra «rifiuto», «sottoprodotto» e «biomassa» ovvero i limiti di una questione mal posta. Rivista di Diritto Agrario 4: 685-705.

Barilla Center for Food & Nutrition (2012). Food waste: causes, impacts and proposals. Available at: http://tinyurl.com/q8pnuzf.

Edwards, V. (2013). A review of the Court of Justice's Case Law in relation to waste and environmental impact assessment: 1992-2011. Journal of Environmental Law, 25: 515-530.

European Commission BIOIS (2010). Preparatory study on food waste across EU 27. Available at: http://tinyurl.com/o3l6svx.

European Commission BIOIS (2011). Guidelines on the preparation of food waste prevention programmes. Available at: http://ec.europa.eu/environment/waste/prevention/guidelines.htm.

European Commission (2012). Guidance on the interpretation of key provisions of Directive 2008/98/EC on waste. Available at: http://tinyurl.com/lknxoy9.

FAO (2014). Food Waste Footprint. Full-cost accounting. Rome. Available at: http://www.fao.org/3/a-i3991e.pdf.

FAO (2013a). E-Forum on Full-Cost Accounting of Food Wastage. Rome, FAO. Available at: http://www.fao.org/nr/sustainability/food-loss-and-waste/food-wastage-forum/en/.

FAO (2013b). Toolkit: Food Wastage Footprint. Rome, FAO. Available at: http://www.fao.org/docrep/018/i3342e/i3342e.pdf.

Lattanzi, P. (2014). Gli ostacoli di ordine giuridico alla riduzione dello spreco alimentare. Rivista di Diritto Agrario 3: 300-325.

Parfit, J. Barthel, M., Macnaughton, S. (2010). Food waste within food supply chains: quantification and potential for change to 2050. Philosophical Transaction of the Royal Society 365: 3065-3081.

Schneider, F. (2013). Review of food waste prevention on an International level, Waste and Resource Management, 166: 187-203.

Waarts, Y., Eppink, M., Oosterkamp, E.B., Hiller, S.R.C.H., van der Sluis, A.A., Timmermans, T. (2011). Reducing food waste. Obstacles experienced in legislation and regulations. Wageningen University, the Netherlands, 128 pp.

7. From food waste to wealth: valuing excess food in France and the USA

M. Mourad
Centre de sociologie des organisations, Sciences Po Paris, 19 Rue Amélie, 75007 Paris, France;
marie.mourad@sciencespo.fr

Abstract

How have recent public policies, corporate initiatives, and mobilizations around 'food waste' impacted? How *excess* food is valued and subsequently managed in France and the USA? This research draws on more than 150 qualitative interviews, observations and document analysis of an emerging 'food waste movement' in Europe and the USA in 2013-2015, bringing together actors involved in a wide range of organizations in the food and waste system. I find that different actors and organizations view 'waste' in different ways depending on their divergent interests in the production, (re)distribution, consumption and disposal of food. Despite endorsing a hierarchy of preferable solutions – prevention, recovery, and then recycling – in practice these actors establish three hierarchies based on environmental, social and economic values. These hierarchies create competition both within and between distinct solutions that are not necessarily compatible. I show that recycling and recovery are dominant solutions because they offer clear equivalences with economic goals, often overlooking long-term environmental and social impacts. Although current initiatives engage in only 'weak' prevention, and thus only create marginal changes in how food is valued and managed, I argue that these mobilizations nonetheless have the opportunity to create more structural changes in food and waste systems.

Keywords: sustainability, surplus, food system, recovery, recycling

1. Introduction

Over the last few years, a wide range of organizations in developed countries have embarked on efforts to address the economic, environmental and social impacts of 'food waste', which has been denounced by European and American advocates (Bloom, 2010; Stuart, 2009) along with international organizations and national agencies (FAO, 2013; Gustavsson *et al.*, 2011; WRAP, 2013). Many private companies and community organizations have started recycling, recovering and preventing what they qualify – in a variable way – as 'food waste'. This article examines how these recent mobilizations impact the way *excess* food is valued and subsequently managed, with respect to sustainable production, distribution and consumption.

The most common framework adopted in efforts against food waste consists of a single 'food recovery hierarchy' that classifies the most appropriate actions to be implemented: prevention (optimizing processes and adapting production to needs), recovery (redistributing food to people who need and/ or want it), and lastly, recycling (feeding animals, using in industrial production, creating energy or compost) (EU Commission, 2008; US EPA, 2011). Yet, I show that actors and organizations value waste differently and establish three hierarchies of 'solutions' based on environmental, social and

Leire Escajedo San-Epifanio and Mertxe De Renobales Scheifler (eds.) *Envisioning a future without food waste and food poverty*

71
DOI 10.3920/978-90-8686-820-9_7, © Wageningen Academic Publishers 2015

economic goals. Solutions are not necessarily compatible and compete both within and between these hierarchies. Dominant approaches are generally focused on the management of existing surplus through recovery and recycling and often overlook long-term, structural environmental and social impacts. Drawing on the concept of 'strong' sustainable production and consumption (Lorek and Fuchs, 2013), I argue that a switch from recycling, recovery and 'weak' prevention to 'stronger' prevention is necessary to achieve incremental change toward sustainability.

2. Theoretical framework: a holistic approach to surplus and waste

Studying food surplus and waste offers the opportunity to extend commodity chains analysis to include the stage where food is no longer a commodity as such, and either enters circuits of free redistribution or becomes another organic resource, before the stage of disposal. The emerging 'movement' addressing food waste also transcends typical dichotomies in the food system such as public vs private and NGO or social movements vs companies (Holt Giménez and Shattuck, 2011) and offers an interesting case to analyse the tension between fundamental and superficial changes in how food systems are governed.

In most developed countries, including France and the United States, food surplus and waste is most prevalent at the stage of consumption (Gustavsson et al., 2011). Yet, a large part of the consumer waste stream is generated by practices upstream, such as packaging, promotional offers, and restaurant portions sizes, as well as structurally-produced consumption habits in society (Evans, 2014). If food waste is often considered as a consumption problem, it is important to look at the chain systematically beyond the dichotomy 'producers' vs 'consumers'. 'Strong' prevention is not restricted to consumer behaviour and requires deeper analysis of political and societal structures.

Existing studies on food waste generally focus individually on environmental, social or economic impacts. As Papargyropoulou (2014) acknowledge, 'the waste hierarchy, as a framework, primarily focuses on delivering the best environmental option (p.110)'. Existing frameworks do not account for potential competition between different sets of values and hierarchies. Besides, most studies on 'food waste' focus on what is actually the bottom of the hierarchy: recycling. A few recent studies have explored the potential of food recovery with regards to social and environmental aspects (Phillips et al., 2011; Reynolds et al., 2015). But there is no macro-analysis of prevention beyond individual behavioural changes. Moreover, advocates generally promote recycling and recovery without analysing the scale and number of intermediaries involved in their implementation. For example, should municipalities implement large-scale composting facilities or promote backyard composting? Should food be recovered informally by grassroots organization or through formal partnerships with national food bank networks?

3. Data and methods: 'diving' into the food waste movement

I collected qualitative data from a wide range of experts, policy makers, corporate representatives, workers, community leaders and activists addressing 'food waste' from 2013 to 2015 in France and in the USA The two countries share similar characteristics related to the organization of food production, distribution and consumption, but they differ in terms of hunger-relief policies and waste management regulations, which allows for comparative analyses. I carried out 68 semi-structured interviews in France and 57 in the USA, along with 19 and 29 exploratory or informal discussions, respectively. I complemented the interviews with more than 80 observations such as conferences and on-site visits of farms, processing facilities, food banks and composting facilities and exploratory 'dives' into dumpsters. In each country, I engaged in regular volunteer work for non-profit grassroots organizations and advocacy groups and participated in national studies and multi-stakeholders initiatives. I regularly read media articles, reports and working documents on 'food waste' and followed news and social media around the world. Based on fieldwork data, I identified how both individual and organizational actors understood excess food and 'waste', enacted solutions to food waste, and related to a broader food waste 'movement.'

4. Results: multiple visions of food waste and potential solutions

A food waste 'movement' with multiple roots

Actors and organizations taking part in the recently emerged food waste movement – such as activists, suppliers, retailers or policy makers – come from diverse fields and often have divergent interests in the organization of food production, distribution and consumption and waste management. For many organizations that use the framing of 'food waste', 'food surplus' was initially part of broader concerns and activities without being the main focus. They were initially rooted in other interests such as anti-capitalism, food justice and sovereignty; local and sustainable food systems; 'zero waste' (focused on all waste streams), environmental protection; charity and social responsibility (through food banking); agriculture and food security (generally food availability at a national level); and industrial efficiency.

Since 2008-2009, an increasing number of actors have been associating their activities with the expression 'food waste', or 'gaspillage alimentaire' (a more normative and usually pejorative French word, which describes a sub-optimal use of food). Significantly, the appearance of the term in French media increased 40-fold between 2010 and 2014 (database Europresse; europresse.com). The term 'food waste' starts to appear on national and local governments' agendas. At the same time, food redistribution organizations relate their donations to avoiding 'waste' and benefiting the planet. In France and the US, more than 30 start-ups and non-profits have been created since 2010 with the stated mission of 'fighting food waste'. Although no national regulations have been enacted at the time of writing, French policy makers released proposals for a 'national policy to fight food waste' in April 2015 (Berkenkamp and Mourad, 2015).

Multiple visions and values of excess food

If most groups converge on a similar framing of food waste as an environmental, social and economic problem, the very definitions of 'food waste' and adequate 'solutions' diverge. Beyond language subtleties, the meanings of 'losses' or 'avoidable' or 'unavoidable' waste vary between regions and organizations, despite efforts to build common measurements (HLPE, 2014; WRI, 2014). For example, a representative of the French wholesale markets includes an incinerator project – supplying energy to a low-income neighbourhood with digested food scraps – in her organization's 'fighting food waste' policy, whereas advocates of food recovery considered burning food still 'waste' (interview, France, May 2015). From a profit-oriented perspective, food is 'wasted' if it is donated for free instead of being sold. From an environmental standpoint, food is 'wasted' if it goes to feed worms in composting systems instead of other animals or, even better, humans.

Behind an overarching hierarchy based on prevention, recovery and recycling, then, I identify three environmental, economic and social hierarchies of actions. Measurement of outcomes differs according to the prioritisation of these factors: tons of waste, CO_2 emissions, or soil and water use for the environmental one; numbers of meals or calories for the social, and monetary profits for the economic. Not only because surplus food is a material perishable good but also because funding and human resources are finite, solutions cannot all be carried out at the same time and are not always compatible, both within and between different hierarchies. In this context, I analyse what solutions are the most *dominant* – which appear the most during interviews, in public instances and media articles, or which are the most concretely promoted, funded and implemented.

Recycling organic resources: an environmental and economic 'win-win'

Recycling entails using 'food scraps' or 'organic resources' to feed animals, create energy or compost. An increasing number of companies develop capital-intensive waste-to-energy infrastructures. For example, a project endorsed by the Innovation Center for US Dairy plans to put 1,300 methane digesters on dairy farms by 2020 to create biogas. Recycling may be after prevention and recovery in all hierarchies, but it is often promoted as the first solution to food waste by municipalities and companies, especially in the USA. The push for recycling, especially compared to prevention, is the need to show directly measurable results (keeping waste away from landfills) to comply with future generation, and provide economic gains.

Tensions arise when recycling happens at an increasing scale through the commoditization of what was formerly waste. As an owner of a family farm illustrates: 'I used to go to a big bakery but now they donate everything to a big guy [a company] ... before they would give bread to 20 people [different farmers]. Also I would get oil from restaurants; now they sell to companies that make biofuels ... it has become a commodity' (interview, California, 2015). Beyond these tensions, recycling risks distracting from prevention and recovery efforts, despite the latter being higher up in the food waste hierarchy. In San Francisco, a communication campaign for composting in 2014 depicted pizza leftovers in a cardboard box with the caption 'California Gold'. But, what about advocating against throwing away leftovers in the first place?

Recovering food surplus: a 'win win win'

Food recovery involves rescuing 'extra' or 'wholesome food' – rarely called 'waste' – in order to bring it to people who need or want it. Producers and businesses see surplus food as 'necessary' because of seasonality, variability of prices and unpredictability of demand. People working for local governments or social enterprises use a similar rhetoric: entrepreneurship, innovative technologies, and better logistics can 'solve the problem of food waste' by connecting necessary surplus food to necessarily 'hungry' people, in a social, economic and environmental 'win-win-win'. Large corporations gain tax incentives for donations – in France, up to 60% of the food stock value – and save on disposal costs while improving their public image. Financial incentives help transform environmental or social benefits into economic benefits. While food recovery was historically mainly carried on by non-profit organizations, a few start-ups now develop for-profit business models. For example, Zero Percent, a mobile application created in 2014 that connects American restaurants with local charities, charges businesses a fee proportional to their tax deductions.

One of the main challenges of food recovery is that food donations are not necessarily suited to the needs of hunger-relief organizations. Many food banks have too much of certain foods such as bread and pastries, particularly in the USA. In one instance, a food bank gave large share of donations to bears and not humans (interview, Arizona, December 2014). In France, the founder of a hunger-relief organization denounced that hunger-relief policies would not question the quality of the food and dignity of access: 'they count in kilograms per poor!' (interview, Paris, February 2014). In that system, some companies are encouraged to donate 'heavy' foods that are not necessarily nutritious, like soda. Activists use the label 'charity washing' (based on the notion of 'green washing') to denounce dumping unneeded products.

In the meantime, food recovery becomes more and more institutionalized. Donations are associated with formalized procedures to protect from liability, ensure compliance with safety rules, and calculate tax incentives. But food recovery does not question overproduction in the first place. More than 3,500 Kcal per capita are produced daily in the USA according to the USDA, while a normal human should consume only around 2,000-2,500 Kcal. Estimates show that less than 10% of surplus food is currently redistributed in the USA (FWRA, 2014), and few ask what would happen if all the food was redistributed.

'Weak' prevention: marginal economic optimization

Businesses and environmental organizations promote optimization to reduce the 'cost of doing business' and 'externalities' of waste. 'Weak' prevention relies on the belief that improved processes and technologies – without a fundamental change in business models – are enough to prevent waste. One large-scale produce farmer argues that 'in about 20-30 years if we are smart, use new technology (including GMOs), we won't hear the word 'waste' anymore' (interview, California, 2015). This vision of optimization as a solution to food waste is supported by many social entrepreneurs in both countries and also relies on raising 'awareness' among consumers. Campaigns focus on drivers of waste at an individual level, such as poor planning, storing and cooking practices or the use of expiration dates, without questioning broader structures in which consumers are led to waste food. There is generally no assessment of what would happen if consumers stopped buying 30-40% of the food that is currently wasted, and no fear that consumer waste would decrease enough to reduce sales. The director of sustainability at a large French retail chain confesses 'when sales decrease because people make more jam at home, pigs will fly!' (interview, France, January 2014). Competitive industries adopt voluntary agreements – such as the industry-led Food Waste Reduction Alliance in the USA – in order to maintain certain market characteristics or preempt government regulations. 'Weak' prevention-as-optimization does not discuss many aspects of food commodity chains that may lead to surplus and waste, such as availability (full shelves at any time), convenience (easy shopping and disposing), or safety (expiration dates).

Toward reducing waste through structural changes

'Strong' prevention, on the other hand, calls for producing and consuming less or at least differently. Advocates of strong prevention see waste as 'nonsense' and challenge 'over-industrialization' and 'homogenization' of food production. They call for more seasonality, proximity to the land, valuing farm work and sharing food through closer social links. If focusing on individual consumers is criticized as a 'weak' approach, 'strong' prevention would still imply deep changes in consumption patterns – less choice or availability, vegetarianism, more time spent on food, less convenience, potentially more risk. Recent initiatives around 'ugly fruit and vegetables' – produce that does not reach markets because of their shape or colour – offer an interesting case of potentially 'strong' prevention. After a French retail store led a very successful marketing campaign to sell carrots that are 'ugly' but 'good in the inside' at a discounted price, many companies in Europe and the USA started exploring similar initiatives. Changing aesthetic criteria may 'strongly' modify long-standing social and cultural expectations on what good produce is and push for less standardization. Farmers could be paid for a larger part of their production and at the same time make fresh produce more affordable to low-income communities. Still, long-term social, environmental and economic impacts are unclear. Farmers show concerns about an overall decrease in prices while retailers benefit from higher margins. In a 'weak' scenario, changing standards may not question who sets standards, nor the power of supermarkets to reject food.

Despite being the only sustainable 'win-win-win', 'strong' prevention is the least implemented solution and mostly appears in marginal social movements or as a non-official discourse. Interviewees often mention individual practices that do prevent waste, but doubt they will be extended. One member of the French retail organization comments that expiration dates are not relevant – 'you can eat a yogurt a month after!' – but makes it clear that he would not say this on behalf of his organization (interview, Paris, January 2014). On the ground, store managers and producers often violate corporate policies and allow informal redistribution of food – letting people access their dumpsters or just giving it to employees – for moral or religious reasons. In the end, 'strong prevention' is still incompatible with current economic paradigms because its social and environmental values do not easily translate into economic ones, unlike recovery and 'weak prevention.'

5. As a conclusion: impact on the food system?

We need to better analyse social, environmental and economic impacts of competing solutions to what is defined as 'food waste'. Sustainable ones require systemic changes in the overall governance of the food system. Yet, dominant approaches to 'food waste' often focus more narrowly on the management of existing surplus in large-scale (over)production systems. Many small-scale systems are being disrupted in favour of supposedly more efficient ones, such as industrial instead of backyard composting or food bank networks instead of local organizations. Increasing the length of commodity chains requires more administrative procedures and more safety measures, with less direct inter-personal relationships, that may generate more surplus and waste. Despite a growing number of studies on 'food waste', very few of them dig into the most appropriate scale at which solutions should be implemented.

Nonetheless, the food waste movement holds potential for incremental change, with individual practices altering business practices and regulation in the long run. I observe the most significant potential for change through a network of actors – notably young professionals with business degrees or specialized trainings in environmental studies – who share ideas across a wide range of companies, public agencies and NGOs. A woman working at the sustainability direction of one of the main French grocery stores says she chose to change things 'from the inside' through 'internal lobbying', adding 'it may be less naïve [than more radical engagements] ... or *more* naïve!' (interview, Paris, May 2015). This 'reformist' or 'neoliberal' approach aims at changing the food system from within (Holt Giménez and Shattuck, 2011). Once individual and business practices evolve, regulations can set higher standards and transform markets by aligning financial goals and productive logics to sustainability. For example, France is now considering a law making supermarket donations mandatory. Further research on the implementation of various solutions – especially prevention, the most needed one – would be useful. By seeing existing opportunities, advocates and policy makers will hopefully seize them and, beyond marginal adaptations, will work on 'strong' changes toward sustainable production and consumption.

References

Berkenkamp, J., Mourad, M. (2015). Food waste inspiration: The French make a bold proposal. Available at: http://tinyurl.com/pf6v3wy.

Bloom, J. (2010). American wasteland: How America throws away nearly half of its food. Da Capo Lifelong Books, Cambridge, MA, USA.

EU Commission (2008). Directive 2008/98/EC on waste (Waste Framework Directive). Official Journal L 312, 22.11.2008, 3-30.

Evans, D. (2014). Food waste: home consumption, material culture and everyday life. Bloomsbury Academic, UK.

Food and Agricultural Organization. (2013). Food wastage footprint. United Nations.

Food Waste Reduction Alliance (FWRA). (2014). Analysis of US food waste among food manufacturers, retailers, and restaurants. Food Waste Reduction Alliance, USA.

Gustavsson, J., Cederberg, C., Sonesson, U., van Otterdijk, R., Meybeck, A. (2011). Global food losses and food waste. UN Food and Agricultural Organization, Rome, Italy.

High Level Panel of Experts (HLPE). (2014). Food losses and waste in the context of sustainable food systems (No. 8). Rome: High Level Panel of Experts on Food Security and Nutrition of the Committee on World Food Security. Available at: http://www.fao.org/3/a-i3901e.pdf.

Holt Giménez, E., Shattuck, A. (2011). Food crises, food regimes and food movements. Journal of Peasant Studies, 38(1), 109-144.

Lorek, S., Fuchs, D. (2013). Strong sustainable consumption governance – precondition for a degrowth path? Journal of Cleaner Production, 38, 36-43.

Papargyropoulou, E., Lozano, R., Steinberger, J.K., Wright, N., Bin Ujang, Z. (2014). The food waste hierarchy as a framework for the management of food surplus and food waste. Journal of Cleaner Production, 76, 106-115.

Phillips, C., Hoenigman, R., Higbee, B., Reed, T. (2011). Understanding the sustainability of retail food recovery. PLoS ONE, 8(10), e75530.

Reynolds, C. J., Piantadosi, J., Boland, J. (2015). Rescuing food from the organics waste stream to feed the food insecure. Sustainability, 7(4), 4707-4726.

Stuart, T. (2009). Waste: uncovering the global food scandal. W. W. Norton & Company, New York, NY.

US Environmental Protection Agency (EPA) (2011). Food Recovery Hierarchy. Available at: http://www.epa.gov/foodrecovery/.

Worldwide Responsible Accredited Production (WRAP) (2013). Household food and drink waste in the UK 2012. Waste & Resources Action Program. Available at: http://tinyurl.com/na8cbo6.

World Resources Institute (WRI) (2014). Food loss & waste (FLW) Protocol. World Research Institute. Available at: http://tinyurl.com/qyd7wzg.

8. Surplus food redistribution for social purposes: analysis of critical success factors

S. Sert*, P. Garrone, M. Melacini and A. Perego
Department of Management, Economics and Industrial Engineering, Via Lambruschini 4/b, 20156 Milan, Italy; sedef.sert@polimi.it

Abstract

The issue of surplus food management has been raised in several articles but has not been analysed deeply, so far, in a satisfactory way. In particular, redistribution of surplus food is highly recommended but the critical success factors to save food in upstream stages of food supply chain are not studied. The purpose of this paper is to explore main critical factors for managing surplus food, in particular for redistributing food for human consumption. In order to achieve it, multiple case study methodology is used. Questionnaires for semi structured interviews are prepared based on an extensive literature review and adapted after expert opinions. The interviews are conducted mostly with the supply chain and logistics managers of food manufacturing companies in food sector operating in northern Italy. The interviews show that reusing options are various such as selling in secondary markets, donation to non-profit organizations, marketing and sponsoring actions. They also highlight the key role played by the adoption of measurement and control systems and the careful design of partnerships with non-profit organizations.

Keywords: surplus food, reuse, redistribution, donation, food insecurity

1. Introduction

According to a report of United Nations' Food and Agriculture Organization (FAO), one third of the food produced for human consumption is lost or wasted globally, i.e. 1.3 billion tons each year (Gustavsson *et al.*, 2011). From agriculture and fishing stage to final household consumption food is wasted throughout the whole supply chain, together with water, cropland and fertilizers used to produce losses and waste (Kummu, *et al.*, 2012; Lunqvist *et al.*, 2008). On the other hand, almost 800 million people were estimated to be undernourished globally (FAO, 2015). It is an important issue even in developed countries; in Europe, nearly 43.6 million people were estimated to be food insecure (Gentilini, 2013) and at the same time, European countries are reported to generate 179 kg per capita of food waste per year (O'Connor, 2013).

The main objective of this paper is to understand critical success factors that can improve the process of surplus food redistribution for human consumption in particular by donation to non-profit organizations. In Section 2, theoretical background is explained. In Section 3, the research methodology is summarized. In Section 4, results emerged from interviews and two confirmatory cases are presented. In Section 5, main conclusions are discussed and a potential route for future study is given.

2. Theoretical background

Surplus food, differently from food waste, can be defined as 'the edible food that is produced, manufactured, retailed or served but for various reasons is not sold to or consumed by the intended customer' (Garrone *et al.*, 2014a). The main causes of surplus food generation (Kaipa *et al.*, 2013; Kantor *et al.*, 1997, Mena *et al.*, 2011) and the possible management options (Garrone *et al.*, 2014a; Griffin *et al.*, 2009; Papargyropoulou *et al.*, 2014) are analysed in the literature. The potential of surplus food donation and its high priority has been emphasized (Gentilini, 2013; Schneider, 2013; Tarasuk and Eakin, 2005; Thang, 2009).

The European Union Directive 2008/98/EC in waste management, applied as Food Waste Hierarchy (Papargyropoulou *et al.*, 2014), sets the priorities by suggesting companies to prevent surplus food generation at source. Even if the firms consider all relevant aspects in order to prevent, it is inevitable in some cases. Exceeding internal sell-by-date (Garrone *et al.*, 2014a), forecasting errors (Darlington and Rahimifard, 2009), processing errors, poor handling or packaging failures such as wrong labels and damaged boxes (Kantor *et al.*, 1997) are the most common reasons of surplus food generation in manufacturing stage together with failed promotions (Kaipa *et al.*, 2013) and failed introductions of new products (Buzby and Hyman, 2012).

Once surplus food is generated, it has to be managed efficiently and effectively within the responsibility of the companies. Reusing surplus food, converting the safe but 'un-saleable' food into consumable food (Kantor *et al.*, 1997), is the second strategy suggested by the Food Waste Hierarchy (Papargyropoulou *et al.*, 2014). This can be done through selling it in secondary markets, donating it to non-profit organizations or using it for marketing and sponsoring actions. With the donation to non-profit organization, surplus food is redistributed to disfavoured population in different modes such as providing food and grocery products for home preparation and consumption or serving meals for on-site consumption. Besides some special cases, the company itself cannot directly reach the people who need to receive food aid. Therefore, the presence of non-profit organization is required to connect both sides of the chain, 'supplier' and 'receiver' (Tarasuk and Eakin, 2005).

According to the Food Waste Hierarchy, if the reuse option is not possible that means the food is not suitable anymore for human consumption, and it should be recycled (EU, 2008) as animal feeding and composting (Papargyropoulou *et al.*, 2014). Finally, before disposal of food waste, it is suggested to recover energy and other valuable materials to reduce the negative environmental impact (EU, 2008).

3. Research methodology

In order to understand management methods applied by the food supply chain companies, explorative case study methodologies are applied (Yin, 2003). The interview questionnaire was developed in three stages. A first draft was prepared based on literature review. This draft was revised by a panel of nine academics and four practitioners. The first draft was adapted based on their feedback and was piloted in two interviews, after which it was decided to include only one additional question.

Four explorative cases are performed with the companies selected after a discussion with supply responsible and the director of Food Bank. The companies are selected based on Food Bank's observations on surplus food management as best cases and worst cases. Two of them are regular donors for a long period of time. Instead, one of them donates from time to time and the last one is a company that in the past was donating its surplus food however the management has recently got a decision of not donating anymore.

Since surplus food management is considered to be sensitive by many organizations, it was decided to offer a confidentiality agreement to all companies interviewed. Interviewees were selected as middle or senior managers with the responsibility of logistics management and/or surplus food management across their organizations.

Each interview lasted around 2 hours and was conducted by three researchers. For confidentiality reasons it was decided not to record the interviews and this meant the role of interviewees was critical. One researcher was responsible for asking the questions while the other two were taking notes. Visits and company reports were used as secondary methods of data collection.

The interviews were analysed independently, data for each of them were coded and put into a standard template and then cross case comparison was performed in order to extract the key information. After that, 2 additional interviews were performed in order to test the conceptual framework.

4. Results and discussion

Four key areas are identified differentiating the companies donating regularly from others. The following section is dedicated to introduce those four concepts.

Surplus food measurement

It has been identified that there are some differences between companies starting from the very beginning: the measurement of the surplus food. While some companies measure surplus food as an important parameter in their management control system, other companies are not aware of the issue.

It has been seen that the company may measure surplus food, only if its generation is extremely high due to some extraordinary events like failed promotions or failed new product introductions (ad hoc measurements). On the other hand, measuring and monitoring surplus food generation regularly in company management control system makes companies aware of surplus food generation on time and facilitate the arrangement of reuse as sales to secondary markets, internal distribution, using it for marketing activities such as sponsoring events and donation to non-profit organizations.

The surplus food generated may also be measured and monitored regularly but different corporate functions may monitor and manage different types of surplus food (periodic – fragmented measurements). For instance marketing function manages commercial returns, while management control function monitors the products sent to the landfill.

In a more structured system, the surplus food measurement indicators should be unified and should be communicated regularly with all functions (periodic – structured measurements). This may increase the probability of allocating surplus food for human consumption, preventing it to become waste.

Degree of process formalization

It has been found that different management methods exist depending on the reason of surplus food generation. For instance, if the reason of surplus food generation is production errors the company may utilize the option of remanufacturing. Similarly, repackaging may be used in case of labelling and packaging errors.

When the reason of surplus food generation is the risk of reaching to internal sell by date, the company may try to sell it to its customers in the primary market at a price below the standard. When the surplus food is not able to be sold anymore with primary channels, the company may put in practice the secondary channels which are alternatives like distributors specialized in surplus products. In some cases, the company may utilize the surplus food sponsoring specific organizations or arranging product tasting events or they may even choose to distribute the products between its employees internally in the organization. The surplus food may be given to a non-profit organization, such as a food bank, food pantry, soup kitchen, etc. that redistributes it to disfavoured population.

Depending on the product type, the surplus food may be sold to companies that produce animal feed, fertilizers, etc. In some cases, the company can obtain small revenue from this channel. The final option is to confer the surplus food (at this stage the surplus food is already transformed to food waste) to third parties for disposal, with or without energy recovery. Typically, the company works with a specialized waste management agency and does not control waste management directly. When the surplus food becomes waste the company contacts a waste management agency to manage the food waste by paying a fee.

If the company doesn't have a procedure based on different causes of surplus food generation, it can be said that the management process is not structured (not structured). On the other hand, the company may have a structured approach for some kind of reasons of surplus food generation such as exceeding internal sell by date (structured for some cases) or may have a structured approach for all kind of reasons of surplus food generation (structured for all cases).

Coordination between functions

The process of managing surplus food involves various business functions (planning, logistics, commercial, CSR, operating staff, etc.). The companies which have a structured surplus food management have stated that different company functions interact with each other regularly in order to decide how to manage surplus food. The coordination between functions can be seen as an indicator of the importance given to surplus food management in a company organization.

In order to manage surplus food in a company, different company functions may not interact with each other for the decision making and one function can manage the whole process without the consultation of others (no coordination), the functions may be in contact through emails, phone calls and informal meetings when it is seemed necessary for the decision making (informal coordination) or the surplus food may be managed through regular meetings involving different functions (formal coordination).

Relationship with non-profit organizations

It has been identified that the relationship between a company and non-profit organizations is an important parameter that facilitates the decision making towards donation. The company may manage the process proactively or may stay reactive to the demand of a non-profit organization. In the same way, the donation process can be held with a regular or variable frequency.

When there is no structured agreement between the company and the non-profit organization in terms of type of products and donated quantity, the non-profit organization contacts the company to check if there is something available for donation and organizes the transport of the surplus food (reactive management – sporadic donation). On the other hand the company may act proactively to manage its surplus food. When the frequency of donation is not established, the company may contact the non-profit organization having large quantities of surplus food to be donated (proactive management – sporadic donation), or the company may define the products to be donated and the relative quantity, together with a detailed schedule concerning the collection (proactive management – regular donation).

Preliminary confirmation

Two companies that produce yogurt as their main stream product are interviewed. In company A, the main reason of surplus food generation is the noncompliance with the market standards. After setting up a new flavour, the first batch of the products is mixed with the previous flavour, the product produced is not suitable for market but it is still suitable for human consumption. The surplus food generated is measured constantly and a fixed procedure exists for distribution employees and donation to non-profit organizations. The relationship is proactive and constant and all the activities are well coordinated.

In company B, the most important reason of surplus food generation is caused by reaching the internal sell by date, due to errors in forecasting demand, overestimation of the performance of certain promotions and picking errors. The company uses alternatives such us promotions, discounts, and selling in secondary markets. When those channels are not available, marketing staff is involved to understand if surplus food can be used as samples to taste in trade shows. Finally, the products can be given to employees. The company donates some surplus food only when the non-profit organization calls and asks, hence, there is not any structured relationship between non-profit organization and the company.

The percentage of surplus food used for human consumption is 100% in company A and 8% in company B. The huge difference in management structure in terms of measurement, coordination and process formalization as well as the relationship with non-profit organization creates the difference in the amount of surplus food saved for human consumption.

5. Conclusions

This paper presents surplus food management in food manufacturing companies and identifies critical success factors that can help increase the amount of surplus food distributed for social purposes and reduce the food waste generated.

Four critical success factors have been identified by analysing explorative cases. Once the surplus food is measured with appropriate performance indicators, the management considers managing it efficiently and effectively by structured decision making approach and good coordination between different functions. The relationship with non-profit organizations is also important to facilitate the donation process and to help the manufacturer to decide towards a donation option.

This paper presents factors to improve surplus food management in food manufacturing companies with a particular focus on redistribution for human consumption aiming to help firms improving the current practices. Understanding the position of donators will help non-profit organizations to collaborate more easily with the food donators and to collect a greater amount of food to distribute to the people in need. Policy makers will be able to use this information in order to innovate current policies related to surplus food redistribution to reduce food insecurity in their community.

Further research should be directed to understand generalizability of the results obtained in this study by enhancing the cross case analysis. Understanding the perspectives of different stages in food supply chain such as retailers and food service operators would also provide a bigger picture to enlighten different perspectives.

Acknowledgements

The authors would like to express their gratitude to the interviewees and colleagues for their contribution to this research, as well as to Foodsaving project, funded by Cariplo Foundation and Lombardy Region.

References

Buzby, J., Hyman, J. (2012). Total and per capita value of food loss in the United States. Food Policy 44, 561-570.

Darlington, R., Staikos, T. and Rahimifard, S. (2009). Analytical method for waste minimization in the convenience food industry. Waste Management 29, 1274-1281.

European Union (2008). Directive 2008/98/EC of the European Parliament and of the Council of 19 November 2008 on waste and repealing certain Directives. Official Journal L 312, 22.11.2008, 3-30.

FAO (2015). Suite of Food Security Indicators. http://faostat3.fao.org/browse/D/FS/E.

Garrone, P., Melacini, M., Perego, A. (2014). Opening the black box of food waste reduction. Food Policy 46: 129-139.

Garrone, P., Melacini, M.,Perego, A. (2014b). Surplus food recovery and donation: the upstream process. British Food Journal 116: 1460-1477.

Gentilini, U. (2013). Banking on food: the state of food banks in high-income countries. Brighton, UK: Institute of Development Studies.

Griffin, M., Sobal, J., Lyson, T. (2009). An analysis of a community food waste stream. Agriculture and Human Values, pp. 67-81.

Gustavsson, J., Cederberg, C. and Sonesson, U. (2011). Global food losses and food waste. Rome, Italy: Food and Agriculture Organization of United Nations.

Kaipa, R., Dukovska-Popovska, I., Loikkanen, L. (2013). Creating sustainable fresh food supply chains through waste reduction. International Journal of Physical Distribution & Logistics Management 43 No. 3: 262-276.

Kantor, L., Lipton, K., Manchester, A., Oliviera, V. (1997). Estimating and addressing america's food losses. Food Review 1997 (Jan-April): 2-12.

Kummu, M., De Moel, H., Porkka, M., Siebert, S., Varis, O., Ward, P.J. (2012). Lost food, wasted resources: global food supply chain losses and their impacts on freshwater, cropland, and fertiliser use. Science of Total Environment 438: 477-489.

Lunqvist, J., Fraiture, C., Molden, D. (2008). Saving water: from field to fork – curbing losses and wastage in the food chain. Stockholm, Sweden: Stockholm International Water Institute.

Mena, C., Adenso-Diaz, B., Yurt, O. (2011). The causes of food waste in the supplier – retailer interface: evidences from the UK and Spain. Resources, Conservation and Recycling, 55: 648-658.

O'Connor, C. (2013). Quantification of food waste in the EU. OECD Food Chain Network http://www.oecd.org/site/agrfcn/Session%201_ClementineOConnor.pdf (accessed 25 May 2015).

Papargyropoulou, E., Lozano, R., Steinberger, J.K., Wright, N., 2014. The food waste hierarchy as a framework for the management of food surplus and food waste. Journal of Cleaner Production 76, 106-115.

Schneider, F. (2013). The evolution of food donation with respect to waste prevention. Waste Management, 33: 755-763.

Tarasuk, V., Eakin, J. (2005). Food assistance through 'surplus' food: insights from an ethnographic study of Food Bank work. Agriculture and Human Values 22, 177-186.

Thang, H. (2009). Food reclamation as an approach to hunger and waste: a conceptual analysis of the charitable food sector in Toronto. Journal of Hunger and Poverty 1, 1-10.

Yin, R. (2003). Case study research design and methods. London, UK: Sage Publications.

Part 3. Improving the efficiency of food chains

9. Vertical integration contracts in agriculture: fair trade and efficiency of the food chain

A. Albanese

Doctoral School on the Agro-food System, Università Cattolica del Sacro Cuore, via Emilia Parmense 84, Piacenza, Italy; albanese.antonio@unicatt.it

Abstract

A complex society such as ours, with its typical consumption, implies on the one hand an agro-food system characterized by a strong specialization and labour division and on the other hand an ever-closer coordination among production, processing, and distribution. In this respect, vertical integration contracts play a key role in the regulation of the relationships in the supply chain, requiring farmers to comply with specific criteria and techniques of cultivation and breeding and to sell all the products to the other party; this, in turn, is obliged to purchase them, after having verified the compliance with all the agreed standards. This approach not only benefits individual contractors, but also increases the efficiency of the whole food system: both in terms of improving the quality of production, which can conform more to the demand of the end markets, and in terms of harmonizing the quantity between supply and demand, thus avoiding the risk of surplus items that remain unsold. However, this integrated system of production exposes farmers to the hazard of abuse because of their position of economic dependence when, in order to adjust their production to the other party's demand, they make specific investments difficult to switch. Indeed, when the contract expires, the buyer can take advantage of the renegotiation by imposing unfair contractual terms and conditions on the farmer. As a consequence, the need exists to develop, even at the interpretation level, legal instruments which, in accordance with the free market, could prevent and correct the imbalances produced by abuses of bargaining power. Fair trade, in fact, cannot deny the role of private autonomy by imposing on the parties a contractual balance determined in advance, but must be achieved by ensuring the weaker party an effective exercise of his contractual freedom. A system of contractual relationships, that, through the meeting of demand and supply, fairly allocates the profits of the food chain, is indeed a necessary prerequisite not only to increase the quantity and quality of goods, but also to ensure the lasting sustainability of agriculture production and its capability to fulfil people's needs.

Keywords: supply chains, sustainable agriculture, agricultural contracts

A complex society such as ours, with its typical consumption's habits, implies on the one hand, an agrifood system characterized by a strong specialization and labour division and, on the other hand, an ever-closer coordination of production, processing, and distribution. In this respect, vertical integration contracts play a key role in regulating the relationships in the supply chain, requiring farmers to comply with specific criteria and techniques of cultivation and breeding and to sell all their products to the other party. The other party, in turn, is obliged to purchase the products after having verified the compliance with all the agreed standards. This approach not only benefits individual parties, but also increases the efficiency of the whole food system: both in terms of improving the quality of production, which can comply better with the needs of the end markets, and in terms of

Leire Escajedo San-Epifanio and Mertxe De Renobales Scheifler (eds.) *Envisioning a future without food waste and food poverty*
DOI 10.3920/978-90-8686-820-9_9, © Wageningen Academic Publishers 2015

harmonizing the quantity between supply and demand, thus avoiding the risk of *surplus* items that remain unsold.

However, this integrated system of production exposes farmers to the hazard of abuse because of their position of economic dependence when, in order to adjust their production to the other party's request, they make specific investments difficult to switch. Indeed, when the contract expires, the buyer can take advantage of the renegotiation by imposing unfair contractual terms and conditions on the farmer. As a consequence, it's necessary to develop legal instruments which, in accordance with the free market, could prevent and correct the imbalances produced by abuses of bargaining power.

In achieving these objectives, mandatory rules play a key role. However the 'myth' of a liberal market – able to regulate itself according to a spontaneous order and to naturally increase the societal overall well-being – is more and more illusory. The very same economic sciences show that, without adequate controls and corrections, the distortions of competition and market imperfections may cause this idealized market's failure, as documented by experience in the agrifood sector. Even from the single food production perspective, there is the risk that free competition, in the absence of legal rules, may comply with economic criteria but violate fundamental human rights. Therefore, protection of higher interests cannot be left to the spontaneous functioning of the market, but calls for a choice of legal policies resulting in a mandatory discipline of contractual relationships, coherent not only with the economic, but also ethical function of the market. Indeed, recognizing the artificiality of the market as an institution governed by legal rules does not necessarily mean to agree with a merely positivistic model of the law, namely with the arbitrary will of the law imposed on the market.

In constitutional systems that identify the foundation of law in ethical values and inalienable rights incorporated in the Constitution, it would be an unacceptable reading of the rules governing economic relations as exclusively aimed to promote market efficiency. However, regulation is legitimate when aimed at ensuring that the private economic initiative in agrifood does not take place to the detriment of values, such as the individual right to healthy and sufficient food, preservation of the land, dignity of human labour in agriculture and, last but not least, fairness in commercial relationships.

To this end, a central role must be recognized to farmers, who not only supply products for the secondary sector, but are themselves recipients of services provided by companies similarly working in the field of primary production. In fact, the supply chain is characterized by the link between several markets, located upstream and downstream agricultural production. The market is not exclusively a physical place, but is essentially a set of contractual relations, which give content to homogeneous and serial exchanges.

Because of this close link the legal regulation of contracts inevitably affects the proper and efficient functioning of the market. Similarly, the rules concerning contracts in the food industry should aim not only at increasing the quantity and quality of products, by offering consumers better goods at lower prices, but also at providing a more equitable distribution of the profits of the entire chain of those active within it.

A system of contractual relationships that fairly allocates the profits of the food chain is indeed a necessary prerequisite not only to increase the quantity and quality of goods, but also to ensure the lasting sustainability of agriculture production and its capability to fulfil people's needs. Moreover, public intervention, although necessary, cannot adopt authoritarian forms of government control and economic planning, but should respect the market as a social institution and ensure that its functioning is not only efficient, but also correct. Therefore, respect for fundamental values, placed at the top of the legal system, has to be combined with an economic system that, through the meeting of supply and demand, favours social progress and the full development of the human personality.

Determining the content of negotiation is not the goal of the law, but that of free negotiation among private parties. Fair trade, in fact, cannot deny the role of private autonomy by imposing a predetermined contractual balance on the parties, but must be achieved by ensuring effective contractual freedom to the each party. This logic does not require a complete and organic regulation of individual contracts in the food chain: it is sufficient to provide regulatory measures, applicable to an indefinite number of different kinds of agreements, in order to restore balance in specific situations of unequal bargaining power. Even with the transition to a technologically advanced and industrialized agriculture, the conditions of structural weakness of the agricultural producer (farmer or breeder) still remain.

In analogy with provisions in labour law, a means to overcome these inconveniences is collective bargaining between producers' organizations and organizations representing companies involved in processing, distribution, and marketing of agricultural products. In particular, we have to consider the 'Framework Agreements' (FA), by which representative bodies of various professional groups determine the content of farming contracts with which their members will agree. The content of these contracts is not predefined by the law, but is determined by collective autonomy through the preparation of a standard contract, i.e. a contract model designed to regulate future contractual relationships between individual parties. Even for contracts in the food supply chain, similarly to employment contracts, collective agreements are functional to balance unequal bargaining power.

Access to FA is free, but the law can facilitate it by allocating public funding for innovations, favouring companies that have signed individual agreements for cultivation, breeding and supply in accordance with FA. The law can also enhance and extend the effectiveness of such agreements in relation to other subjects, stating that the terms of these contracts cannot be derogated by individual contracts damaging farmers' interests and entitling them to ask for it, even when they are not members of organization that signed the agreement. Instead, the parties other than the farmers are not bound by the FA, if they did not join the organizations that have signed it.

Even though applicable, FA cannot regulate all the essential elements of future individual relationships. In fact, in compliance with antitrust laws, they can only predict the procedures and criteria for price differentiation in relation to the production process followed and the quality characteristics of products; they cannot intervene on the supply's prices, that can only be established by the parties to individual contracts. Nonetheless, these restrictions to collective bargaining, if compatible with the principles of the free market, may induce the farmer to accept unfair and not sufficiently profitable remuneration and can affect the functionality of the market as a means for efficient resource allocation.

In order to overcome this and other situations of unequal bargaining power, which may occur in relationships within the supply chain, regulatory intervention is also appropriate at the level of individual bargaining. This intervention may imply prohibiting improper conduct of negotiations in order to ensure an effective freedom of contract to the weaker party, identified in the farmer selling his products to wholesalers, supermarket chains, or to the processing industry. It can also be appropriate to impose some formal requirements and a minimum mandatory content for the contract, in order to establish more transparent contractual relationship and prevent buyer's requests for undue benefits.

In this logic, binding the parties to determine the price of products at the time of signing the contract, fight the common practice of setting the price of the sale of future agricultural products, by reference to price lists and mercurial at harvest. In this way the negative effects of price volatility in the market of agricultural products are attenuated and the farmer can plan investments according to more rational and efficient criteria.

For the same reason the contractual freedom of the parties may be limited by requiring them to determine the duration of the supply and by excluding the eligibility of relations of indefinite duration. The farmer could, then, plan his economic activity, safe from sudden cancellations often instrumental to the renegotiation of the conditions that may be unfavourable for him.

Another important protection is aimed at countering the delays in the payment for the supplies of food products, which should take place within a maximum peremptorily required by law, providing for default interest as an effective deterrent against late payment.

All these regulatory measures share a common concern toward minimizing unequal bargaining power, as a prerequisite for a legitimately balanced contract according to a criterion of justice, operating not in opposition to parties' autonomy, but simply ensuring effective freedom of contract for everyone. This discipline complies with a specific trend in modern legal systems. In fact, following the demands of the post-liberal society, the principle of freedom of contract is here replaced by judicial review of the content of the agreement, based on a plurality of elements such as unequal bargaining power and unfair contract terms.

In this approach both formal and substantive justice are involved. In fact, in line with the idea of procedural justice, the law does not impose authoritatively a contractual settlement, considered as an *a priori* right; however, it corrects the imbalances generated by abuse of bargaining power that would limit the freedom of the economically weaker party, forcing it to accept unfair conditions. Indeed, the direct intervention of the State in socio-economic relations, also in terms of control of trade negotiations, does not adequately cope with the complexity of a technological society; this also depends on the insufficient cognitive competence of the legislative bodies and on the difficulty of founding legal choices on the majority principle as a criterion of truth.

The overcoming of natural law theories, as they have been developed before Kantian critical philosophy, and the failure of the historical experiences of legal positivism, raise the need for new criteria for legitimizing the law. Legitimization cannot be represented any more either as conformity to abstract, self-evident values or as the arbitrary will of a political majority. The issue of the ethical foundation of positive law, instead, can be framed in terms of social consensus on the founding values of the legal system and its being compliant with them.

According to this neo-institutional view of the legal system, the government should leave more space to the market, as an instrument of social self-regulation, and should entrust the task of providing concrete answers to people's needs to private autonomy, as defined by the law. Differently from the neo-liberal model, public intervention is not limited to establishing organizational and procedural rules, with reference to a principle of formal equality of contracting parties. Instead, it aims at removing social and economic barriers that, by limiting the freedom and equality of the parties, prevent the full development of the human person even in the context of contractual relationships.

References

Albanese, A. (2008). Contratto mercato responsabilità. Giuffré Editore Milano, 274 pp.

Balekjian, W.H. (1986). Legal Reasoning and Economic Reasoning. Rechtstheorie. Vernuft und Erfahrung im Rechtsdenken der Gegenwart. Duncker & Humblot, pp. 375-377.

Birthal, P., Gulati, A., Joshi, P.K. (2005). Vertical Coordination in High-Value Food Commodities: Implications for Smallholders. MTID Discussion Paper No. 85. International Food Policy Research Institute, Washington DC, USA.

Brennan, G., Buchanan, J.M. (2008). The Reason of Rules: Constitutional Political Economy. Cambridge University Press, 168 pp.

Castronovo, C. (2005). Autonomia private e costituzione europea. Europa e diritto privato. Giuffré Editore Milano, pp. 29-50.

Collins, H. (2003). The Law of Contract[2]. Cambridge University Press. 441 pp.

Galizzi, G., Venturini, L. (1999). Vertical relationship and coordination in the food system. Physica-Verlag Heidelberg, pp. 61-92.

Joskow, P.L. (1998), Asset specifity and vertical integration. The New Palgrave Dictionary of Law and Economics I. Macmillan Oxford, pp. 107-108.

Klein, B., Crawford, R.G., Alchian, A.A. (1978). Vertical integration, appropriable rents and the competitive contracting process. The Journal of Law and Economics, XXI, 297-326.

Kötz, H., Flessner, A. (1997). European Contract law. Clarendon Press Oxford, 286 pp.

Mengoni, L. (2011). La questione del diritto giusto nella società post-liberale. In: Castronovo C., Albanese, A., Nicolussi, A. (eds.) Metodo e Teoria giuridica, Giuffré Editore, Milano, pp. 55-71.

Nogler, L. (1997), Saggio sull'efficacia regolativa del contratto collettivo. Cedam Padova, 251 pp.

Pietrobelli, C., Saliola, F. (2008). Power relationships along the value chain: multinational firms, global buyers and performance of local suppliers. Cambridge Journal of Economics 32, 947-962.

Polinsky, A.M. (2011), An introduction to law and economics. Wolters Kluwer Law & Business, 195 pp.

Raiser, L. (1977). Die Aufgabe des Privatrechts. Athenäum-Verlag, 263 pp.

Schwartz, A., Scott, R.E. (2003). Contract theory and limits of contract law, 113 Yale Law Journal, pp. 541-620.

Trebilcock, M.J. (1993). The limits of freedom of contract. Harvard University Press, 310 pp.

Unger, R.M. (1976). Law in modern society. toward a criticism of social theory. The Free Press New York, 315 pp.

Williamson, O.E. (1985). The economic institution of capitalism: firms, markets, relational contracting. The Free Press New York, 450 pp.

10. Improving sustainability of fruit and vegetable processing industry by sub-product transformation

T. Dietrich[1*], J. Wildner[2], F. D'urso[3], R. Virto[4], C. Velazquez[5], C. Sacramento Santos Pais[6], B. Sommer Ferreira[7], A. Carolas[7], L. Prado-Barragan[8], M.P. De Castro[9] and S. Verstichel[10]

[1]TECNALIA, Leonardo Da Vinci, 11, 01510 Miñano, Spain; [2]ttz, An der Karlstadt 10, 27568 Bremerhaven, Germany; [3]BIOESPLORA, Via Della Repubblica 51, 72018 San Michele Salentino, Italy; [4]CNTA, NA-134, km 50, 31570 San Adrian, Spain; [5]UCR, Rodrigo Facio, 2060 Pedro, Costa Rica; [6]UMinho, Largo do Paco, 4704553 Braga, Portugal; [7]BIOTREND, Biocant Park – Nucleo 4, Lote 2, 3060-197 Cantanhede, Portugal; [8]UAM-I, Prolongacion Canal de Miramontes 3855, 14387 Mexico D.F., Mexico; [9]PROTEOS, Calle Almansa 14, 02006 Albacete, Spain; [10]OWS, Dok Noord 5, 9000 Gent, Belgium; thomas.dietrich@tecnalia.com

Abstract

Few years ago, the prices of bulk agricultural raw materials such as wheat, rice or corn have been low because of increasing agricultural yields. This tendency has drastically changed, with competition between biomass for food vs for chemicals or biofuels, becoming a societal issue. According to OECD/FAO the demand for agricultural products is expected to remain firm and crop prices to remain above pre-2008 period. On the other hand up to 25% of fruit and vegetable production is lost or wasted during agriculture and post-harvest activities in Europe. Therefore new approaches are needed to reduce the environmental impact of waste and to produce feedstock's for industry. The implementation of the innovative cascading concept of the European project TRANSBIO for valorisation of sub-products from fruit and vegetable processing industry will improve sustainability of fruit and vegetable production as well as reduce competition between biomass for food vs industrial applications. TRANSBIO characterized and selected appropriate sub-products followed by appropriate pre-treatment and hydrolysis strategies to produce value-added compounds such as food ingredient and platform chemical succinic acid, biopolymer polyhydroxybutyrate (PHB) and enzymes for detergents. TRANSBIO investigated three fermentation strategies; submerged fermentation (bacteria, yeasts) and solid state fermentation (filamentous fungi). Molecular screening procedures for selection of PHB producing bacterial candidate strains were established and PHB production on sub-product hydrolysates was implemented. From by-products 692 yeast isolates could be obtained, including two yeasts able to produce higher amounts of succinic acid without any fermentation optimization. Afterwards fermentation optimization and down-stream processing implementation took place. Solid state fermentation using filamentous fungi could be implemented and enzymes for detergents obtained. Potential of by-products for biogas production could be proven.

Keywords: by-products, valorisation, biopolymer, enzymes, chemical building block

1. Introduction

All over the planet land is converted to human use, primarily to agricultural land. In 2012, more than 40% of available lands in EU27 were utilized as agricultural area (Eurostat, 2015). The transformation of forests, grasslands, wetlands and other vegetation types to agricultural land is the driving force

Leire Escajedo San-Epifanio and Mertxe De Renobales Scheifler (eds.) *Envisioning a future without food waste and food poverty*

DOI 10.3920/978-90-8686-820-9_10, © Wageningen Academic Publishers 2015

95

behind the serious reductions in biodiversity, impacts in water flows as well as biogeochemical cycling of carbon, nitrogen, phosphorous and other important elements. Nitrogen and phosphorus are both essential elements for plant growth, so fertilizer production and application is the main concern (Rockström *et al.*, 2009). Beside others, anthropogenic perturbation levels of land-system change and biochemical flows exceed already proposed planetary boundaries (Steffen *et al.*, 2015). Therefore no further transformation from land to agricultural area is recommended. In order to ensure feeding of population as well as providing feedstock's for growing biofuel and biomaterial industry, new approaches are needed to use existing raw materials in a sustainable manner.

2. Fruit and vegetable by-products

Continuous use of petroleum sourced raw materials as industrial feedstock is nowadays widely recognized as a major barrier towards sustainable development resulting in growing appreciation that the management and utilization of natural resources need to be improved. Developing a sustainable bio-based economy that uses eco-efficient bio-processes (enzymatic, microbial) and renewable raw materials (e.g. by-products from fruit and vegetable transforming industry) is one key strategic challenge for the 21st century. Improved understanding of biotechnology is making it possible to utilize biomass and organic waste materials in a highly efficient and sustainable manner. Kind and amount of by-products generated during fruit and vegetable processing and preservation depends strongly on the raw material. Nevertheless, close to 50% of all fruits and vegetables in the European Union go to waste says a report by the Food and Agriculture Organization (FAO, 2011). These losses occur throughout the entire food chain, from 'farm to fork', but the predominant losses are linked to agriculture, post-harvest and processing. A huge quantity of fruit and vegetable wastes and by-products from the processing industry are available throughout the world. A large proportion of these wastes are dumped in landfills or rivers, causing environmental hazards (FAO, 2013). Due to extensive characterization of available waste and by-products from endive, green bean, lettuce, pea, sweet corn, potato, chard, cardoon, broccoli as well as banana and pineapple processing could be suitable to be used as fermentation feedstocks. After establishing appropriate pre-treatment and enzymatic hydrolysis procedures fermentable hydrolysates could be obtained for by-products shown in Table 1.

Table 1. Sugar content in enzymatic hydrolysates of different fruit and vegetable by-products.

By-products	Sugar content in enzymatic hydrolysates (g/l)						
	Sucrose	Glucose	Fructose	Mannose	Galactose	Xylose	Arabinose
Sweet corn	16.7-32.4	52.4-61.5	20.7-29.6	–	–	2.5-7.1	1.5-7.5
Cardoon	n.a.	152.8	n.a.	n.a.	n.a.	n.a.	n.a.
Bean	n.a.	210.4	n.a.	n.a.	n.a.	n.a.	n.a.
Endive	–	32.1-47.6	40.2-54.8	–	–	5.6-9.9	0.9-6.7
Potato	–	163-182	–	–	–	–	–
Banana pulp	0.76	92.34	81.93	0.21	0.14	0.33	0.24

3. Bacterial production of biopolymer polyhydroxybutyrate

Around 90 million tons of food wastes are generated in the EU each year. The largest fraction of EU food waste among the four sectors (manufacturing, households, wholesale/retail, food service/ catering) considered, produces household at about 42% of the total amount (European Commission DG ENV, 2010). In this context, packaging is playing a vital role in containing and protecting food as it moves through the supply chain to the consumer. It already reduces food waste in transport and storage, and innovations in packaging materials such as plastics will provide new opportunities to improve efficiencies (Verghese *et al.*, 2013). Plastics are widely used in many applications such as in packaging, building materials and commodities as well as in hygiene products (Abou-Zeid *et al.*, 2001). However, environmental pollution caused by the indiscriminate dumping of plastic waste has created a global problem. Conventional plastics that are synthetically derived from petroleum are not readily biodegradable and are considered as environmentally harmful waste (Kumar *et al.*, 2004).

Plastic waste generation is set to continue growing and the development of new materials continues apace. In 2008, total generation of post-consumer plastic waste in EU-27, Norway and Switzerland was 24.9 Mt. Packaging is by far the largest contributor to plastic waste at 63% (European Commission DG ENV, 2011). Therefore new environmentally friendly plastics for food packaging solutions must be developed. Polyhydroxybutyrate (PHB) is a bio-based and biodegradable polyester with highly attractive qualities for thermo-processing applications. PHB is produced via fermentation of renewable feedstock's (carbon substrates) within microorganisms. PHB accumulates as granules within the cytoplasm of cells and serves as a microbial energy reserve material (US Congress, Office of Technology Assessment, 1993).

At present, the raw material costs account for as much as 40 to 50% of the total production cost for polyhydroxyalkanoates (PHA) (Shen *et al.*, 2009). Use of lower cost carbon sources, recombinant *Escherichia coli* or genetically engineered plants should all lead to reductions in production cost (Suriyamongkol *et al.*, 2007). The European project TRANSBIO (www.transbio.eu) is supporting these approaches by screening and selecting for appropriate wild-type strains of *Cupriavidus necator* (*Ralstonia eutropha)* and other strains able to produce PHB. Additional a genetically engineered *E. coli* was designed containing the phaCAB operon responsible for PHB production derived from *Cupriavidus necator* (*Ralstonia eutropha*). The recombinant microorganism expresses the three genes *phaC* (PHA synthase), *phaA* (3-ketothiolase) and phaB (NADPH-dependent acetoacetyl-CoA-reductase). The recombinant pET24-CAB *E. coli* accumulate PHB up to 12% of the Cell Dry Weight.

Currently on-going are fermentation process development and up-scaling test trials using hydrolysates obtained from fruit and vegetable by-products as carbon source. Additional, green solvents and enzyme cocktail extraction are compared with classical chloroform extraction. Green solvents and enzyme cocktail extraction showed promising extraction capability with purities reached of 82.45±1.91% (anisole) and 76.2±3.54% (enzyme cocktail extraction).

4. Succinic acid – nutraceutical and platform chemical

Succinic acid (SA) is a key building block for a wide range of secondary chemicals and finds use in chemical, pharmaceutical, food and agricultural industries. Market for SA is projected to witness a compound annual growth rate of 22.6% by volume between 2014 and 2019, and is expected to generate a global market value of $ 486.7 million by 2019 (MarketsandMarkets, 2015). Currently petroleum-based SA is the commonly used succinic acid globally. Due to the negative impact of the petroleum-based SA, emphasis is laid on production and consumption of bio-based succinic acid. Therefore the market for conventional SA is estimated to decline by 2019. Succinic acid is produced biochemically from glucose using metabolic engineering of the natural succinate overproducers (like *Actinobacillus succinogenes*, *Anaerobiosprirrilium succiniciprudens*, and *Mannheimia succiniciprodudens*) as well as metabolic engineered overproducers like *E. coli*, *Corynebacterium glutamicum* and *Saccharomyces cerevisiae* (Cheng *et al.*, 2013). Nevertheless, the use of metabolic engineered organisms for the production of food grade succinic acid is at least questionable. Therefore the TRANSBIO partners performed huge screening activities using culture dependent methodology and metagenomic analysis to identify new natural yeast strains which may be used for succinic acid production. As the strains should be able to grow on the by-product hydrolysates, by-products from fruit (strawberry, green beans, peach, apple, pear, pepper) and vegetable (cardoon, mix of salad, asparagus, artichoke, fresh chard, processed chard) were used for screening activities. A total of 692 yeast isolates were obtained, 450 from fruit by-products and 242 from by-products from vegetable processing.

Nevertheless, the major technical hurdle for the development of succinic acid as a building block includes the development of very low cost fermentation routes. Therefore the partners performed extensive test trials starting with 96 deep-well plates used at the partner Biotrend for high-throughput screening followed by test trials in controlled bioreactors. Based on these test trials two different strains were selected, which were able to produce higher amounts of succinic acid than the reference strains used (Figure 1). It could be shown that naturally occurring yeast strains are able to produce similar amounts of succinic acid than genetically modified yeast strains (Raab *et al.*, 2010; Otero *et al.* 2013).

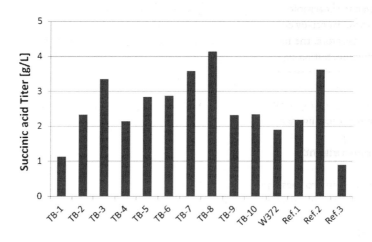

Figure 1. Succinic acid titer for different TRANSBIO approaches with Saccharomyces cerevisiae TB-1 – TB-10 (W372 = commercial strain as tested in TRANSBIO and used in Ref.1 (De Klerk, 2010); Ref.2 (Raab et al., 2010); Ref.3 (Otero et al., 2013).

5. Enzyme production by solid state fermentation

Another innovative approach for the valorisation of by-product is enzyme production by solid state fermentation. Use of an aerobic submerged culture in a stirred-tank reactor is the typical industrial process for enzyme production involving a microorganism that produces mostly an extracellular enzyme. Filamentous fungi are the preferred source of industrial enzymes because of their excellent capacity for extracellular protein production (Jun *et al.*, 2011). As the production of metabolites such as enzymes depend on the fermentation conditions, solid state fermentation (SSF) represents a unique alternative as SSF simulates the natural environment/microenvironment for filamentous fungi. As detergent industry represents the largest single market for enzymes with the 25-30% of total sales, TRANSBIO focused on the production of lipases and proteases for detergent application. After extensive screening activities, 14 thermo-stable lipolytic fungi strains showed qualitative lipase activity. Four of these lipolytic fungi strains were found to produce lipase under laundry detergent components. Nine thermo-stable proteolytic fungi strains showed qualitative protease activity. Six proteolytic fungi strains were found to produce proteases under laundry detergent components. These strains are currently used for SSF development in technical scale. Enzyme extraction procedures were established and first extracts are used for detergent development.

6. Biogas from fruit and vegetable by-products

As global population continues to grow and will need to be fed and will expect to live an improved life style, consuming more energy questions are raised related to 'food versus fuel'; how much land and other resources are available, how should they be used and what are the priorities? (European Biofuels Technology Platform, 2015). Not all biofuels can be considered sustainable to the same extent. There are significant differences between first and second generation biofuels as well as between biofuels of the same generation. First generation biofuels are made from food crop feedstock while second generation biofuels are made from agriculture and forestry waste. Current biomass resources used for anaerobic biogas production are manure or sewage, municipal waste, green waste or energy crops. Especially energy crops are using agricultural area and/or could be also used in the food chain. Therefore, the main focus of TRANSBIO is to enlarge the base of raw materials for biogas production, testing by-products from fruit and vegetable transforming industry as well as sub-products after fermentation approaches in a sophisticated cascading approach. During the continuous lab-scale anaerobic digestion tests, a stable methane content of 60% was measured in the biogas. This corresponds to a low heating value of 21.6 MJ/Nm^3. Relatively low levels of H_2S and NH_3 were detected, indicating that no large problems are expected for biogas valorisation of by-products.

Acknowledgements

The research leading to these results has received funding from the European Union's Seventh Framework Programme (FP7/2007-2013) under grant agreement No. 289603.

References

Abou-Zeid D.M., Muller R.J., Deckwer W.D. (2001). Degradation of natural and synthetic polyesters under anaerobic conditions, Journal of Biotechnology, 86(2): 113-126.

Cheng, K.-K., Wang, G.-Y., Zeng, J., Zhang, J.-A. (2013). Improved Succinate Production by Metabolic Engineering. BioMed Research International, Volume 2013: 538790.

De Klerk, J.-L. (2010). Succinic acid production by wine yeasts. MSc Thesis, Stellenbosch University, Department of Viticulture and Oenology, Faculty of AgriSciences, South Africa.

European Biofuels Technology Platform (2015). Food versus fuel. Available at: http://biofuelstp.eu/food-vs-fuel.html.

European Commission DG ENV (2010). Preparatory study on food waste. Technical Report – 2010-054. Available at: http://tinyurl.com/4424obc.

European Commission DG Environment (2011). Plastic waste in the Environment – Final Report. Specific contract 07.0307/2009/545281/ETU/G2 under Framework contract ENV.G.4/FRA/2008/0112. Available at: http://tinyurl.com/6fc59w3.

Eurostat (2015). Land use by NUTS 2 regions. Available at: http://ec.europa.eu/eurostat/data/database.

FAO (2011). Global food losses and food waste – Extent, causes and prevention. Rome, Italy.

Suriyamongkol, P., Weselake, R., Narine, S., Moloney, M., Shah, S. (2007). Biotechnological approaches for the production of polyhydroxyalkanoates in microorganisms and plants – A review. Biotechnology Advances 25(2) 148-175.

Jun H., Kieselbach T., Jönsson L.J. (2011). Enzyme production by filamentous fungi: analysis of the secretome of Trichoderma reesei grown on unconventional carbon source. Microbial Cell Factories 10: 68.

Kumar M.S., Mudliar S.N., Reddy K.M.K., Chakrabarti T. (2004). Production of biodegradable plastics from activated sludge generated from a food processing industrial wastewater treatment plant, Bioresoure Technology, 95(3): 327-330.

MarketsandMarkets (2015). Succinic acid market by source (petroleum and bio-based), application (1,4 BDO, Polyurethane, Food, Resins, Coatings, Pigments, Plasticizers, Pharmacy, De-icer solutions, PBS, PBST, Solvents and Lubricants, Cosmetics, Others) – Global Trends and Forecast to 2019. Report Code: CH 2917. Available at: www.marketsandmarkets.com.

OECD/Food and Agriculture Organization of the United Nations (2014). OECD-FAO Agricultural Outlook 2014, OECD Publishing.

Otero, J.M., Cimini, D., Patil, K.R., Poulsen, S.G., Olsson, L., Nielsen, J. (2013). Industrial systems biology of *Saccharomyces cerevisiae* enables novel succinic acid cell factory. PLOS ONE 8(1): e54144.

Shen L., Haufe J., Patel M.K. (2009). Product overview and market projection of emerging bio-based plastics – PRO-BIP 2009 – Final Report, Universiteit Utrecht, the Netherlands.

Raab A.M., Gebhardt G., Bolotina N., Weuster-Botz D., Lang C. (2010). Metabolic engineering of *Saccharomyces cerevisiae* for the biotechnological production of succinic acid. Metabolic Engineering, 12(6): 518-525.

Rockström, J., Steffen W., Noone K., Persson Å., Chapin F. S., Lambin III, E., Lenton T. M., Scheffer M., Folke C., Schellnhuber H., Nykvist B., De Wit C. A., Hughes T., van der Leeuw S., Rodhe H., Sörlin S., Snyder P. K., Costanza R., Svedin U., Falkenmark M., Karlberg L., Corell R. W., Fabry V. J., Hansen J., Walker B., Liverman D., Richardson K., Crutzen P., Foley J. (2009). Planetary boundaries: exploring the safe operating space for humanity. Ecology and Society 14(2): 32.

Steffen, W., Richardson, K., Rockström, J., Cornell, S. E., Fetzer, I., Bennett, E. M., Biggs, R., Carpenter, S. R., de Vries, W., de Wit, C.A., Folke, C., Gerten, D., Heinke, J., Mace, G.M., Persson, L.M., Ramanathan, V., Reyers, B., Sörlin, S. (2015). Planetary boundaries: Guiding human development on a changing planet. Science, 347(6223): 1259855.

US Congress, Office of Technology Assessment (1993). Biopolymers: making materials nature's way-background paper, OTA-BP-E-102 (Washington, DC: US Government Printing Office, September 1993). http://ota.fas.org/reports/9313.pdf.

Vandamme E.J. (2009). Chapter 1 – agro-industrial residue utilization for industrial biotechnology products. In: P. Singh nee' Nigam, A. Pandey (eds.) Biotechnology for agro-industrial residues utilisation – utilisation of agro-residues. Springer Science + Business Media B.V., Berlin, Germany, pp. 3-11.

Verghese K., Lewis H., Lockrey S., Williams H. (2013). Final Report: The role of packaging in minimizing food waste in the supply chain of the future. RMIT University, Centre for Design.

11. Food waste in Kenya: uncovering food waste in the horticultural export supply chain

E. Colbert

Feedback, 18 Ashwin Street, London E8 3DL, United Kingdom; edd@feedbackglobal.org

Abstract

Feedback is an environmental organisation based in the UK who campaigns globally to put a stop to food waste. Feedback have held mass public feasts around the world using food that would have only gone to waste to feed people instead of bins. Alongside these events and its other two campaigns, The Gleaning Network and The Pig Idea, Feedback also delivers hard-hitting research from the field in order to shine a light on the social impact food waste has on farmers across the globe. Feedback's most recent investigations into Kenya's horticultural export supply chain have uncovered systemic issues relating to imbalances of power and unfair trading practices that have a significant impact on food waste levels, as well as farmer livelihoods and by extension food security. Last minute alteration or cancelation of demand forecasts, unnecessarily strict cosmetic specifications and unpredictable fluctuations in demand and price from European retailers and their buyers mean that farmers are left unable to sell large amounts of produce, as secondary markets are not responsive or lucrative enough to absorb this produce. This research demonstrates how farmers in Kenya are wasting up to 50% of their harvests due to cosmetic standards, whilst others are forced to dump entire crops as a result of order cancellations. Not only do these issues result in wasted resources such as land, energy, agro-chemicals and fuel; they also cause severe financial loss to exporters, farmers and farm workers. Financial risk is transferred down the supply chain to the weakest actors reducing living standards and forcing many into debt cycles. There are an evolving number of initiatives concerned with 'food losses' in the global south. These initiatives generally refer to 'post-harvest' losses (PHL), focusing on infrastructural issues, poor harvesting methods and inadequate storage of crops. This research, based on qualitative interviews with Kenyan exporters and farmers, intends to extend this discussion further to better understand food waste in developing countries, as opposed to food losses. As this research demonstrates, there is a great need for reducing food waste that occurs independently from improvements in PHL reduction. Food waste reduction can be achieved with limited public investment, compared to PHL reduction, instead requiring innovations in business practices to avoid unfair trading practices that force farmers to waste their produce.

Keywords: Africa, agriculture, supply chains, supermarkets, vegetables

1. Introduction

Through its ongoing research into local and export horticultural supply chains, Feedback has uncovered systemic issues related to imbalances of power and unfair trading practices in these supply chains that have a significant impact on food waste levels, as well as farmer livelihoods and by extension food security at the international level.

Last minute alteration or cancellation of demand forecasts, unnecessarily strict cosmetic specifications and unpredictable fluctuations in demand and price from retail buyers often mean that farmers are left with large amounts of produce that they cannot sell, as secondary markets are not responsive enough or lucrative enough to absorb this produce. Whilst some of this surplus is fed to livestock the majority of the food is simply dumped or at best composted.

2. Defining food waste versus food losses

There is an evolving number of initiatives concerned with 'food losses' in the Global South. These initiatives generally refer to what has come to be known as 'post-harvest' losses (PHL), focusing on infrastructural issues, poor harvesting methods and inadequate storage of crops. Feedback's research extends this discussion further to better understand food *waste* in developing countries, as opposed to food losses.

Food waste should be understood as any food intended for human consumption that is discarded or left to spoil as a result of actions and decisions taken by stakeholders across the supply chain (farmers, brokers, exporters, importers, retailers, and consumers) and which relate to the way that the market is structured. Food waste reduction can be achieved with low cost solutions – such as the Tesco green bean case study mentioned above – compared to PHL reduction which often requires investment in relatively costly infrastructure, and therefore can have a disproportionately large positive impact in terms of food waste prevention, increasing the wellbeing of local communities and improving food security by increasing food availability where it is most urgently needed.

3. Summary of Feedback's recent research in Kenya

Since 2013 Feedback has focused part of its investigations into supplier level food waste in Kenya, one of the largest exporters to the UK and Europe of luxury horticultural produce such as fine green beans, baby corn, sugar snap peas and 'mangetout'. Feedback's investigations revealed the case of a grower in Kenya supplying EU and UK supermarkets, who was forced to regularly waste up to 40 tonnes of green beans and other vegetables every week, amounting to 40% of what he grows. This would have been enough to feed over 250,000 people every week in a country where 3 million people are dependent on food aid. Some of this unwanted produce was sold on the local market or donated to those in need, but a large proportion of it was either left to rot or fed to livestock at best.

A survey of approximately 20 Kenyan farmers which we conducted in early 2014 found that over 70% of them believed that wastage and spoilage of their produce was a direct result of the imposition of cosmetic standards and/or changes to forecast orders by UK and EU retailers or middlemen. Feedback recently revisited to Kenya for two weeks to conduct further research into the Kenyan horticultural supply chain in order to gain more insights into the causes of food waste affecting Kenyan producers and exporters as well as stakeholders in similar supply chains across the African continent and beyond.

We discovered that this offloading of risk is widespread in the Kenyan agricultural sector producing food for European markets. Not only do these issues result in high levels of wasted resources such as land, waste, energy, agri-chemicals and fuel; they also cause severe financial loss to exporters, farmers and farm workers. Financial risk is transferred down the supply chain to the weakest actors forcing many into debt cycles and reduced living standards.

As demand for luxury horticultural products is rapidly increasing, it is likely that greater numbers of suppliers will be affected by these unfair trade practices. Around Africa, farmers are following Kenya's lead and seeking export markets in Europe and the US. Similarly, our initial investigations in Peru, Ecuador and Guatemala have revealed that the growth of luxury horticultural product sectors including asparagus, green peas and artichokes are resulting in similar practices.

Here, we summarise some of the findings of the research:
- All of the farmers and exporters interviewed for this research expressed having experienced problems with food being rejected on the grounds of cosmetic specifications, i.e. produce that was perfectly good to eat but that was the wrong shape, size or colour to be exported.
- All farmers interviewed had experienced financial loss as a result of the rejections caused by cosmetic specifications.
- The average amount of food rejected at farm-level reported by farmers was 32.9%.
- Exporters reported an average of 44.5% of produce being rejected before being exported, inclusive of farm level waste.
- Farmers reported having up to 50% of their produce rejected because of cosmetic standards or order forecast amendments.
- Farmers reported being forced into cycles of debt as a result of uncompensated order cancellations.
- Exporters waste 30% of French beans through the practice of 'topping and tailing' for cosmetic specifications.

4. Feedback's impacts

Feedback is an environmental organisation that campaigns to end food waste at every level of the food system. We catalyse action on eliminating food waste globally, working with governments, international institutions, businesses, NGOs, grassroots organisations and the public to change society's attitude toward wasting food.

Feedback has a strong track record in catalysing concrete changes in the policies of retailers in the UK, Europe and recently in the US. Following our investigation into waste of green beans in Kenya, Feedback met with Tesco to challenge them to stop the practice of topping and tailing French beans that leads to a third of produce being wasted. As a result Tesco changed their buying policy, instead opting for just topped beans. During the research process, an exporter was interviewed who supplies Tesco and therefore had become a beneficiary of this change in purchasing policy. She said:

> When we were doing the top and tail we were basing our yield calculations on 67%. This means that out of what was delivered here ... we would provide for 33% waste just from top and tailing ... When our customer made the switch from cutting both sides of the bean ... the yield is 77%.

The exporter, now only having to trim one end of the bean, had reduced their waste by a third. This reduction led to annual savings of seven million shillings (approximately £ 50,000). This saving also had a knock-on effect for farmers. As the exporter paid their farmer per packability, the farmer could expect a higher price as more of their produce was being exported.

Since this initial challenge, at least three major retailers are now only trimming one end of their French beans rather than both. Tesco has also renegotiated its contracts with farmers, basing its orders on anticipated crop yields to limit last minute cancellations. Feedback is now continuing campaigning efforts to end the unfair trading practices highlighted in this research and to engage retailers in changing their cosmetic specifications to ensure less food is wasted.

5. Conclusions and recommendations

Two areas of recommendations addressing solutions to food waste have emerged from this research: *food waste* reduction, and *food* redistribution.

Food waste reduction

Measures to avoid food waste should always be at the heart of any intervention or wider strategy aiming to tackle the problem. This report has identified that food waste in the Kenyan horticultural export sector is being caused to a large extent by systemic issues related to the patterns of behaviour and actions of stakeholders in these supply chains and should therefore be tackled from these root causes. These actions primarily originate from the top of the supply chain, i.e. European retailers and importers, and are an expression of the imbalances of power that dominate these supply chains. It is therefore at this level that most of the opportunities for interventions lie, which can achieve significant and measurable reductions in food waste, reinforce the livelihoods of local farmers and improve access to food where it is needed most.

European retailers have a great deal of leverage and control over their supply chains and therefore have the opportunity to adopt a proactive role in changing wasteful patterns of behaviour in their relationships with direct and indirect suppliers. This report concludes with the following two recommendations with regards to retailer action:

- Relaxation of cosmetic specifications.
  Retailers should relax unnecessarily strict cosmetic specifications, and ideally gradually abolish these standards in due course, to allow farmers to sell a larger percentage of the produce grown for export markets which would in turn reduce the need to systematically overproduce in order to ensure there is a sufficient buffer to meet order quantities. The examples given of Tesco changing its French bean trimming policy demonstrates the relative ease in which small changes in the buying policies of such retailers can have huge impacts on farmers and exporters. Cosmetically imperfect produce can be sold as grade two produce, or used in added value and processed food production lines.
- Improving forecasting accuracy and spreading the risks of demand fluctuations.
  It is important for retailers to work directly with their suppliers to ensure that farmers are not disproportionately affected by fluctuations in demand for certain products. This could be ensured for example by improving forecasting methods and models to increase accuracy with the direct input of their suppliers; changing the structure of their supply chains (for example by creating a more direct relationship with primary farmers); guaranteeing the purchase of a certain percentage

of their suppliers crop or fully compensating their suppliers for last minute order adjustments; and helping their farmers access local or secondary markets for their excess produce either by relationship brokering or by investing in relatively low cost initiatives that can extend the 'shelf life' and add value to the rejected produce by repurposing it.

There are examples of existing initiatives by UK retailers who have already taken positive steps to this direction (as has been demonstrated by the Tesco banana example). Such interventions are easily replicable and their replication should be encouraged as they have immediate and significant gains in terms of food waste reduction and improving the efficiency and fairness of retail supply chains.

Where voluntary commitments are not upheld, legislation in British law states that UK supermarkets should compensate direct suppliers in cases of order forecast amendments or cancellations (Groceries Supply Code of Practice; Section 10). The practice of order cancellations and amendments are therefore illegal under British law and the British Government should be investing time and resources into investigating potential breaches of this legislation. As the Groceries Supply Code of Practice does not currently protect indirect suppliers more research should be considered into the relationships between supermarkets and their direct suppliers who procure food from other countries.

When interviewing farmers and exporters, Feedback found that not a single person had heard of the Groceries Supply Code of Practice or the Adjudicator. Information about this regulatory body should be shared with the various stakeholders in Kenya's horticultural industry to increase awareness of this office.

Two further recommendations regarding food waste reduction relate to the development of local markets for products currently grown solely for the export market:
- Development of local market for horticultural products usually grown for export.
  There is scope to increase local demand in Kenya for products currently grown for the export market only. Farmers and exporters interviewed for this report both expressed a desire to be able to sell their rejected yet good quality produce to both the export and local markets. However, they identified a lack of demand in both of these markets as a challenge to selling their food outside of the conventional export supply chains.
- Development of domestic value addition processing industry.
  There are a number of produce types that form part of the typical Kenyan diet that are currently rejected with no secondary market to be sold to. This report has identified a range of different initiatives that not only add value to such produce but also extend the shelf life of otherwise quickly perishable foodstuffs. Processing produce via the methods suggested in this report not only reduces waste but also generates greater incomes for people involved in the industry.

Redistribution

In a country where millions of people are without adequate food and nutrition, infrastructure should be put in place to ensure surplus food is redistributed to those who need it.

Small quantities of food are currently given to schools, children's homes, street children centres and medical centres around Nairobi from the export industry. However, there are infrastructural challenges blocking a) more food being redistributed and b) food being redistributed outside of Nairobi in the rural areas of Kenya. Farmers and exporters claimed that it was generally not economically viable for them to redistribute food themselves due to the cost of labour and transportation.

Centralised collection or redistribution points may provide a solution for this problem, streamlining the process of delivering large quantities of produce to numerous social organisations. Exporters interviewed by Feedback showed enthusiasm towards the idea of an independent redistribution system if it could overcome the aforementioned challenges. The majority of the exporters are based around the airport, either in private warehouses; in government-run pack houses such as the Horticultural Crop Development Authority; or within export processing zones. Establishing a redistribution network within this export area would provide a centralised location for surplus food to be collected at low cost. Further research should be conducted in this field to understand how such a scheme could operate without affecting local markets. Logistics present a further challenge to such a scheme, especially as the produce in question is highly perishable and may not survive long distance journeys if not refrigerated.

It is important to note that, in line with the food waste pyramid, avoidance and reduction should be the principle goal of any initiative addressing food waste. There is no doubt that the redistribution of surplus food has a number of substantial yet short term social benefits. Ultimately, the overproduction of food leading to high levels of food waste must be stemmed in order to provide longer-term social, environmental and economic development globally.

12. Traditional knowledge and sustainable agriculture: the strategy to cope with climate change

V. Sandhya, G.P. Reddy, S.Z. Ali and P. Kumar K.L.*
Agri Biotech Foundation, Rajendranagar, Hyderabad-500030, Telangana State, India; sandhyarao28@gmail.com

Abstract

Indigenous Traditional Knowledge (ITK) is developed over a period of time through accumulation of experiences and intimate understanding of environment. This knowledge is gradually vanishing due to population pressure and industrial development. Traditional crops and cropping practices are becoming extinct, of which modern farming practices are major contributors. Further climate change is a significant change in the statistical distribution of weather patterns. The Intergovernmental Panel on Climate Change Third Assessment Report 2001 concluded that current knowledge on adaptation to climate change is limited and emphasized the need for research on viable adaptation measures. India is vulnerable to climate change as the majority of its population depends on agriculture, which is climate sensitive. Agriculture in India is involved in domestic food supply, employment and cash income. Recognizing the challenges to mitigate climate variability promotes more sustainable and resilient agriculture. The traditional knowledge for ecosystem management and use of natural resources is gaining credence as a key weapon to fight against climate change. Understanding the ITK will help sustain farming practices preventing plant genetic erosion and environmental deterioration. It contributes to sustainable food security and conservation of variety and variability of animals, plants and soil properties. To document indigenous and traditional knowledge on coping with climate change in agriculture in India, Agri Biotech Foundation conducted case studies in five villages under distinct rainfall zones with severe drought (<50%), moderate drought (50 to 75%), mild drought (75%). The study was undertaken under six major themes like indigenous practices for water and soil conservation, plant protection, unique agricultural practices, cropping systems and community level adaptation. This paper presents field based case studies and explains the relevance of these experiences for sustainable agriculture. There is a need to strengthen dissemination of indigenous knowledge and integrating with modern approaches in climate change resilience.

Keywords: indigenous people's knowledge, coping strategies, food security

1. Introduction

Traditional knowledge is the knowledge, innovations and practices of indigenous and local communities. Developed from experience gained over the centuries and adapted to the local culture, traditional knowledge is transferred from generation to generation. It becomes collectively owned and takes the form of stories, songs, folklore, proverbs, cultural values, beliefs, rituals, community laws, local language, and agricultural practices, including the development of plant species and animal breeds. Traditional knowledge is mainly of a practical nature, in fields such as agriculture, fisheries, health, horticulture, forestry and environmental management (Cruz and Ramos, 2006).

Leire Escajedo San-Epifanio and Mertxe De Renobales Scheifler (eds.) *Envisioning a future without food waste and food poverty*
DOI 10.3920/978-90-8686-820-9_12, © Wageningen Academic Publishers 2015

This traditional knowledge of indigenous peoples and local communities for ecosystem management and sustainable use of natural resources is gaining credence as a key weapon in the fight against climate change. The Intergovernmental Panel on Climate Change has highlighted the role of indigenous knowledge and crop varieties in adaptation (Parry *et al.*, 2007) and the Institute of Advanced Studies at the UN University recently identified more than 400 examples of indigenous peoples' roles in climate change monitoring, adaptation and mitigation (McLean, 2010). Further in 2010, UN Framework Convention on Climate Change adopted an 'enhanced action on adaptation' that identified the need on traditional and indigenous knowledge.

There are at least five types (Table 1) of traditional knowledge useful for adaptation in agriculture (Swiderska *et al.*, 2011b).

Despite this growing recognition, the role of traditional knowledge in adapting agriculture to climate change remains largely undervalued. Decision makers tend to focus on intensifying production through modern agriculture as the key to adaptation and food security. But for many local communities it is not modern agriculture but traditional knowledge that has enabled them to cope with extreme weather and environmental change over centuries (Swiderska *et al.*, 2011a).

This knowledge has given rise to thousands of traditional crop species and varieties that local farmers have domesticated, improved and conserved over generations. The communities of the Potato Park in Cusco, Peru, hold more than a quarter of the 4,000 or so potato varieties found in the country through traditional knowledge (ANDES, the Potato Park Communities, 2011). The Karst Mountains of southwest China, the Bolivian Andes and coastal Kenya farmers are severely impacted by changes in climate. Adopting more resistant and flexible local crop varieties, using native plants for biocontrol,

Table 1 Traditional knowledge for adaptation in agriculture (adapted from Swiderska et al., 2011).

Traditional knowledge	Adaptation in agriculture
Climate forecasting	Traditional knowledge help forecast local weather
	Predict extreme events
	Farmers can monitor climate change in and fill the gap of scientific models
Resilient properties	Traditional farmers often live on marginal land where climate change impacts and selection pressures are greater. This enables them to identify resilient crop species and varieties for adaptation
Plant breeding	Farmers, particularly women and the old are active plant breeders, conserving local landraces and selecting seeds for preferred and adaptive characteristics over generations
Wild crop relatives	Local communities often draw on wild areas around farms for crop improvement and domestication as well as to supplement their diet and provide food when crops fail
Farming practices	Traditional farming practices from water, soil or pest management to erosion control and land restoration conserve key resources for resilience and adaptation, such as biodiversity, water, soil and nutrients

planting and conserving resilient local landraces, participatory plant breeding, using modern and traditional crops and knowledge, community seed production, planting resilient traditional varieties, sharing seeds, sharing animals proved vital in adapting to the changes. Farmers in all study sites choose traditional crop varieties over modern ones because they are better adapted to local conditions and more likely to survive environmental stress and climatic variability (Swiderska *et al.*, 2011a).

More than half of households in China still use local landraces of maize and rice as they taste good, are better adapted to mountainous and barren land, and have drought and dislodging resistance. Evidence from Guangxi province shows that most farmer-improved landraces survived the big spring drought in 2010 (Swiderska *et al.*, 2011b). In coastal Kenya, many farmers are going back to using traditional maize varieties because they are hardy and able to cope with unpredictable weather conditions and local pests. And in the Bolivian Andes, farmers are using local potato variety (*Doble* H), in response to new pests and water shortages. Similarly sharing seeds and knowledge is found across most traditional farming communities. In Southwest China, annual seed fairs and local culture activities organized by women farmers have revived seed and knowledge sharing to meet emerging economic and climatic challenges. Women's groups produce their own seeds selecting local varieties, improving them and sharing them with other villages. The idea of sharing does not only apply to seeds, during periods of drought in Kenya, livestock farmers share their animals with different sources of water and pasture to minimize chances of losing all their stock (IIED, 2011). Among the indigenous communities of North East India, forests are maintained collectively by the villagers for the forestry needs of the villagers. Any villager is prohibited from clearing the villages or cutting down woods without prior permission of the village durbar/ village council. Community forestry empowers the people over decision making through participation in planning and management (Varte, 2012).

Indigenous communities have come up with many resource management practices that build on detailed knowledge of their environment. Sacred groves like in Meghalaya or community forests common among indigenous communities in North East India fulfil many critical ecosystem functions such as providing seed banks for local species, providing habitat for seed dispersing animals and predators on local pests (Varte, 2012). Due to effective community management by many indigenous communities, there is evidence of significant improvement in the conservation of natural resources and enhanced soil and water management (Varte, 2012).

In developed and developing countries, farmers and indigenous and local communities have traditional knowledge, expertise, skills and practices related to food security in a wide range of ecosystems, including fragile and harsh ones. Such practices can be applied in innovative ways to help tackle today's problems. The application of traditional knowledge for climate change resilience for landscape and water management, soil conservation, biological control, ecological agriculture and livestock practices, and plant and animal breeding often enhances food security. Many rural peoples have generated traditional knowledge related to indigenous crop and plant varieties, animal breeds, landraces and wild species that they use as food, medicine and other products to ensure food and livelihood.

Of the 10,000 to 15,000 plants that are known to be edible, about 7,000 species have been used in agriculture for food and fodder throughout human history. Of the 7,000 animal breeds developed by traditional herders, 1000 have become extinct over the last century. Food security is threatened when these breeds/varieties are affected by pests and diseases and by climate changes. (FAO, 2009). Climate change, extreme weather events are having increasing impacts threatening agricultural production on which people depend for their subsistence. Many rural communities have vast knowledge of previous variations in climate and weather and have developed mitigation and adaptation strategies for ensuring food security. Communities apply traditional knowledge in early warning systems to calculate risks or detect extreme weather events, droughts or floods. They use it in adapting subsistence strategies for agriculture, fishing, and forestry, improving water and resource management, selecting which resources to use to mitigate or adapt to climate change effects. The loss of such knowledge results in increased food insecurity, while livelihoods decline and biodiversity disappears.

Climate change and drought are both an environment and a development issue. Nowhere is this more critical than in India. One of the key climate-related vulnerabilities of India's economy is its heavy dependence on the monsoons. Monsoon analysis reveals that some part or the other of the Indian subcontinent has been hit by drought almost every two years.

The disastrous effect of feeble or failed monsoons has been particularly acute in the state of Andhra Pradesh (Figure 1), where more than 70% of the people depend on agriculture for their livelihood. The human and social costs of these droughts are devastating and wide-ranging, resulting in crop-yield failure, unemployment, erosion of assets, decrease in income, reduction in living conditions, impoverished nutrition and health, and increased vulnerability to other shocks.

2. Traditional Knowledge to cope with climate change: some case studies from India

Based on existing literature on indigenous and traditional knowledge on coping mechanisms to climate change in agriculture a systematic literature review process was adopted and a model was developed to study six themes like water and soil conservation, plant protection, unique agricultural

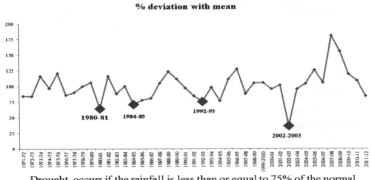

% deviation with mean

Drought occurs if the rainfall is less than or equal to 75% of the normal rainfall, severe drought would prevail if it is less than or equal to 50% of the normal rainfall (Indian Meteorological Department (IMD)

Figure 1. Rainfall data of Anantapur of Andhra Pradesh State in India.

practices, cropping systems and community level adaptation of coping strategies which either specifically or generally, represented the views of traditional and indigenous people in respect with climate change in agriculture.

Based on the stratified random sampling technique five villages, Atmakur, Thopudurthi, East-Narsapuram, Basapuram and Mallapuram, of Anantapur district under distinct rainfall zones with severe drought (<50%), moderate drought (50 to 75%), mild drought (75%) were selected for case studies. Pretested case study schedule was used to collect primary data from the respondents of the five villages. The selection of respondents was done based on the purposive sampling method. Farmers aged more than 50 years were selected for the study, as they could provide information on the pre and post green revolution agricultural practices and could assess those changes in present agro-climatic context. Here we present key findings during the case studies on traditional tank management practices traditional soil management practices, and traditional cropping system and plant protection.

Tank irrigation

Management of traditional irrigation system called 'Neeruganti System' adopted by Guvvalagutlapally village community at the time of conceptualization of tank water resource utilization and is continued till today (Figure 2). A committee comprising of the landholders of the command area takes care of tank management activities. The tank committee acts as an apex body in dealing with activities relating to tank management. An elder person called Pinna Pedda elected as chairman of the committee

Figure 2. Traditional tank irrigation systems in Guvvalagutlapally Village. (A) Nagaraju Tank – traditional water tank. (B) Water supply channel to field. (C) Village committee for tank management. (D) Rice cultivation under the tank.

discharges his authority under the directives of 'Tank Committee'. Pinna Pedda recruits people to help in monitoring the discharge canals and in rationalizing the water usage over the fields. They are referred as 'Neeruganti', one who safeguards the water resource.

Neeruganti acts as a water distributing agent across the command area. To ensure adequate irrigation to each landholding in the command area he follows methods like 'Gaging Method' and 'Grading Method'. In 'Gaging Method' Neeruganti fixes various sizes of gages at the mouth of individual field canal to ensure equitable allocation of tank water. In 'Grading Method' prior to the allocation of water to the field, Neeruganti categorizes the entire command area into different grades based on the water retention ability of the land. Neeruganti estimates level of irrigation required for crop cultivation in each grade and reports it to the tank committee. Tank committee decides the amount of water to be distributed to various grades. The duties of Neeruganti is to ensure a uniform supply of water to all the fields in the command area, to inform the farmers in case the crops are affected by pests or diseases, to ensure proper maintenance of the tank outlets.

The Neeruganti follows the decisions taken by the committee regarding the use of water and the maintenance of the tank. If anybody violates the rules by disturbing Neeruganti, the committee reserves an authority to punish them. Expenses relating to tank maintenance and Neeruganti wages are met from the money/grains collected from each landholder of the command area. Further one person from each landholding family participates in desilting acidities during tank maintenance.

Soil management

Farmers in Anantapur district followed traditional soil management practices that were complementary to nature. In relation with soil texture management, four practices like spreading of tank silt in the fields of black soil, ant hill soil in horticulture crop fields (musambi, orange and mango), mixing of alluvial rivulets sediments and spreading of canal silt in the fields are practiced. Texture of the soil changed for cultivation of new crops and to increase water retention or percolation ability for transforming the soil to climatic variations.

Farmers in these study villages used barren lands with less rock and trees as agricultural fields, they used farm implements called 'Gora' (Figure 3) to level the shallows and applied cattle dung during the tilling process. In Thopudurthi and Basapuram villages, crops like castor, sesame, and jowar were preferred as first crop in the newly developed agricultural field.

For improving the fertility of the soil, farmers followed practices such as manuring the field prior to the sowing, applying humus collected from forest or green manuring or by cattle dung every year. Green manure/compost are prepared in pits by using wild plant residues which were crushed using cattle.

Figure 3. Gora.

These are effective and time-tested traditional practices used by the farmers in arresting the soil erosion in the field. Farmers built bunds with soil or semi-solid material by including grasses and farm wastes. They also used small rocks to strengthen these bunds. These bunds are also made to control the flood water current. Near the gully areas in the field, 'Sappillu' is built with loose stones to drain out excess water from fields. Farmers developed a non-intensive tilling process to protect the soil traction against wind and rain forces (Figure 4).

In the case of soil moisture preservation and to increase the percolation ability of the soil to rain water practices like strip tillage, night irrigation, critical irrigation, avoiding weed removal during scanty rainfall, retaining the weed residues on the field were followed.

Cropping systems

Traditionally mixed cropping system was followed across the Anantapur district. Pulse crops like red gram, horse gram, black gram, green gram were sown along with millets like jowar and bajra. Pulses were also sown as inter crop with oilseed crops like groundnut and sesamum. They used mixed cropping system to overcome persistent droughts in the district. Farmers traditionally practiced crop rotation method to improve the soil fertility and to prevent the spread of pests and diseases. During drought periods crops like jowar, horse gram and ragi and cotton were preferred over millets, pulses and other commercial crops. They traditionally cultivated crops like castor and red gram along with groundnut or green gram or cow pea, during recent decades, to withstand extended dry spells.

Figure 4. Traditional soil management practices. (A) Application of tank/canal silt in the field. (B) Soil bund around the field. (C) Sappillu in gully areas.

Traditional knowledge and crop varieties are not only linked, each depends on the other. For example, maintaining and transmitting traditional knowledge relies on the use of diverse biological resources, both wild and domesticated, while the reintroduction of traditional crop varieties has revived related traditional knowledge and practices (IIED, 2009).

Plant protection

During the case study in Anantapur and Bellary districts a wide range of traditional plant protection practices were explored. The practices like sprinkling of cow urine across the borders of the field, cultivation of crops using mixed cropping, cultivation of border crops, crop rotation are the predominant farming cultural practices observed in the study regions.

Another category of traditional practices that promote healthy growth of crops to tolerate pests and diseases under extreme weather includes fortifying the field with cattle dung and forest humus, using locally available wild plants, leaves, grasses as green manure, use of oil cake extracted from neem, castor, groundnut, sesamum (Figure 5).

In the case of preventive measures for pests and diseases, practices like cultivation of local cultivars, crop rotation of pulses with millets or oil seeds, cultivating trap crops (like jowar cultivation in groundnut, marry gold cultivation in chilli fields) with in the field or across the borders, application of neem extract on fruit crops, use of lime water and *erra matti* (oxidised red sediments) to make marks on the trunks of the trees up to three feet from ground level to prevent the pest and disease.

Further to mitigate the impacts of pest and disease set-offs on the yields and income losses, the practices like broadcasting of potter-ash, sprinkling of cattle extracts like urine, dung water mixture on infected plants, sprinkling of buttermilk on infected rice fields, sprinkling of turmeric and lime water on vegetable crops, sprinkling of horse gram boiled juice on infected ground nut and rice crop, spraying of safflower oil on infected fields of groundnut, usage of food traps made with boiled rice bran and jaggery and neem oil across the borders of the field to control the pest transfer from one filed to another, broadcasting of boiled jowar mixed with pottery ash to attract the preys to feed on

Figure 5. Plant Protection practices. (A) Mixing of green manure. (B) Application of farm yard manure.

the pest present in the field are the predominant practices followed to mitigate the impact of pest and disease on yield and net income of the crop production.

In spite of proven abilities of traditional agricultural system, doubts still exist about their efficacy. Further refinement and convergence of traditional practices with modern scientific knowledge are the need of the time to realize climate resilient agriculture.

3. Conclusions

Climate variations are associated with the manifestation of floods, drought and other natural calamities. Traditional knowledge and farming practices offer huge potential for building resilience and adapting agriculture to climate change. The indigenous communities have succeeded to develop coping strategies and survived for centuries against livelihood demising threats like floods and droughts. It is important to realize the application of indigenous knowledge on climate change coping strategies. The importance of traditional knowledge in adapting to climate change issues should be addressed as a priority in national adaptation actions and global climate negotiations. Further, recommendations should be made on integrating indigenous knowledge with modern approaches to optimize the advantages in climate change adaptation and resilience.

Acknowledgements

We thank Central Research Institute for Dryland Agriculture (CRIDA) Hyderabad for sponsoring the project on 'Documentation of Indigenous coping strategies to climate change in different areas of agriculture with special reference to drought prone areas of Andhra Pradesh and Karnataka' under, National Initiative for Climate Resilience Agriculture (NICRA) network project of Indian Council of Agricultural Research (ICAR).

References

ANDES (Peru), the Potato Park Communities, International Institute for Environment and Development (IIED) (2011). Community biocultural protocols: Building mechanisms for access and benefit-sharing among the communities of the Potato Park based on customary Quechua norms. IIED, London.

Cruz, P.D., Ramos, A.G. (2006) Indigenous health knowledge systems in the Philippines: a literature survey Paper presented at the XIIIth CONSAL Conference, Manila, Philippines, pp 1-20.

Food and Agriculture Organization of the United Nations (FAO). (2009). FAO and Traditional Knowledge: the Linkages with Sustainability, Food Security and Climate Change Impacts. Available at: http://tinyurl.com/ncozv23.

International Institute for Environment and Development (IIED), ANDES, FDY, Ecoserve, CCAP, ICIPE, KEFRI. (2009). Protecting community rights over traditional knowledge: implications of customary laws and practices. IIED, London.

McLean, G.K. (2010). Advance guard: climate change impacts, adaptation, mitigation and indigenous peoples. A compendium of case studies. UNU-IAS.

Swiderska, K.Y., Song, J., Li, Reid, H., Mutta D (2011a) Adapting agriculture with traditional Knowledge), Briefing, The International Institute for Environment and Development (IIED). Available at: http://tinyurl.com/q7qgu8w.

Swiderska, K., Reid, H., Song, Y., Li, J., Mutta, D., Ongogu, P., Mohamed, P., Oros, R., Barriga, S. (2011b).The role of traditional knowledge and crop varieties in adaptation to climate change and food security in SW China, Bolivian Andes and coastal Kenya. Paper prepared for UNU-IAS workshop Indigenous Peoples, Marginalised Populations and Climate Change: Vulnerability, Adaptation and Traditional Knowledge, Mexico.

Varte I.Z. (2012). Role of indigenous traditional knowledge in sustainable resource management (with special reference to North East India). Two day regional seminar on Environmental Issues in North east India, Churachandpur, Manipur, 12-13 October.

13. Use of permethrin coated nets to protect stored grain from pests infestations

M. Anaclerio[1], M. Pellizzoni[2]*, V. Todeschini[1], M. Trevisan[2] and G. Bertoni[3]
[1]Department of Sustainable Crop Production, Faculty of Agriculture, Food and Environmental Sciences, Università Cattolica del Sacro Cuore, 29122 Piacenza, Italy; [2]Institute of Agricultural and Environmental Chemistry, Faculty of Agriculture, Food and Environmental Sciences, Università Cattolica del Sacro Cuore, 29122 Piacenza, Italy; [3]Institute of Zootechnics, Faculty of Agriculture, Food and Environmental Sciences, Università Cattolica del Sacro Cuore, 29122 Piacenza, Italy; marco.pellizzoni@unicatt.it

Abstract

Surveys carried out in some villages of Democratic Republic of the Congo showed that insect pests are a major cause of food loss for rural populations. Cereal grains are often compromised by pests in the field and during storage. Among cereals cultivated for family consumption in these areas of Africa, mainly maize and rice are both seriously infested by Coleopteran pests belonging to the genus *Sitophilus*. Any approach currently adopted to solve this problem, as the manual removal of visible insects, is often inadequate or insufficient. Due to the limited availability of chemical insecticides and their high price in these countries, our aim is to study a simple and cheap way to protect grain after harvest. The technology derives from anti-malaria mosquito nets, also known as insecticide treated nets or bednets, modified for the new purpose. Nets consist in fine-mesh polyester coated with permethrin (2.0%), a synthetic second generation pyrethroid that gives repellent and insecticide effect, while ensuring a high human safety. The idea is to use these nets to wrap jute bags or other containers in which grain is commonly stored in poor countries. Preliminary bioassays with adults of *Sitophilus oryzae* showed that knock-down values for 50 and 95% of exposed insects (KD_{50} and KD_{95}, respectively) were equal to 4.8 and 9.5 min. We are currently carrying out further *in vivo* bioassays (in Italy) and field studies (in Congo) to test the efficacy of the treated net. The coating of structures for grain preservation could be a viable sustainable strategy for limiting food losses in countries that lack resources and technologies for pest control.

Keywords: *Sitophilus*, weevils, pyrethroids, food protection

1. Introduction

Postharvest losses are recognized as a major constraint in Africa with reports of annual losses averaging 30% of durable stored grains (Golob and Webley, 1980). Food availability and accessibility can be improved by increasing production and reducing losses in order to satisfy the future food demand and ensure the global food security (Alexandratos *et al.*, 2012).

In the developing countries, the greatest losses during storage to cereals and grain legumes are caused by insect pests (Obeng-Ofori, 2010). In fact insects may cause severe losses of stored food for their ability to infest raw materials or processed products during transportation, storage in warehouses and retail stores (Germinara *et al.*, 2010).

Stored food insect pests can be classified into 'penetrators' when they are able to actively infest food by perforating packages or 'invaders' if they can only enter through existing holes (Hou *et al.*, 2004) caused by other pests or accidental damage (Mullen and Mowery, 2000). Insect pests are able to find food using their sense of olfaction and to move from one package to another (Allahvaisi *et al.*, 2011).

Among all the post-harvest insect pests, the cosmopolitan weevils belonging to the genus *Sitophilus* (Coleoptera: Dryophthoridae) are particularly problematic in the developing countries and they are well known for their ability to perforate several different packaging materials (Fleurat-Lessard and Serrano, 1990). The most important species are the granary or wheat weevil (*Sitophilus granarius*), the rice weevil (*Sitophilus oryzae*) and the maize weevil (*Sitophilus zeamais*) which feed on several kinds of cereal grains, beans, nuts and dried fruits. Females lay eggs in holes they bore in the seeds, usually one per grain, then the newborn larvae feed on the seed endosperm until they emerge as new adults, leaving the seed completely hollow (Plarre, 2010). In African rural areas, cereals cultivated for family consumption, mainly maize, sorghum, rice, wheat and millet, are seriously infested by these weevils during storage and from surveys carried out in some villages of Democratic Republic of the Congo emerged that insect pests are a major cause of food loss for rural populations.

Some basic techniques are currently adopted to solve this problem, as the manual removal of visible insects, but they are often inadequate or insufficient and the chemicals are almost unused, both for their poor availability and their high costs. At the Faculty of Agriculture, Food and Environmental Sciences of Università Cattolica del Sacro Cuore (Piacenza, Italy) we are currently testing a new method to protect stored grain in a simple and cheap way.

The idea and the technology derive from anti-malaria mosquito nets, also known as insecticide treated nets (ITNs) or bednets, which have been modified to wrap jute bags, polypropylene bags or other containers in which grain is commonly stored in developing countries. These 'long lasting insecticidal nets' are guaranteed to maintain their efficacy for at least five years and are sold for under $ 5 in Africa (approximately 5-6 m^2). The quick knockdown effect given by pyrethroids, in contrast to the slower action of other groups of compounds (Wickham *et al.*, 1974), could contribute to the protection of stored products from pests attacks.

The present study intended to evaluate the efficacy of ITNs in the laboratory, and then transfer the acquired knowledge to the reality of poor countries. This technique, if promoted by the international community, could be submitted in the context of integrated pest management because of its economic and ecological sustainability.

2. Materials and methods

The modified nets consist in a fine-mesh (reduced compared to that employed for mosquitoes) polyester coated with permethrin (2.0%), a synthetic second generation pyrethroid which acts as repellent and insecticide, while ensuring a high human safety. Permethrin is commonly used for public health treatments against lice and other parasites.

Preliminary bioassays have been performed with adults of *Sitophilus oryzae* exposed to the treated nets; insects come from colonies reared at the Università Cattolica del Sacro Cuore facilities for many years.

In the first bioassay the weevils have been exposed to treated net following the WHO cone bioassay procedures (WHO, 2013) and the knocked down individuals were counted minute by minute to estimate the knock-down values for 50% and 95% of exposed insects (KD_{50} and KD_{95}, respectively).

In the following bioassay, five groups of 10 weevils, in five replicates, were exposed to the treated net in plastic boxes, forcing the contact for 1, 3, 5, 10 and 30 min and then transferred into clean plastic boxes with rice for feeding. An untreated check was kept as control. Since knockdown following application of an insecticide consists in a state of intoxication and partial paralysis which usually, but not necessarily, precedes death, affected weevils were recorded after 1, 24 and 48 h.

3. Results and discussion

The cone bioassay showed that KD_{50} was equal to 4.8 min and KD_{95} was equal to 9.5 min, as shown in Figure 1. Results of the other bioassay are shown in Figure 2. All the *S. oryzae* adults in the untreated control were alive at each time interval. The slightly decreased number of affected insects after 24 and 48 h compared to 1 h check is due to the fact that the immediate knockdown effect of pyrethroids is often temporary at sub-lethal rates or times of exposure.

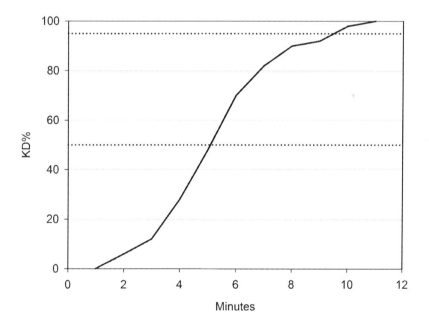

Figure 1. Knockdown curve for Sitophilus oryzae.

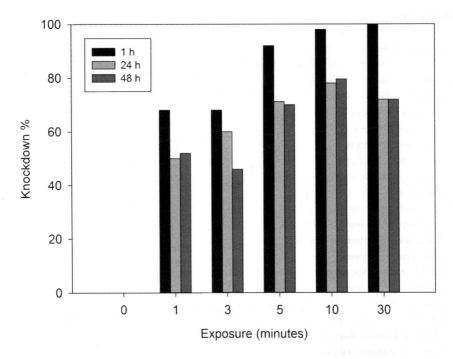

Figure 2. Effect of different exposure times (0 = control).

We can deduce that the simple wrapping of a jute or polypropylene bag with treated nets could increase efficiently the protection from pests because a very short time of contact with the insecticide is sufficient to affect insects.

To evaluate the possible application of these nets it is necessary to better assess insect capacity and speed of penetration within the bags used for cereals storage. For this reason we are currently carrying out further *in vivo* bioassays to test the efficacy of the treated nets, comparing the penetration of *S. oryzae* inside different kind of sacks (jute and polypropylene) coated with treated or untreated nets. Field test are currently ongoing in Congo as well. Preliminary results, not presented in this work, confirm that the modified net coated with permethrin 2% is able to ensure a total protection from pests attack. Field studies, however, underlined that grains are often stored already infested; in this case the ITNs protection can only reduce the entity of the infestation avoiding additional entry from outside.

Even if permethrin is safe for humans and treated nets never get into contact with the stored grain, some studies are currently in progress to evaluate any possible translocation of insecticide active ingredient from net to grain. Given the low costs and the easy application, the treated coating of sacks or other containers for grain preservation could be a viable sustainable strategy for limiting food losses in countries that lack resources and technologies for pest control.

References

Alexandratos, N., Bruinsma, J. (2012). World agriculture towards 2030/2050: the saving water. From Field to Fork-Curbing Losses and Wastage in the Food Chain 2012 revision. Working paper: FAO: ESA No. 12-03, p.4.

Allahvaisi, S., Maroufpoor, M., Abdolmaleki, A., Hoseini, S., Ghasemzadeh, S. (2011). The effect of plant oils for reducing contamination of stored packaged-foodstuffs. Journal of Plant Protection Research 51: 82-86.

Fleurat-Lessard, F., Serrano, B. (1990). Resistance des films de matieres plastiques aux perforations par les insectes nuisibles aux denrees. 2. Etude methodologique sur les tests *Sitophilus oryzae* (L.) et Prostephalus truncatus Horn. Sciences des Aliments 10, 521-532.

Germinara, G.S., Conte, A., Lecce, L., Di Palma, A., Del Nobile, M.A. (2010). Propionic acid in bio-based packaging to prevent *Sitophilus granarius* (L.) (*Coleoptera, Dryophthoridae*) infestation in cereal products. Innovative Food Science and Emerging Technologies 11: 498-502.

Golob, P., Webley, D.J. (1980). The use of plants and minerals as traditional protectants of stored products. Report Tropical Product Institute, G138.

Hou, X.W., Fields, P.G., Taylor, W. (2004). The effect of repellents on penetration into packaging by stored product insects. Journal of Stored Products Research 40: 47-54.

Mullen, M.A., Mowery, S.V. (2000). Insect-resistant packaging. International Food Hygiene: 11, 13-14.

Obeng-Ofori, D. (2010). Residual insecticides, inert dusts and botanicals for the protection of durable stored products against pest infestation in developing countries. Julius-Kühn-Archiv 425: 774.

Plarre, R. (2010). An attempt to reconstruct the natural and cultural history of the granary weevil, *Sitophilus granarius* (*Coleoptera*: *Curculionidae*). European journal of Entomology: 107, 1-11.

Wickham, J.C., Chadwick, P.R., Stewart, D.C. (1974). Factors which influence the knockdown effect of insecticide products. Pesticide Science 5(5): 657-664.

World Health Organisation (2013). Guidelines for laboratory and field testing of long-lasting insecticidal mosquito nets WHO/HTM/NTD/WHOPES/2013.1. Geneva.

14. Rural development plays a central role in food wastage reduction in developing countries

A. Minardi[1*], *V. Tabaglio*[2], *A. Ndereyimana*[1], *M. Fiorani*[1], *C. Ganimede*[2], *S. Rossi*[1] and *G. Bertoni*[1]

[1]*Institute of Zootechnics, Università Cattolica del Sacro Cuore, Via Emilia Parmense 84, 29122 Piacenza, Italy;* [2]*Department of Sustainable Crop Production, Area Agronomy and Plant Biotechnologies, Università Cattolica del Sacro Cuore, Via Emilia Parmense 84, 29122 Piacenza, Italy; andrea.minardi@unicatt.it*

Abstract

According to the Food and Agriculture Organization of the United Nations (FAO), roughly one-third of food produced for human consumption is lost throughout the Food Supply Chain. This Food Wastage (FWa) could be divided in Food Loss (decrease in mass or in nutritional value of food for human consumption), and Food Waste (food discarded because left to spoil or expired). FAO suggests to start analysis of FWa from harvesting, but several authors suggest to start from sowing because FWa can also occur between sowing and harvesting. For this reason, to limit FWa it is important to adopt better agricultural practices (seed quality, appropriate organic and synthetic fertilizers, type of rotation, multiple cropping, pruning, and pesticide control). Each item of FWa – according to FAO including only edible part losses – could be further divided in *unavoidable*, *not economically avoidable* and *economically avoidable*. Besides the efforts to reduce last one, it is also important trying to recycle *unavoidable* and *not economically avoidable* losses (i.e. as animal feeds or composting). In all these aspects there are great differences between Developed Countries and Developing Countries, mainly justified by available structures and infrastructures. Therefore we have carried out some experiences on FWa reduction in Dem. Rep. of Congo and India. Within established women groups, the following activities have been made: (1) planting improved seeds of rice in India has almost doubled the grain yield without any other change; (2) constructing warehouse in common to contain hand-crafted silos for cereals, in order to avoid insect attack, resulted feasible and very effective; and (3) drying or frying with palm oil fresh vegetables allows prolonged storage. To conclude, acting to reduce FWa to obtain maximum potential yield, as well as reducing the post-harvest losses, recycling FWa as animal feed, composting and anaerobic digestion, limits the need for food production rise. However, uncertainties remains about the actual extent of FWa; difficult is therefore the estimating the amount of its reduction.

Keywords: pre-harvest losses, post-harvest losses, developing countries, rural development, food wastage, food wastage recycling

1. Introduction

Reducing Food Losses (FL) and Food Waste (FW) is a topic that is receiving much attention because of its close link with two major issues of the moment: guaranteeing access to food for a rapidly growing world population and the correct management of waste, in order to also minimize the environmental impact of agricultural production (Bagherzadeh *et al.*, 2014). According to the Food and Agriculture Organization of the United Nations (FAO) estimates (FAO, 2011), one third of the food produced in 2009, the equivalent to 1.3 billion tons of food, was lost along the Food Supply Chain (FSC), with

Leire Escajedo San-Epifanio and Mertxe De Renobales Scheifler (eds.) *Envisioning a future without food waste and food poverty*

125

DOI 10.3920/978-90-8686-820-9_14, © Wageningen Academic Publishers 2015

the FSC being defined as all the steps from harvest to consumption. This estimate differs from that of Didde (2013) and the World Wide Fund for Nature (WWF, 2013), which include agricultural production losses in their estimates, in agreement with the recommendations of the European Union (EC, 2011). The FAO study (FAO, 2011), as well as disregarding the production phase, also presents a series of uncertainties that will be discussed in detail hereafter. However, having said that, FAO estimates on FL and FW are believed to be the best estimates currently available (Lipinski *et al.*, 2013).

2. Defining losses and waste

Food losses and food waste

A single definition of what is meant by FL and FW is not currently available (Bagherzadeh *et al.*, 2014). In accordance with the FAO (FAO, 2011) FL are defined as the losses that occur from harvest through to distribution, while FW is what is wasted during distribution and consumption. Buzby and Hyman (2012) agree with these distinctions and they justify the exclusion of the pre-harvest phases because of the difficulty to obtain a trustworthy dataset. We, on the other hand, in accordance with Didde (2013) and WWF (2013), believe that pre-harvest losses (those that occur during production) are just as important as post-harvest losses. Another contested issue is whether losses and wastes should include everything that is lost up until consumption, regardless of the reason, or whether only 'edible' losses and wastes should be accounted for (thereby excluding fruit peel and pits, bones, hulls and bran, etc. from FL and FW estimates). The FAO (FAO, 2011), Buzby and Hyman (2012) argue for the latter, Bagherzadeh *et al.* (2014) believe the issue is irrelevant, and Didde (2013) believes that losses and wastes should include everything. Regardless of the above definitions, FL and FW can be further defined by the technological, behavioural and economical limitations from which they are generated. In particular, FL and FW can be distinguished as being unavoidable or avoidable (Edjabou *et al.*, 2014); production processes include risk factors, among which only some are controllable, and the extent to which they can be controlled depends on the tools that are available. If the entire FSC is taken into consideration, risk factors include adverse climatic conditions, illness and inadequate means of transport, food processing and food preparation etc., which are usually avoidable in the developed countries only. Whether FL and FW can be defined as unavoidable or avoidable also depends on the decisions made along the FSC, decisions that are also driven by ethical goals (i.e. organic farming).

An additional issue related to the uncertainty of their definition is how the losses and wastes are quantified; in literature their quantities are expressed in kg, kcal and monetary value (usually US dollars) (Edjabou *et al.*, 2014), thereby rendering some of the data incomparable (Dahlén and Lagerkvist, 2008) and limiting its relevance. A better definition of losses and wastes would facilitate the publication of more reliable estimates.

Pre-harvest losses

Though often not considered, pre-harvest losses are all the losses that occur during production in the field and are the losses that constitute the first reduction in the available amount of food. These losses can be caused by a range of factors, such as seed choice and crop and pest management, but they also include choices such as opting for a reduced yield in order to achieve a supreme quality; for example, choosing an antique rather than an innovative plant variety for a misunderstood attempt to return to better savours and more natural and safe agriculture. These losses can again be divided into avoidable and unavoidable: avoidable losses are those related to choice as a function of market demand (e.g. seed type and quality), while unavoidable are the losses due to the weather (heavy rain, hailstones, drought, etc.) or to either a lack of technological solutions or the decision not to use them (e.g. GMO). Although the losses referred to above are difficult to evaluate, they represent an essential phase, particularly important for the subsistence agriculture because decisions made in this phase can determine the losses that occur successively (e.g. the chosen variety may be more susceptible to contamination by insects, mould, and bacteria post-harvest).

Post-harvest losses

These are the losses that occur during harvest, processing and preservation because the produce is left un-harvested or mechanically damaged during the harvest, lost or damaged during transport from the field to the factory, lost or damaged during processing and preservation and lost or damaged during distribution and commercialization. Losses can be of a physical nature (weight and volume) and of a qualitative nature (loss of vitamins, other nutritional compounds, scent or taste), and once again they can be defined as avoidable and unavoidable, always remembering however that the products themselves can be perishable and 'non-perishable'. A reduced loss of perishable produce can only be achieved via appropriate technological means (refrigeration or thermal treatments); nonetheless, loss of perishable produce is inevitable. Inevitable losses can be greatly reduced when dealing with 'non-perishable' produce (e.g. dried grains and legumes) because simple treatments can protect them against insects and rodents.

Distribution and consumption losses (waste)

In developed countries 66% of FW is represented by waste along the retail chain; the other 34% is domestic waste (Bernstad Saraiva Schott and Andresson, 2014). Distribution wastes occur when foods become substandard (e.g. stale bread, out of date products), or when fresh produce is damaged and aesthetically unappealing. Not all food categories are subject to the same amounts of waste; according to Schneider and Wassermann (2004) the waste generated by Austrian supermarkets is 50% vegetables, 30% fruit, 9% cereals and the remainder is animal products. In domestic circumstances FW is mostly due to mismanagement and incorrect conservation of the purchased foods (Salhoefer *et al.*, 2007), though there is also some misunderstanding of the best before dates on food labels (and although care should be taken, there is often no immediate risk in consuming the food in the subsequent days). A further cause of FW is represented by impulsive food shopping and special promotions by supermarkets, both of which encourage an excessive accumulation of foods and increase the risk of food going out of date before being consumed. If non-edible domestic waste is also taken into consideration, FW would include losses during food preparation (e.g. fruit and vegetable peels, eggshells and bones) or left-overs.

3. Technically unavoidable losses and wastes

Among the causes of today's unavoidable waste many can be tackled in the future (if research solves some of the inherent problems and developing countries will be enabled to install the necessary structures and infrastructures); nevertheless it is worthwhile to reflect on the losses that may occur when the technical means which are needed to avoid them, are still unavailable or too costly. One such example of the first scenario is waste due to adverse climatic conditions (drought, excessive rain, wind, etc.); an example of the latter scenario is the loss of fresh fruit, vegetables, meat and fish due to a lack of machinery, storage, refrigeration, etc. in poor countries (APO and FAO, 2006). All these factors limit the agricultural yield in developing countries, particularly in Africa; yields are much lower than they could potentially be. In a study carried out in 10 countries around the Greater Lakes region of East Africa the yields seemed to be uniformly low, with only 15% to 30% of the agricultural potential being achieved (La Ferrara and Milazzo, 2015) because of a lack of proper seeds and of tools and agrochemicals.

Other unavoidable losses – if more than just the edible losses are accounted for – are the FL and FW that come about during processing and preparation of food/drink (cassava peel, bones, eggshells, pineapple peel, leaves and core, external lettuce leaves, etc.). Processing wastes include tomato skins, bran, oilseed meals and are often utilized as by-products in animal feeding.

4. Recycling food losses and food waste: perspectives

As it has already been said, large quantities of FL and FW can be generated along the FSC, and both the FL and FW may be unavoidable (inedible material, perishable produce no longer suitable for consumption), but nonetheless FL and FW may be useful in agronomic and economic terms, as well as in terms of environmental sustainability. The considerable quantity of nutrients in FL and FW could be recycled to feed animals, which are a key aspect of poor-agriculture: economic, social and food security. Such 'recycling' is probably as old as agriculture itself, with domestic animals having always eaten human food wastes, and although its practice in developed countries has gradually been lost (Mopaté *et al.*, 2011), its praxis in developing countries continues, particularly because the feed supply does not meet the demand (Foley *et al.*, 2011). Recently the FAO (FAO, 2013a) indicated various applications that can exploit this type of waste, and giving them value as animal feed. Further emblematic examples of this application (Bertoni, 2015) are: rice husk fed to pigs in India, fermented cereal residues from beer production, fed to pigs in DR Congo, non-edible parts of fruits and vegetables fed to chickens and pigs. That which cannot be used as animal feed can be composted.

Developed countries have a contrasting situation; only some of the aforementioned FL and FW is used as animal feed, and the rest is added to the already huge volumes of litter that needs to be burnt, causing environmental pollution. Fortunately, however, when organic litter is collected separately it is possible to activate the more virtuous processes of composting or anaerobic digestion, both of which reduce the aforementioned disposal and pollution problems. Paradoxically, in the urban areas of underdeveloped countries, the recycling of FL and FW became more difficult and therefore are burned or simply downloaded in the landfill or in the rivers with environmental pollution problems (FAO, 2013a).

5. Examples of loss-reduction in developing countries

In developing countries, losses are of particular relevance in primary production processes and in the successive phases of processing and preservation; here below are details of our experience in

India (Meghalaya) and the Democratic Republic of the Congo (Eastern Kasai), within the project 'Production of appropriate food: sufficient, safe and sustainable' funded by R. and E. Invernizzi Foundation.

Pre-harvest losses

Regarding production losses on the field, the best possibility to reduce losses is to have access to locally ameliorated seeds that are better adapted to the environment, because agrochemicals cannot be made available at an affordable price. For this reason, in India we compared an improved rice variety with the local variety – (the improved variety was supplied by the KVK Centre, Krishi Vigyan Kendra – ICAR Research Complex for NEH Region). A two-year trial indicated that it is possible to achieve higher yields; paddy rice yields increased from 2.0-2.5 to 3.5-4.5 Mg/ha. Therefore, it would be sufficient to shift from traditional varieties to the improved varieties produced in local research centres, as a first step to improving the living conditions of the people (freeing land so it can be used for other crops and increasing their income and giving them the potential to make investments, etc.).

In DR Congo the pre-harvest losses (represented by lower production) are a consequence of difficulty to optimize cultivation techniques due to lack of fertilizers, irrigation and agrochemicals; in some cases the inputs are not available, and in other cases the elevated costs prevent their use. Within the framework of sustainable intensification of DR Congo's traditional agricultural systems, a field trial was set up in 2012/2013 (approx. 1.5 ha) to compare 5 local cassava ecotypes being planted in 3 different ways. Results showed that one tested ecotype (i.e. Kakwanga) was more resistant than others to the mosaic virus and it was more productive. The identification of varieties that are inherently more resistant to local adversities could be a first step towards rural development.

Post-harvest losses: processing and/or preservation

After proper drying, cereals and legumes can be lost due to infestation by insects or rodents. The FAO (2008) demonstrated that the use of small metal silos (family-sized) enables the dry grains and seeds to be kept for more than six months. Analogous structures have therefore been installed at a reasonable price in DR Congo, using flattened corrugated iron tiles, moulded to form a silo by applying the techniques used for wrought iron; since both the corrugated iron and the necessary skills are readily available, the construction of such silos could generate useful non-agriculture jobs.

This conservative strategy has been seen to have numerous advantages: the possibility to adapt to the local circumstances; the effectiveness of the silos to protect the grains and seeds from insect and rodent infestation for more than six months; and a longer life of the new structures with respect to what was used traditionally.

Fresh produce, like fruit and vegetables, is readily perishable post-harvest because of the hot climates and the lack of structures and infrastructures for transport, storage and cooling (Rolle, 2006). Therefore, in accordance with the FAO (FAO, 2013b) the authors decided to develop two 'models' that increase food security and are compatible with the structures already available and with the modest quantities of produce needing to be managed. The first model uses sun-drying to conserve the excess fruit; the second model exploits home-preservation, via, for example, the production of plantain chips. Home-drying of foods rich in sugars (fruits) requires optimal climatic conditions and a system to protect the food from flies, but both are not always available. Frying foods in palm oil, on the other hand, is a promising option.

6. Conclusions

Fresh foods are often produced in environments that cannot be controlled, and the fresh produce itself can be characterized according to its perishability. It follows, therefore, that some loss of fresh produce along the FSC is inevitable, starting from its field production until consumption. In order to facilitate a reciprocal understanding of workers and stakeholders it would be useful if the following aspects were more clearly defined:

- If the term 'losses' should be applied to that which occurs prior to distribution and the terms 'waste' to that which occurs successively. From our point of view the distinction is not entirely precise, but it is acceptable.
- If that which occurs in the field prior to harvest is included in 'losses' (but in some cases is included as 'waste'), as we would wish it to be, because they reduce the amount of food made available.
- If the portion of non-edible food parts (seeds, peels, bones, etc.) is excluded from what is considered waste – as we would wish it to be – but remembering that such food parts have always been used in animal feeding and/or recycling.
- The relevance of interventions to reduce FL and FW must be also considered, because sometimes their economic and environmental cost can be higher than the cost of producing what was lost.

Clarity on the issues listed above would facilitate the estimate of the losses and wastes and their clear positioning within the FSC, and it would help identifying tools for reducing them (where this is possible and appropriate). With particular reference to developing countries, where the highest FL are pre- and post-harvest, the technological and economical limitations of the interventions done to avoid them are of primary importance. It follows that the developed countries are asked to supply their own know-how and financial support, but above all, their technical assistance. The latter is essential in order to successfully implement the suggested changes (e.g. better adapted seeds, simple tools); nevertheless, the technicians from developed countries must simply support a local staff that better interacts with population and acquires competence. Finally, it must not be forgotten that the use of agrochemicals may also be among the suggestions; while their use brings more food into the balance, there are also some food safety and sustainability issues related to their use.

References

Asian Productivity Organization and Food and Agricultural Organization. (2006). Postharvest Management of Fruits and Vegetables in the Asia-Pacific Region. Rome, Italy.

Bagherzadeh, M., Inamura, M., Jeong, H. (2014). Food Waste Along the Food Chain. OECD Food, Agriculture and Fishering Papers, No 71. OECD Publishing.

Bernstad Saraiva Schott, A., Andresson, T. (2014). Food waste minimization from a life-cycle perspective. Journal of Environmental Management 147: 219-226.

Bertoni, G. (eds.) (2015). Produzione ed uso del cibo. Sufficienza, sicurezza e sostenibilità. EGEA, Italy. (in press).

Buzby, J.C., Hyman, J. (2012). Total and per capita value of food loss in the United States. Food Policy 37: 561-570.

Dahlén, L., Lagerkvist, A. (2008). Methods for household waste composition studies. Waste Management 28: 1100-1112.

Didde, R. (2014). Taking care of the harvest. Wageningen World 3, pp. 20-25.

Edjabou, M.E., Jensen M.B., Götze, R., Pivinenko, K., Petersen, P., Sheutz, C., Asturp, T.F. (2014). Municipal solid waste composition: Sampling, methodology, statistical analysis, and case study evaluation. Waste Management 36: 12-23.

European Community (2011). Preparatory study on food waste across EU 27. Available at: http://tinyurl.com/4424obc.

Foley, J.A., Ramankutty, N., Brauman, K.A., Cassidy, E.S., Gerber, J.S., Johnston, M., Mueller, N.D., O'Connell, C., Ray, D.K., West, P.C., Balzer, C., Bennett, E.M., Carpenter, S.R., Hill, J., Monfreda, C., Polasky, S., Rockström, J., Sheehan, J., Siebert, S., Tilman D., Zaks, D.P.M. (2011). Solution for a cultivated planet. Nature 478: 337-342.

Food and Agriculture Organization (2008). Household metal silos key allies in FAO's fight against hunger. Rome, Italy.

Food and Agriculture Organization (2011). Global food losses and food waste – Extent, causes and prevention. Rome, Italy

Food and Agriculture Organization (2013a). Toolkit. Reducing the food wastage footprint. Rome, Italy.

Food and Agriculture Organization (2013b). Utilization of fruit and vegetable wastes as livestock feed and as substrates for generation of other value added products. Rome, Italy.

La Ferrara, E., Milazzo, A. (2015). Sviluppo e Nutrizione. Il ruolo delle norme sociali EGEA, Milano, Italy, 129 pp.

Lipinski, B., Hanson, C., Lomax, J., Kitinoja, L., Waite, R., Searchinger, T. (2013). Reducing food loss and waste. World Research Institute. Washington, DC.

Mopaté, L.G., Kaboré-Zoungrana, C.L., Facho, B. (2011). Disponibilités et valeurs alimentaires des sons de riz, maïs et sorgho dans l'alimentation des porcs à N'Djaména (Tchad). Journal of Applied Bioscience 41: 2757-2764.

Parfitt, J., Barthel, M., Macnaughton, S. (2010). Food waste within food supply chains: quantification and potential for change to 2050. Philosophical Transcritions of the Royal Society 365: 3065-3081.

Salhofer, S., Obersteiner, G., Shneider, F., Lebersorger, S. (2007). Potential for the prevention of municipal solid waste. Waste Management 28: 245-259.

Schneider, F., Wassermann, G. (2004). Social transfer of products in retail. On behalf of the Initiative 'Abfallvermeidung in Wien', Vienna, Austria.

World Wide Fund for Nature (2013). Quanta natura sprechiamo? Le pressioni ambientali degli sprechi alimentari in Italia. WWF, Rome, Italy.

15. The use of web-based technology as an emerging option for food waste reduction

C. Corbo[1] and F. Fraticelli[2,3]*
[1]*Università Cattolica del Sacro Cuore, Largo A. Gemelli, 1, 20123 Milan, Italy;* [2]*Università Politecnica delle Marche, Piazzale R. Martelli 8, 60121 Ancona, Italy;* [3]*Università della Svizzera Italiana, Via Giuseppe Buffi 13, 6900 Lugano, Switzerland; f.fraticelli@univpm.it*

Abstract

In recent years web technologies have been enabling new ways for reducing the food waste by enhancing unexplored connections between donors and beneficiaries of food commodities. In this context, the aim of this paper is to introduce a framework that can serve as a basis for identifying best technology practices for food waste management. As an explorative example, the framework is applied to 8 Italian IT platforms. This paper fosters the understanding of the implication of technology in pushing for emerging models in the food chain organizations, in terms of roles and policies.

Keywords: innovation, food chain, food banks, web platforms, efficiency of food chains

1. Introduction

Every year, whilst about four billion metric tonnes of food per annum is produced, about 30-50% of this is lost along the overall food supply chain (Institution of Mechanical Engineers, 2013), from the production to the consumption stage. In addition, significant quantities of resources (as land, water, energy, soil and organic matter) are lost in the production of food that will be never consumed, as well as the huge amount of economic resources invested in food production (estimated in USD 1 trillion each year; Beretta *et al.*, 2013; FAO, 2014) and consumption (Buzby *et al.*, 2012). Food losses have significant environmental burden also because of landfill disposal and consequent greenhouse gases emissions (Adhikari *et al.*, 2006). Food poverty and food waste are symbols of inequalities and inefficiencies of our contemporary food systems (Midgley, 2014): with the world's population anticipated to reach 8 billion people by 2025, this quantity of wastage is unacceptable for moral, ethical, health, environmental, social and economic reasons (Williams *et al.*, 2015).

Generally, the term 'food loss' refers to a decrease in mass or nutrition value of food that was originally intended for human consumption (Grolleaud, 2002). Food losses can occur at different points in the food supply chain, from the raw materials production to household consumption (Figure 1). Food losses occurring at the end of the food chain are rather call 'food waste': they are related to retailers' and consumers' behaviours (Gustavsson *et al.*, 2011; Parfitt *et al.*, 2010). Although food losses in industrialized countries are as high as in developing countries, in developing countries food losses occur mainly at post-harvest and processing levels (due to lack of proper technologies for food storage and logistic in the developing countries, often in combination with extreme climatic conditions), while in industrialized countries, more than 40% of the food losses occur at retail and consumer levels (Gustavvson *et al.*, 2011).

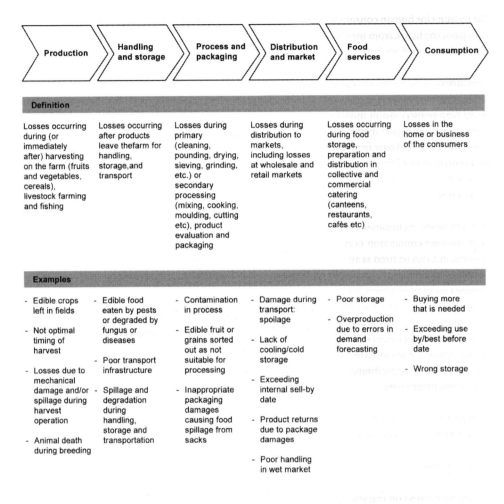

Figure 1. Food chain steps and players (adapted from Beretta et al., 2013; Garrone et al., 2014; Lipinski et al., 2013).

How food waste can be reduced? According to the 'food waste hierarchy' (Papargyroupoulou *et al.*, 2014), the first strategy to reduce food waste should be obviously the prevention of food surplus creation (for example, improving the demand planning and retailing, using price promotions for foods slightly damaged or nearing expiry). Once food surplus has been created, it can be managed in several ways: it can be re-used for human consumption and, when it is no more possible to recover it for this purpose, it can be used to feed animals or in industrial processing or composting. Finally, food products that are not recovered must be disposed of as waste (incineration of in landfills) (Garrone *et al.*, 2013).

The re-using for human consumption is the preferred option: this generally is done collecting surplus food from the fields, from manufacturing companies, from food service providers for distribution to the needy, through redistribution network and 'food banks' (Garrone *et al.*, 2013).

Particularly, food banks are considered as a key frontline response to the growing problem of food poverty in first world societies through food redistribution. Even before 'organizations', the food banks are a redistribution model of surplus food. In fact, food banks have basically a 'matching' role within the food chain: they reduce food waste by linking charitable organizations to other players of the food chain. Apart from donation to food banks, however, few studies provide a quantitative representation of different alternatives to re-use food for human consumption. Some authors (e.g. Beretta *et al.*, 2013) suggest, for example, donations between consumers when an excessive quantity of food has been bought or prepared.

In recent years, technology has been enabling new ways for reducing the food waste by enhancing a peer-to-peer connection between givers and beneficiaries of food. Social networks, mobile apps, websites, etc. can be used as an efficient way to reduce waste and fight hunger. And they actually are: in the last three years, both in Europe and abroad, it has been increasing the number of technology solutions for food donation, exchange or selling between consumers, farmers and retailers. Smartphone applications as *LeftoverSwap*, online communities as *CropMobster* and websites as foodsharing.de are increasingly connecting consumers, farmers, shops, restaurants and consumers to share, exchange or buy (at a special price) surplus food, with the main goal to reduce food waste and, in some cases, to fight hunger. Also in Italy this emerging trend can be recognised, with several mobile apps and websites allowing or facilitating food sharing, food donation and, in some cases, the sale of products close to the expiry date.

Since this phenomenon is still at an emerging phase, we decided to use an explorative methodology to address the topic and figure out the main occurring trends.

2. Methods

This study is based on the analysis of web and mobile platforms for food waste reduction currently operating in the Italian market. Following the principle of theoretical sampling (Mayan, 2009; Santos and Eisenhardt, 2005), we have selected a portion of the full set of Italian platforms that, according to our analysis, represent the 'best' examples in technology for food waste reduction.

The sample was built by using a snowball approach (Goodman, 1961; Noy, 2008). Since technology for food waste reduction is a recent and innovative topic, we started to build our sample by analysing the start-ups listed in 'WHOISWHO' registry held by ItaliaStartup Association (http://www. italiastartup.it/whoiswho/), a permanent observatory on the Italian innovation ecosystem sustained by the Italian Ministry of Economic Development.

At the time of the analysis (6 May 2015), the total number of listed start-ups was 3,481. Among these, 664 were involved in IT services and had an operating website suitable to be used in order to gather valuable information for our research. Since none of these start-ups resulted to be involved into the development of web or mobile platforms for food waste reduction, we used a search engine to obtain a list of popular technology currently recognized as valuable for our research. Thanks to this approach, we finally came to a list of 8 different web and mobile platforms suitable to be included in our analysis.

The data for our analysis were therefore collected from the official websites of each platform. They were collected in the primary language of the website: only 1 website was analysed in English, while the rest in Italian. In the end we downloaded, read and analysed 38 pages from the official websites (incl. texts, images, snapshots and links to embedded videos, downloadable documents, etc.).

In order to address the impact of web and mobile technology on food waste management, we wanted to understand how each company addresses and explains their value proposition in their own words (Thompson, 1997), namely which categories they use 'naturally' when they tell their own story to the stakeholders.

In the analysis of the secondary data (every type of the contents incl. texts, images, videos on official websites) we used a qualitative approach to obtain a thick description of the platforms. The data were analysed using latent and manifest content analysis (Mayan, 2009) via selective and open coding (Strauss and Corbin, 1990).

3. Empirical application: the Italian example

As an empirical illustration, we tested a framework aimed at describing the main characteristics of eight web-sites and mobile technologies (namely, IT platforms) enabling food waste reduction in the distribution and consumption stage. These IT platforms are delivered by eight different organizations (private or non-profit) and are well established (in terms of general awareness) in the Italian market. The list of products was retrieved by using the above-mentioned methodology and is summarized in Table 1. This data set is obviously limited and the analysis should therefore be considered as an illustration rather than as a test of the validity of the framework. In this paragraph, we first explore the framework dimensions separately and then describe a number of categories falling out of the data when we combine all the dimensions.

With regard to 'transaction approach', we studied the food transfer criteria mentioned in each platform website. Six platforms (Ratatouille, Fame Zero, Breading, Bring The Food, IFoodShare and Scambia Cibo) offered full functionalities to let users communicate the food surplus they have, in order to let other users take that food without giving back anything. In other terms, these platforms had a clear 'donation' approach. One platform (Last Minute Market sotto casa) let point of sales to use a SMS to communicate their surplus to end-users that signed up in the platform. These users went to the shop and bought food at a discounted price. This platform was clearly sale-oriented.

One platform (MyFoody) was a marketplace in which point of sales can sell products at a discounted price. Consumers could buy food at a low price and take it directly from the reseller. This platform was considered a 'second market' and, therefore, it was profit and sale oriented under this perspective. However, at the same time, this platform let users donate part of their shop to non-profit organizations (NPOs), and was therefore partially donation-oriented. We considered this platform as an 'hybrid' one: mostly oriented to a sale approach but – at the same time – ready to host end-users with a selfless spirit.

Table 1. Analysed samples.

Platform	Short description
Ratatouille	Food-share app. Users are able to upload a description and pictures of the unwanted perishables, allowing residents meet up with others in the community to pick it up. Type: app (iOS) http://www.ratatouille-app.com/
MyFoody	E-commerce geolocalized platform, allowing retailers and producers to sell products that risk to be wasted (closed to the expiry date, slightly damaged or with aesthetic defects). Consumers can book and buy products, that are taken at the selling point. Type: website http://www.myfoody.it/
FameZero	App and website aiming to facilitate food donations from people, organizations or public institutions to non-profit organizations (NPOs) working in the field of food collection and redistribution. Between the 'Donors' and the 'Food collecting point', the 'Angels' work in the name of and on behalf of FameZero, to collect food. Type: mobile app (iOS and Android) www.famezero.com
Last Minute sotto casa	'Last-minute' marketplace for surplus products, connecting groceries and shops with consumers. An 'alert' is received on the consumer's smartphone every time a shop is selling food at a special price near the consumer's area. Type: website www.lastminutesottocasa.it
Breading	A digital platform that allows the redistribution of surplus food, connecting food services (restaurants, cafés, take-always and bakeries) with NPOs working in the field of food redistribution. The NPOs, thanks to a geolocation service integrated in the mobile app, can monitor 'alerts' launched by retailers. Type: website and mobile app http://breading.foundation/
Bring The Food	Crowdsourcing web/mobile application that allows donors to seamlessly publish offers and easily coordinate collections. Type: mobile app (Android) http://www.bringfood.org/public/landing?locale=it
iFoodShare	Web platform allowing the food redistribution and share. Consumers, retailers, shops, groceries, bakeries and farmers can use the platform to distribute surplus food. Type: website http://ifoodshare.org/
S-Cambia Cibo	Web-platform allowing the food donation between consumers, food services and organizations. The food is photographed and posted on the website by the donor, hence it appears on the map; keepers can privately contact the donor to arrange the appointment. Type: website http://www.scambiacibo.it/

Surprisingly, we did not find any platform expressly aimed at a food exchange (food for food).

With regards to the type of actors involved into the food transaction, we analysed the type of users that are mainly targeted by each platform. Being originated by the idea of 'matching', all these IT solutions need two main types of actors in order to be successful: 'givers' on one side and 'takers' on the other one. While takers are only placed in the consumption stage of the food chain, according to our analysis, givers can be players positioned both at the consumption stage, at the distribution stage and at the food services stage.

In order to simplify the analysis, we divided givers into two main clusters: 'business givers' (distributors or food service providers) and 'customers givers' (households).

Three out of eight platforms (MyFoody, Last Minute Market sotto casa and Breading) are designed in order to let business actors to give food surplus to external customers. In these cases, players are provided with specific functions aimed at mass communicating their surplus, in order to redistribute it among different players of the 'consumption stage'.

Three platforms (Ratatouille, Bring the food and Scambiacibo) are clearly targeted to 'customers givers'. These platforms are really similar to each other: in the first case, users can share their extra food with other users and give out their surplus, while in the second case, households can share some amount of food surplus (in terms of quantity and type) with other users. Two platforms (FameZero and IFood Share) are mostly oriented to Business Givers (municipalities or institutions, restaurants and shops) as well as to households.

The last dimension that we took into account in our analysis was the existence of a 'mediator' involved into the food transaction. As introduced in the first paragraph, food chains are including new types of actors that are aimed at redistributing food within players that rely at different stages of the chain. Some examples of these actors are food banks, NPOs and NGOs that take care of food surplus in order to feed indigents. While all the analysed platforms consider customers as takers of food surplus, some of them are aimed at providing mediators with food in order to let them feed customers. In particular, two of them (FameZero and Breading) are designed in order to include mediators (NPOs or other users within the platform) in their own operations.

4. Results and discussion

In the empirical setting of this paper, we successfully implemented a framework for the classification of technologies for food waste reduction. The applied framework aimed at describing how – thanks to technology – food surplus is redistributed among players at different stages of the food chain and is based on three dimensions: (1) 'transaction approach' (free or paid food transaction); (2) 'type of actors' involved as 'givers' in the transaction (business or consumers); and (3) 'existence of a mediator' within the transaction.

The framework was able to correctly and exhaustively describe several platforms, as mentioned in the Table 2.

Table 2. IT platforms for food waste management: the Italian case.

IT Platform	Type of transaction	Type of givers	Type of mediator	Approach[1]
Ratatouille	donation	households	none	2
MyFoody	mostly sale	businesses	none	3
FameZero	donation	hybrid	peers	1 and 2
Last Minute sotto casa	sale	businesses	none	3
Breading	donation	businesses	npos	1
Bring The Food	donation	households	none	2
iFoodShare	donation	hybrid	none	1 and 2
Scambia Cibo	donation	households	none	2

[1] 1 = transaction approach; 2 = type of actors involved as givers in the transaction; 3 = existence of a mediator.

Thanks to the classification of different platforms made through the framework, we are now able to distinguish among different emerging options in food waste reduction.

The first one, is made of the platforms that have a 'donation-oriented' transaction approach and that need a qualified mediator (an NPO) to conclude those transactions. These platform have a potential huge impact on redistribution networks, since they are able to quickly reallocate food surplus among charities listed in the platform: even a small number of business users, therefore, can potentially feed a huge number of indigents through the partnership with NPOs. At the same time, the safety of food is guaranteed by the type of givers (business) and the type of mediators (NPOs) that bring the food to end-customers. The platforms of this category are considerable 'quasi-food-banks'. Even if only for specific types of food, their technology is in fact aimed at linking charity organizations and professional donors.

The second approach arises when givers are households and mediators are absent or peers. In this case, a big dilemma arises: at one hand, micro-transactions of food can happen between a great number of users, and can potentially feed a huge number of indigents of a very different type as well as rich people (we therefore call this approach 'food sharing technology'). At the other hand, there is no control on food safety. Since a qualified mediator is missing, food takers are not completely aware of food quality, and this is potential a threat for them. This issue, is partially reduced when donors are 'businesses', but an attention for food safety is needed.

The third approach is represented by all the platforms that enhance the development of a secondary market for food surplus distribution. In this case, the food safety is guaranteed by the type of giver (a business player) and the food surplus is redistributed in a network that is forbidden to NPOs and food banks. Since the food surplus is transferred in exchange for money, the poorest players among indigents have no access to the system. Under this point of view, those kinds of platform are 'elitary'. They reduce food waste but are not able to increase food fungibility by feeding hungry poor.

The analysis of the distribution of platforms on different categories is very helpful to highlight some implications for this research. First of all, the food unsafety is a concrete issue that must be taken into consideration by authorities for health and hygiene control. Under this point of view there is room for an enrichment of the existing legislation in terms of food preservation. At the same time, it can be argued that redistribution networks and the food chain itself could be potentially affected – in terms of design – by this kind of technology. The degree of centrality and the role of food banks, as well as of NPOs, could be rethought at the light of this technology.

These options, indeed, can be seen as effective tools to prevent food waste, allowing the community to organize a food surplus redistribution model by themselves in a quick and simple way. At the same time, some concerns arise. The more IT platform are oriented to 'sharing', the less people have to prove they are needy (as for food banks access). Of course, people with little money use it more, but this is not a needed condition to use it. This situation is not an issue itself, but affects the degree of equity in food surplus distribution. At the same time, the shift from a 'non-profit' to a 'profit' business model for food surplus management has a deep impact in the efficiency and efficacy of waste management, in terms of food availability, redistribution networks' roles and composition.

5. Conclusions

This paper was a first attempt to offer a framework for a systematic recognition of different technology options for food waste reduction. Our study showed that IT-platforms for food waste prevention are really increasing nowadays, and that a diversified plethora of solutions is already available on the market. Even if they are a concrete and helpful solution to food surplus management and waste prevention, the more these IT platforms are successful, the more potential threats and issues arise: the risk for an elitism in 'technology access' and food transactions poses the problem for effectively hunger facing, while the bypass for traditional mediators (like food banks or NPOs) requires for an adapting of food safety norms and laws.

Under this perspective, this research was helpful to identify a new research agenda on this topic: in order to generalize the evidences of this research, and therefore overcome the limits of this paper, new examples of IT platforms must be analysed in order to test the validity of the presented framework on different cultural context and at different stages of the food chain.

References

Adhikari, B.K., Barrington, S., Martinez, J. (2006). Predicted growth of world urban food waste and methane production. Waste Manag. Res. 24 (5): 421-433.

Beretta, C., Stoessel, F., Baier, U., Hellweg, S. (2013). Quantifying food losses and the potential for reduction in Switzerland. Waste Management 33:764-773.

Buzby, J., Hyman, J. (2012). Total and per capita value of food loss in the United States. Food Policy, 37(5): 561-570.

FAO (2014). Food wastage footprint. Full-cost accounting. Final Report. Available at: http://www.fao.org/3/a-i3991e.pdf.

Garrone, P., Melacini, M., Perego, A. (2013). Feed the hungry. The potential of surplus food recovery. Fondazione per la sussidiarietà, Milano, Italy, 245 pp.

Garrone, P., Melacini, M., Perego, A. (2014). Surplus food recovery and donation in Italy: the upstream process. British Food Journal, 116(9): 460-1477.

Grolleaud, M. (2002) Post-harvest losses: discovering the full story. Overview of the phenomenon of losses during the post-harvest system. Rome, Italy: FAO, Agro Industries and Post-Harvest Management Service.

Goodman, L.A. (1961). Snowball sampling. Annals of Mathematical Statistics 32, 148-170.

Institution of Mechanical Engineers (IMechE) (2013). Global food: waste not, want not. IMechE, London, UK.

Gustavsson, J., Cederberg, C., Sonesson, U., van Otterdijk, R., Meybeck, A. (2011). Global food losses and food waste. extent, causes and prevention. FAO, Rome, Italy. Available at: http://tinyurl.com/kkn52e6.

Lipinski, B., Hanson, C., Lomax, J., Kitinoja, L., Waite, R., Searchinger, T. (2013). reducing food loss and waste. Working Paper, Installment 2 of Creating a Sustainable Food Future. Washington, DC: World Resources Institute. Available at: http://www.worldresourcesreport.org.

Mayan, M.J. (2009) Essentials of qualitative inquiry. Left Coast Press.

Midgley, J.L. (2014). The logics of surplus food redistribution. Journal of Environmental Planning and Management 57 (12): 1872-1892.

Noy, C. (2008). Sampling knowledge: The hermeneutics of snowball sampling in qualitative research. International Journal of social research methodology, 11(4), 327-344.

Papargyropoulou, E., Lozano, R., Steinberger, J.K., Wright, N., Bin Ujang, Z. (2014). The food waste hierarchy as a framework for the management of food surplus and food waste. Journal of Cleaner Production 76:106-115.

Parfitt J., Barthel, M., Macnaughton, S. (2010). Food waste within food supply chain: quantification and potential for change to 2050. Philosophical Transactions of the Royal Society B: Biological Sciences 365(1554): 3065-3081.

Santos, F.M., Kathleen, M.E. (2005). Organizational boundaries and theories of organization. Organization Science 16(5): 491-508.

Thompson, C.J. (1997). Interpreting consumers: a hermeneutical framework for deriving marketing insights from the texts of consumers' consumption stories. Journal of Marketing Research 34: 438-455.

Williams, I.D., Schneider, F., Syversen, F. (2015). The 'food waste challenge' can be solved. Waste Management 41: 1-2.

16. Case study of paprika supply chain efficiency in Malawi Central region

L. Repar[1]*, S. Onakuse[1], J. Bogue[1] and A. Afonso[2]
Food Business and Development Department, University College Cork, Ireland; [2]GESPLAN research group, Technical University of Madrid, Spain; l.repar@umail.ucc.ie

Abstract

Rapid population growth and income increase in countries across the world resulted in rising demand for quality and diverse food. Trade liberalisation and modernisation of production, processing and distribution systems enable agro-food companies to quickly access raw materials directly from farmers. However, the threats of exclusion and unbalanced benefit allocation from high-value markets represent the challenges for vulnerable small farmers affected by the transformation of food industry. The re-emerging contract farming tends to balance market imperfections and include small farmers into modern food chains. This research looks at the supply chain dynamics of paprika sector in Malawi Central region with the purpose to identify key challenges its players face, and propose changes to improve supply chain efficiency. Paprika supply chain was explored using mixed method approach: 428 household questionnaires and eight focus group discussions with small farmers, and five key informant semi-structured interviews with the company and enabling environment representatives. The data were triangulated and analysed using SPSS and NVivo. The result shows that paprika supply chain consists of direct players (farmers and company) and indirect players with significant role in building chain's efficiency (agro-dealers, NGOs, civil societies, Government, financing institutions, vendors). Additionally, for small farmers efficiency implies secured inputs and highest possible selling price. For the company, efficiency is related to minimised transaction costs and obtained planned volumes of appropriate quality. The research finds that communication quality and consistency, together with the trust level among players influence paprika supply chain efficiency. Moreover, the lack of understanding and complying with the contract terms hinders effective performance. The failure to define selling prices in the contract encourages farmers' opportunistic behaviour, e.g. side-selling. The study concludes that companies can increase farmer's loyalty by awarding long-term value through input bonuses, renting equipment and transportation, or subsidising the purchase of mobile devices and applications.

Keywords: contract farming, food chains transformation, loyalty, small farmers

1. Introduction

Characterized as globalised with free flow of goods, services, capital and knowledge – the world is undergoing transformation processes in many sectors and the consequences are far-reaching. Agri-food sector provides an example of existing intertwined systems, which depend and rely on each other. Here, the increased population with raising income demands higher quantities of quality food. Companies adjust their supply chains to meet the consumers' requirements by engaging into partnerships with different suppliers across the globe. Therefore, the agri-food sector is switching from independent markets towards more tightly aligned food supply chains that use different forms

Leire Escajedo San-Epifanio and Mertxe De Renobales Scheifler (eds.) *Envisioning a future without food waste and food poverty*
DOI 10.3920/978-90-8686-820-9_16, © Wageningen Academic Publishers 2015

of vertical coordination, while through trading mechanisms companies are rapidly integrating their operations into global markets (da Silva, 2005; Vavra, 2009). Da Silva (2005) finds that trade liberalisation and mobility of capital flows are affecting the organisation and performance of agri-food sector, pushing it to modernise with more close management of production, processing and distribution activities.

Furthermore, the transformation of agri-food sector stimulated the supermarket revolution (Reardon and Berdegué, 2002), and re-organised procurement practices placing even more emphasis on international supply chains. At the same time, public sector is decreasing its support and services provision, while private grades and standards become driving mechanism of differentiation (da Silva, 2005).

Agri-food transformation affected farmers, too. Vavra (2009) noticed that changes occurred in relatively short period, hence the speed might raise concerns on how actors in food supply chains adapt to new conditions. Moreover, Reardon and Barret (2000) argued that the 'industrialisation' of world agriculture resulted in increased size of farms, but reduced number of farms. In addition, authors point out farmers have to adjust production practices to resemble manufacturing processes, and have to engage in closer relationship with various processors, buyers and other players in the supply chain. Eaton and Shepherd (2001) indicated the threat for small-scale farmers in the age of expanding agribusinesses by questioning their efficient participation on dynamic markets, and concluding that since large farms are increasingly important for the operation of modern agri-food sector small farmers might be marginalised and forced to divert into other sectors, as well into urban areas.

The threats of exclusion and unbalanced benefit allocation from high-value markets represent the challenges for vulnerable small farmers affected by the transformation of food industry. The re-emerging contract farming tends to balance market imperfections and include small farmers into modern food chains. This research looks at the supply chain dynamics of the paprika sector in Malawi Central region with the purpose to identify key challenges its players face, and propose changes to improve supply chain efficiency.

2. Methodology

The study applied mixed-method approach using quantitative (household questionnaire) and qualitative tools (focus group discussion and semi-structured key informant interview) in parallel. Household questionnaire consisted of 78 closed questions and 11 thematic groups as follows: household characteristics, contract details, motivation and satisfaction, future plans in relation to contract, input supply and extension services, payment and meeting the requirements, communication, relation and networking, market and information access, housing and assets, health, education and food security, and farm characteristics. The including criteria for the population considered were set in relation to having signed the contract for paprika export with the company. Therefore, the small farmers eligible for the study purpose had to be under the out-grower contracting scheme, had less than two hectares, and were representing the household head. In order to explore the supply chain efficiency under different circumstances, two geographical locations were selected for conducting the surveys. First area of interest (Nhkotakota district in Central region) was located around Malawi

Lake, with the distance of over 150 km from the capital Lilongwe, and involved irrigated paprika cultivation. Second area of interest (Lilongwe district in Central region) was situated within 80 km of the capital, where the farmers rely on rain-fed paprika cultivation. After the purposive sampling of the area, the study used random sampling in two stages to select the participants: firstly the clusters in the districts were randomly selected, and secondly the farmers in the clusters were randomly selected using mainly the groups of 10 representatives from each cluster. In total, 428 questionnaires were conducted and analysed using SPSS software.

The focus group discussions represented a qualitative tool used simultaneously and complementary with the questionnaires in selected area. The groups consisted of mainly 8-10 farmers (mixed and non-mixed gender groups), who discussed and expanded on the topics related to supply chain dynamics, agronomic practices, motivation for entering the contract, preferences for contract configuration, perceptions in the community, communication and future plans. The discussions lasted averagely for 90 minutes. In total, eight focus group discussions were performed and analysed using NVivo software.

Since the supply chain consists of many players on the market, the study aimed at bringing them together through the semi-structured key informant interviews, which were conducted with the representatives from the sectors that play the key role in chain's efficiency: private sector, government, civil society, NGO and informal channels (vendors). The interviews were analysed in NVivo software in order to triangulate the data collected from household questionnaires and focus group discussions, and clarify the challenges identified in paprika supply chain.

The findings from quantitative and qualitative tools were integrated at the final point of analysis to better reflect and cross-check the relations among the chain players, and to explore the way issues can be addressed to increase the supply chain efficiency by transferring the benefits, such as added value and better prices, directly to the primary producers and the company.

3. Results and discussion

Paprika supply chain in Malawi Central region: direct and indirect players

In Malawi Central region paprika is considered as non-traditional crop, which is commercially mainly grown for the export purposes. The cultivation of paprika demands significant amount of input supply, in most cases seeds, fertilizers, pesticides and equipment. Additionally, most of the primary producers of paprika in Central region are households with less than two hectares, heavily dependent on external support for the inputs, and food and income insecure. Therefore, the supply chain involves many different players whose roles and responsibilities are supposed to fill in the gaps within and around the paprika chain. The study identified direct and indirect players that participate in building the chain's efficiency (Figure 1).

The direct players include the company and the primary producers. In case of studied paprika supply chain, the initial relationship between the two parties is regulated through an out-grower contract with an individual small farmer or a farmer group representative. The contract prescribes terms and conditions for the delivery and purchase of whole pod dried paprika and obligations of each

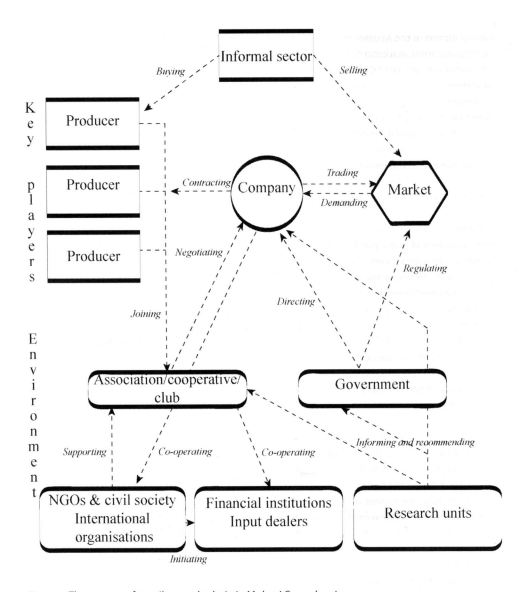

Figure 1. The concept of paprika supply chain in Malawi Central region.

party in doing so. Moreover, the contract specifies the undesired cases of contract breach and the consequences that follow.

Since in explored paprika supply chain the company provides inputs only partially (the seed), other players were found to have a significant role in facilitating efficient paprika cultivation, delivery and marketing. Although informal sector, NGOs, civil societies, international aid organisations, Government, financial institutions, input dealers and research units are formally recognised as indirect players in the Malawian paprika chain, their overall contribution in the chain is meaningful. For example, joint marketing coming from small primary producers is usually supported by NGOs, civil society or international organizations, which enable easier communication and access to companies, financial institutions or input dealers. The government facilitates contracting in terms of creating policies, strategies and recommendations for commodity trade, and directs agribusiness towards including small farmers into the market opportunities. Research units explore the chains and critically identify main challenges that hinder the development, and therefore their endowment is reflected in making recommendations for the informed policy decisions and disseminating them through the whole supply chain.

Efficiency and what does it mean for smallholders, company and enabling environment

While the ultimate aim that both direct players in the chain are trying to achieve is the high-value marketable produce, the study identified differences in players' perception of the efficient supply. In case of small producers, the secured inputs for cultivation, coupled with the highest possible received price represent the efficient system. However, the company strives to obtain the desired quantities of good quality product with the lowest transaction costs involved. For most players in enabling environment, the efficient supply chain is supposed to involve the sustainable concept of commercialised farming, which at the same time supports small producers in fighting the poverty traps and enables the platform for the private agri-business sector to initiate the economic prosperity of Malawi through agricultural production and export.

Factors influencing the efficiency: communication, trust, understanding, compliance and contract configuration

Nevertheless, to achieve the supply chain efficiency, main challenges must be identified and addressed. The study found that communication quality between the company and the small producers suffers mostly due to the considerable geographical distance, which hinders regular contact. Furthermore, the information on the paprika price is communicated late in the season, making small producers hesitate to trust the honest benefits distribution in terms of premiums received for the quality paprika. Additional issue is created with the existence of the informal sector (vendors) that operates in a manner to cover the company's failures in communication and pricing aspect. Specifically, vendors maintain friendly relationships with the small producers throughout the year and visit them often to secure stable supply. In addition, the price offered by vendors exceeds company's price and tempts small producer's loyalty. Therefore, although by selling the paprika to vendors (instead of the company) small producers are able to generate some higher income, the long-term consequences of such side-selling damages the trust and brings into question the producers' basic understanding of and compliance with the contractual obligations. Also, the analysis of the company's contract offered to the small producers exposed the power imbalance manifested in the absence of sufficient information on pricing formula, quantities that the company will purchase, the property rights, and non-existent exit options in the written form of the contract.

4. Conclusions and recommendations

The supply chains often consist of many players, whose roles and relations intertwine at different levels and contribute to the efficiency of the whole chain. Ideally, the supply chain should secure the maximum economic benefit at the minimum costs to the key players involved, with the support of enabling environment in terms of services, finances and policies. The explored paprika supply chain in Malawi Central region involves direct and indirect players, and the efficiency observed is obstructed with the lack of communication and trust, failure in understanding and honouring the contract, and even poor configuration of the contract. Although the responsibility of improving the chain's efficiency lies firstly with the two direct players, the enabling environment has the potential to facilitate upgrading of the relations, and consequently promote more equal benefits and liability allocation.

In the study, the lack of trust and loyalty showed to be one of the factors influencing the side-selling and hence the decrease of chain's efficiency. In order to address the issue, the company might award long-term value by closing the gap which small farmers indicated, for example by distributing input bonuses, renting equipment and transportation, or subsidising the purchase of mobile devices and applications. Another solution is to start cooperation with the informal sector by using the vendors as mediators in communication with the small producers, and thus share the risks, costs and benefits. In any case, without actively involving the enabling environment to the extent where those indirect players enable the development of more efficient and sustainable supply system, the chain will lack valuable sources and chance for upgrading, and it will not reach the potential of alleviating the poverty food insecurity among small farmers while at the same time providing the foundation for economic prosperity of Malawi.

References

Da Silva, C.A.B. (2005). The growing role of contract farming in agri-food systems development: drivers, theory and practice. Rome, FAO, Agricultural Management, Marketing and Finance Service.

Eaton, C. and Shepherd, A.W. (2001). Contract farming; partnerships for growth. FAO Agricultural Services Bulletin. Rome, FAO.

Reardon, T. and Barrett, C.B. (2000). Agroindustrialization, globalization, and international development. An overview of issues, patterns, and determinants. Agricultural Economics 23: 195-205.

Reardon, T. and Berdegué, J.A. (2002). The rapid rise of supermarkets in latin america: challenges and opportunities for development. Development Policy Review 20(4): 371-388.

Vavra, P. (2009), Role, usage and motivation for contracting in agriculture. OECD Food, Agriculture and Fisheries Working Papers, No. 16, OECD Publishing. doi: 10.1787/22503674570.

Part 4. Proposals and strategies which empower consumers

17. Get consumers truly informed of their food choices!

A. Corini

Doctoral School on the Agro-Food System, Università Cattolica del Sacro Cuore, Piacenza campus, Via Emilia Parmense 84, 29122, Piacenza, Italy; antonia.corini@unicatt.it

Abstract

The right of consumers to make informed choices is stated in Art. 8 of Regulation (EC) No. 178/2002. Moreover, the right to receive information is recognized in the Charter of Fundamental Rights, in Art. 11. Regulation (EC) No. 178/2002 mentions the objective of the protection of consumer interests and the importance of their information also in other provisions, for instance, Art. 9, 10 and 16. However, it seems that nowadays consumers are not effectively informed, especially when speaking about future challenges related to food. In fact, sometimes consumers are not duly informed on food waste and loss issues and on the rising number of food poverty situations going alongside with the future increase of the population. If consumers were really informed they could develop their own ideas on which to found their food choices. Moreover, if consumers were aware of the fact that they could contribute, with their daily lives, to the achievement of global challenges, maybe they would make food sustainable choices. Sustainability is a key word for our future and it involves many sectors: food policies, agriculture, environment and energy. Consumer awareness of the active role that they could play in order to build their own future, would be, first and foremost, possible, if they were truly informed. In this context, measures should be taken, for instance information campaigns, 'training' courses and soft law provisions. A real involvement of the civil society is extremely important because the present and future issues related to food are and will continue to be challenges of the society itself.

Keywords: consumer interests protection, participation, labelling, GMOs, global challenges

1. Introduction

Information for consumer interests protection and for their participation. EU law is so devoted to the safeguarding of consumer information that their right to receive it is recognized in the EU Charter of Fundamental Rights, in Art. 11.

Consumers have to be considered as the purchasers of goods or services supplied in the market by public or private entities (Capelli, 2013). Given this broad notion of consumers, it is evident that consumers' rights and interests vary as to the type of goods and services that they buy, pertaining to different market sectors.

2. EU Law on food safety and provisions for the protection of the consumers

In the field of food safety, European legislation has the aim of protecting consumer health, which is a primary objective of public relevance (Capelli, 2013).

Therefore, in the domain of food safety, the right to receive information is derived from the primary objective of consumer health protection. In particular, food consumers have the right to make informed choices, as stated in Art. 8 of Regulation (EC) No. 178/2002 laying down the general principles and requirements of food law, establishing the European Food Safety Authority and laying down procedures in matters of food safety.

These choices may be considered informed if the consumers know the food that they eat (e.g. characteristics, origin, properties, etc.) without being misled by fraudulent or deceptive practices, by food adulteration and any other behaviour that may mislead the consumers. In effect, when the so-called food frauds are committed, the main damage is caused to those consumers whose interests and level of information required are affected.

Consumer interests are protected through several other provisions of the Regulation (EC) No. 178/2002. Art. 9 sets forth the public consultation of consumers during the preparation, evaluation and revision of food law. Then, Art. 10 states the requirement to provide detailed public information when there are reasonable grounds to suspect that a food or a feed may present a risk for human or animal health.

Also, Art. 16 of the same Regulation makes reference to consumer information by stating that this information shall not mislead the consumers. This refers, in particular, to labelling, to advertising, to the presentation of food or feed and, generally speaking, to all the information that is made available about some food or feed through whatever medium.

3. Food information to consumers: labelling and in particular the issue of GMOs

As to the issue of labelling, a very important Regulation has recently entered into force: Regulation (EU) No. 1169/2011. As it has been affirmed, this Regulation *'has meant a further step'* in relation to the food information available to consumers (González Vaqué, 2013).

Always on the theme of labelling, it has to be noted that EU law on GMOs provides accurate rules on this subject. In particular, Regulation (EC) No. 1829/2003 on genetically modified food and feed affirms that '(...)*In addition to other types of information to the public provided for in this Regulation, the labelling of products enables the consumer to make an informed choice and facilitates fairness of transactions between seller and purchaser*' (Recital n 17).

In other provisions, the same Regulation states that labelling also has the function of providing final users with accurate information on the composition and properties of the feed (cf.: Recital 20), and that it shall specify whether the food contains or is made from GMOs (cf.: Recital 21). More in detail, Art. 12 sets forth that the label shall identify if the food contains GMOs or ingredients derived from GMOs (if the 'GMO' content is higher than 0.9%).

Moreover, Art. 9 of the Directive 2001/18/EC, on the deliberate release into the environment of genetically modified organisms, highlights the importance of information and consultation of the public.

These rules on consumer information are extremely important. When speaking about GMOs, there are both epistemological and cultural uncertainties (Stefanini, 2006) so that all the legal, political, economical and ethical aspects that may concern GMOs should be taken into consideration.

As has been stated, *information is the key* (Kolodinsky, 2007). In effect, information and knowledge are the key factors in order to be able to understand the possible impacts and peculiarities that derive from the use of scientific techniques. Having this knowledge, people could be free to choose, whether to consume GMOs or not, being also aware of the consequences of their choices.

However, it frequently occurs that people have a distorted perception of scientific issues, especially when these are much-debated questions such as GMOs. When speaking about GMOs, both negative and positive characteristics are pointed out. Taking into account the negative ones, there could be, for instance, general impacts on the ecosystem as well as risks of 'contamination' (Berti, 2006).

There may be also positive issues; principally: higher resistance of GMO plants, higher quality of products as well as an increase of food production (Berti, 2006). This is obviously important, bearing in mind that the world population is expected to reach 9.6 billion by 2050, according to the United Nations World Population Prospect of 2012.

Even if we all know that these impacts concerning GMOs have not been 'measured' in the long term yet, information and knowledge are the key factors in order to make people free to choose whether or not to consume GMO products.

4. Information and global food challenges

Consumer information is a key element not only to protect consumer interests but also to make them contribute to global food challenges. In effect, consumers, during their daily lives, could do their part by dealing with projects/programmes with the aim of reducing the number of food poverty situations. Also, they could make sustainable choices, taking into consideration the environmental impact of their food choices.

The importance of consumer involvement may be understood by considering the issue of food waste. This represents a very serious problem both because food is wasted while a high number of people in the world go hungry and because many resources have to be invested for the waste disposal process.

Percentages of food waste are shocking. According to a study of the Food and Agriculture Organization of the United Nations (FAO) of 2011 on Global Food Losses and Food Waste – Extent, causes and prevention, the pro capita food loss in Europe and North-America is 280-300 kg/year.

It is evident that these situations could be reduced if every single consumer took part in the fight against food waste. For example, by avoiding to buy more food than the amount that is needed and by not throwing away food that is still fit for human consumption.

Clear legal definitions (food waste, use by date, use before end)

However, an involvement on the part of consumers is not sufficient. A real commitment on the part of the Institutions and the food operators is necessary. In particular, a clear legal definition of what has to be considered as food waste should be given. Nowadays, only a 'general' definition of 'waste' is provided, in Art. 3 point 1 of the Directive 2008/98/EC on waste, that makes reference to the activity of discarding or to the intention or requirement of discarding.

Another explanation is needed on the difference between the indications 'use by date' and 'use before end', which is not always very clear.

Contribution on the part of producers

In addition, also food operators should contribute to the fight against food waste by using sustainable productive processes, adequate packaging and, generally speaking, by providing all the information required on the label.

This commitment is certainly important, in effect, as written in the Communication from the Commission to the European Parliament, the Council, the European Economic and Social Committee and the Committee of the Regions – Roadmap for a Resource Efficient Europe of 2011, *'a combined effort by farmers, the food industry, retailers and consumers through resource-efficient production techniques, sustainable food choices (...) and reduced food waste can contribute to improving resource efficiency and food security at a global level'.*

This combined effort may have really significant consequences in relation to the sustainability of many sectors: food policies, agriculture, environment and energy.

Consumer involvement

The role of consumers is crucial. In order to facilitate their participation, consumers should not only be aware of the rights and interests recognized to them but they should take a step further. They should be active in asking for their involvement, by conducting information campaigns, by creating new tools both for the protection of their rights and interests and for the achievement of global food challenges. In one message consumers should *get political* (Nestle, 2013).

To conclude, in order to achieve the two fundamental objectives of consumer interest protection and consumer participation, many actors are involved and many steps should be taken. However, the unavoidable starting point is represented by a fundamental objective: to get consumers truly informed about their food choices.

References

Berti, M. (2006). La dimensione economica e ambientale. In: Il diritto degli OGM tra possibilità e scelta – Atti del convegno tenuto presso la Facoltà di Giurisprudenza di Trento 26 novembre 2004. In: Casonato, C. and Berti, M. (eds.) Litotipografia Alcione S.r.l., Trento, Italy, pp.77-89.

Capelli, F. (ed.) (2013). La tutela giuridica del consumatore (contraente-debole) in Europa e nel mondo. Editoriale scientifica, Napoli, Italy, pp. 3-32.

European Commission, 2001. Directive 2001/18/EC of the European Parliament and of the Council of 12 March 2001 on the deliberate release into the environment of genetically modified organisms. Offical Journal L 106, 17 April 2001, 1-39.

European Commission, 2002. Regulation (EC) No 178/2002 of the European Parliament and of the Council of 28 January 2002. Laying down the general principles and requirements of food law, establishing the European Food Safety Authority and laying down procedures in matters of food safety. Official Journal L 31, 1 February 2002, 1-24.

European Commission, 2003. Regulation (EC) No 1829/2003 of the European Parliament and of the Council of 22 September 2003 on genetically modified food and feed. Official Journal L 268, 18 October 2003, 1-23.

European Commission, 2008. Directive 2008/98/EC of the European Parliament and of the Council of 19 November 2008 on waste. Official Journal L 312, 22 November 2008, 3-30.

European Commission, 2011a. Communication from the Commission to the European Parliament, the Council, the European Economic and Social Committee and the Committee of the Regions (20 September 2011, COM (2011) 571 final). Roadmap for a Resource Efficient Europe. Brussels, Belgium.

European Commission, 2011b. Regulation (EU) No 1169/2011 of the European Parliament and of the Council of 25 October 2011 on the provision of food information to consumers. Official Journal L 304, 22 November 2011, 18-63.

Food and Agriculture Organization of the United Nations (2011). Global food losses and food waste – extent, causes and prevention. Rome, Italy. Available at: http://www.fao.org/docrep/014/mb060e/mb060e.pdf.

González Vaqué, L. (2013). The new regulation on food labelling: are we ready for the 'D' day on 13 December 2014? European Food and Feed Law Review 3: 158-167.

Kolodinsky, J. (2007). Biotechnology and consumer information. In: Brossard, D., Shanahan, J., Nesbitt, T.C. (eds.) The media, the public and agricultural biotechnology. Wallingford, UK, Cambridge, MA, USA, pp. 161-175.

Nestle, M., (2013). Food politics: how the food industry influences nutrition and health. University of California Press, Berkeley, USA, 510 pp.

Stefanini, E. (2006). Gli OGM e i diritti fondamentali. In: Il diritto degli OGM tra possibilità e scelta – Atti del convegno tenuto presso la Facoltà di Giurisprudenza di Trento 26 novembre 2004. In: Casonato, C. and Berti, M. (eds.) Litotipografia Alcione S.r.l., Trento, Italy, pp. 59-76.

United Nations (2013). Department of Economic and Social Affairs, Population Division. World Population 2012. United Nations Publication, August 2013. Available at: http://tinyurl.com/phz9q7w.

18. A social perspective on food waste: to what extent consumers are aware of their own food waste

R. Díaz-Ruiz[1]*, M. Costa-Font[2] and J.M. Gil[1]

[1]CREDA-UPC-IRTA, C/ Esteve Terrades 8, 08860 Castelldefels, Spain; [2]SRUC Rural College, West Main Road, Edinburgh EH9 3JG, United Kingdom; raquel.diaz.ruiz@estudiant.upc.edu

Abstract

The proliferation of food waste (FW) worldwide at every stage of the food chain exerts various environmental, social and economic consequences. Importantly, the existing literature suggests that most waste is generated in the last stage of the chain, namely food consumption, hence it is crucial to understand how such behaviour is engendered. Consumers' FW behaviour has been related with their attitudes, values, knowledge and behaviour towards food, planning and shopping routines, lifestyle and consumption habits in general and recycling or green behaviour. Based on the state of the art, the general aim of the present paper is to go one step forwards in reaching a better understanding of consumer behaviour in relation to food waste generation. This paper sets out a perceived food waste generation indicators using three groups of questions that confronts respondents with: (1) knowledge about food waste, subjective and objective one; (2) awareness of the problem; and (3) behaviour of food waste generated. In the present case-study, it has been noticed that citizens are not aware about their own FW generation. It is also important to highlight that respondents do have a biased perception as regards to which agents of the food chain are more responsible of overall FW generation. That is, they consider that agriculture and retailers are the main FW generators, while the literature demonstrates the opposite. Despite the lack of awareness, respondents do consider that to waste food is a problem that arouses interest and concern mainly for ethical and economic reasons. Finally, it has been noticed that consumers are confused about FW reduction strategies named recycling and prevention. The present work analyse household FW behaviour in the Barcelona metropolitan area, by means of taking into consideration citizen's opinion. The study employs face to face and online interviews conducted during June-July 2013 in Barcelona to a sample of 220 individuals responsible on their household food purchase.

Keywords: food waste, consumer awareness, consumer behaviour

1. Introduction

The 2008 food price crisis made evident to the international community that food security can be a complex threat to society. It was since that moment that food waste is gaining great international relevance. In order to handle food security we must develop an efficient food management system at both societal and individual levels paying special attention to resource consumption, food consumption and waste management. In this last context, FAO's 2011 food waste report noticed that one-third of all food produced for human consumption is lost or wasted every year. In particular, in EU 27, consumers waste 76 kg of food per person per year, with big variation between countries (Bio Intelligence Service, 2010).

Leire Escajedo San-Epifanio and Mertxe De Renobales Scheifler (eds.) *Envisioning a future without food waste and food poverty*
157

DOI 10.3920/978-90-8686-820-9_18, © Wageningen Academic Publishers 2015

However the data above, it is important to highlight that there is no international consensus on the subject of food waste (FW) concept and quantification despite the spread of initiatives attaining at a proper FW framework definition (FUSIONS, 2014; Gustavsson *et al.*, 2011; HLPE, 2014). There is no official or common definition at the international level and neither at the European level, three notions: food wastage, food loss and food waste are randomly used in the different reports published up to the moment. Since up to the moment an international database on FW do not exist, the different studies use a variety of methodologies to quantify it, such as dietary cooking books, questionnaires, organic waste counting, environmental impacts, and so on.

This lack of homogeneity brings about data collection divergences not allowing an adequate FW cross country comparison. In developed countries the focus must be on industry, retailers, restaurants and households, dealing with behaviours and attitudes towards food (Kosseva, 2013). It is important to highlight that as further on the food chain the waste happens more are the resources misused. Bio Intelligence Service (2010) reported that for the case of Europe, 60% of household FW could be avoided. Therefore, the food waste issue must be considered not just from a waste management perspective, but also from households' food management and dietary planning perspective.

From the previous studies it can be said that the consumers' profile characteristics associated to higher food waste are: lower age individuals, women, one-person households, high income households and households with children. Some work has been done also regarding to the reasons of household food waste such as (Cecere *et al.*, 2014; HISPACOOP *et al.*, 2012; Parfitt *et al.*, 2010; Quested *et al.*, 2013). These studies reported four main groups of reasons for food waste generation: storage problems, insufficient attention of go by dates when purchasing, inefficiencies on food preparation, life style. Up to the moment, investigations considering the extent of the food waste problem measuring its environmental, economic and social impact. However, consumer's behaviour, perceptions, knowledge and attitudes towards food waste are almost not considered in the literature up to now.

The aim of the present work is to analyse Barcelona metropolitan area (as an example of European city) household food waste behaviour, by means of taking into consideration citizen's opinion. The FW behaviour is analysed using three groups of questions that confronts respondents with: (1) knowledge about food waste, subjective and objective one; (2) awareness of the problem; and (3) behaviour of food waste generated.

2. Methodology

The study employs a survey conducted during June-July 2013 in Barcelona metropolitan area to a sample of 220 individuals over 18 years old responsible on their household food purchase (sample error ±5.5% (95%)). The questionnaire was self-administrated, and assistance was provided if necessary. The survey was implemented by quotas selecting representative factors, such as gender, age, are of residence, education, income and children in charge (Table 1). We did a pre-test of eleven questions to 20 individuals and finally, the definitive questionnaire had 38 questions (self-designed and questions from WRAP (2007) survey). It was divided in four blocs: food habits, waste management, environmental awareness and food waste. Descriptive analysis has been done in order to understand individuals' behaviour towards food waste behaviour. The data has been performed using SPSS Statistics 15.0.

Table 1. Characteristics of the respondents (%; n=220) compared to the general population (% between brackets).

Data collection		Gender		Children under 16	
In person	52.3%	male	37.3% (48.5%)[1]	no children	68.2% (65%)[4]
Online	47.7%	female	61.8% (51.5%)[1]	with children	31.8% (35%)[4]
Age >18		**Studies**		**Working status**	
18-34	36.4% (26.9%)[1]	basics	17.3% (36.3%)[2,3]	employee	60.9% (79.8%)[2]
35-49	33.6% (29.6%)[1]	medium	39.5% (32.7%)[2]	entrepreneur	11.8% (10.4%)[2]
50-64	20% (21.5%)[1]	graduate	41% (30.3%)[2]	pensioner	10.9%[b]
>65	9.5% (22.0%)[1]	no answer	2.3%	unemployed	14.5% (15.4%)[2]
				no answer	1.80%
Housing structure		**Household size**		**Rent per capita and year**	
Unipersonal	12.3% (17.8%)[2]	1	11% (17.8%)[2]	Less than € 12,000	26.8%
Couple	23.2% (30.9%)[2]	2	29% (38.5%)[2]	€ 12,001-18,000	15%
Family	55.9% (45.8%)[2]	3	26% (19.6%)[2]	€ 18,001-24,000	20%
Sharing flat	6.8% (5.0%)[2]	4	26% (18.6%)[2]	€ 24,001-36,000	13.6%
No answer	1.8%	≥5	5% (5.5%)[2]	€ >36,001	11.4%
		no answer	3%	no answer	13.2%

[1] Population in the metropolitan area of Barcelona (AMB) 2012 (http://www.idescat.cat/pub/?id=aec&n=250&lang=en).

[2] Population AMB 2011 (http://www.idescat.cat/pub/?id=aec&n=250&lang=en; Idescat/IERMB, 2011).

[3] Without studies + basic studies.

[4] Population Catalonia 2007 (http://www.idescat.cat/pub/?id=aec&n=250&lang=en).

3. Results

The study tries to evaluate three indicators of FW: (1) knowledge about food waste, subjective and objective one; (2) awareness of the problem; and (3) behaviour of food waste generated by means of an auto-evaluation and the description of the causes. Furthermore additional information regarding to food habits, environmental concern and waste management behaviour is provided.

Food waste-knowledge
- *Received information.* Those who has received any information about FW, has received the information from the TV half of the times (49.4%) less than once a month. The survey contained a question showing a list of international and national studies focused on FW. The 63.6% of the sample knew at least one study on the list.
- *Objective knowledge.* Three different definitions where shown to participants in order to evaluate their knowledge about the definition of FW. Two of them were true and the other was false. We

have developed and indicator summing the number of right guess: high knowledge (3), medium (2), low (1) and do not know (0). Almost half of the participants (47.7%) have shown medium knowledge about the definition. Only 9.5% guessed the three definitions and a 15.5% do not know any of the definitions.

- *Subjective knowledge.* It is important to distinguish between objective and subjective knowledge. The former represents what consumers really know, while the latter reveals which consumers believe they know. To evaluate this concept respondents answered their level of agreement in relation to four statements from totally agree to totally disagree. Half of respondents consider themselves informed about food waste, while 79% think that more information is needed on the subject. More than a half (62%) reveals that they are not aware of food waste generated in the metropolitan area. Finally, 70% believe that if they were better informed, they would reduce their amount of food waste generation.

Food waste-awareness

- *Interest.* Nearly half of respondents (48%) say that they think about the waste from time to time. Just a 5.5% claim they are not interest on the topic. Those who are very interested and think every day on it are the 24%. And, finally, 21% think on it once a week.
- *Concerning, reasons.* More than a half (67%) of respondents are worried about wasting food, while a 23% are less or not worried. Those who are worried are mainly due to social and economic reasons: '*It makes me feel guilty, there are starving people*' (78%); '*It's a waste of my money*' (74.5%); '*It's a waste of good food*' (41.3%). The main reason to not be worried of wasting food (n=12) is: '*I don't throw away much food*'; '*Want to eat the freshest food so there's bound to be some waste*'.
- *Food supply valuation.* It is known that food waste occurs across the food value chain. However, the responsibility for this waste is not equally distributed and thus is perceived by participants. The results reveal that the common perception is that supermarkets and shops, as well as catering and restaurants, are those with greater responsibility for the food waste; and in second place, the food industry, followed by domestic consumption. Agriculture producers are perceived as the less responsible. Overall, is the opposite of the data offered in the studies that quantify the volumes along the food supply chain (Gustavsson *et al.*, 2011).

Food waste-behaviour

The quantification of the generation of food waste has been through three questions considering the type of food that is valued as well as situations in which the waste is generated tacking the reference of (WRAP, 2007). The responses are measured on a scale of five levels, among *None* and *A lot*.

The first way to quantify it was for food groups (Table 2). Almost 70% of respondents said they threw a lot or a reasonable amount of unavoidable waste. For the group of fruits, vegetables and salads, 50% have answered *A small amount*. However, for most of the other food groups respondents have declared they do not generate any amount *None of raw meat & fish* (72.3%), *Processed meat & fish* (57.3%), *Ready meals/convenience foods* (65.9%), *Bread and cakes* (42.3%), *Cheese & yogurt* (67.7%). For cooked meat and fish the mean answer has been somewhat majority (36.8%), as well as bread and pasta (40.5%).

Table 2. Food type waste indicator-answers.

	How much of the following food items – whether prepared/cooked or not – have you thrown away into the bin?				
	Quite a lot	A reasonable amount	Some	A small amount	None
Inedible food waste (e.g. peelings, bones)	39%	28%	17%	10%	5%
Fruit, vegetables or salad	2%	5%	15%	50%	28%
Raw meat and fish	0%	1%	2%	22%	72%
Processed meat & fish (e.g. sandwich meats)	0%	0%	3%	37%	57%
Ready meals/convenience foods (e.g. pizza, burgers)	2%	1%	4%	25%	66%
Bread and cakes	2%	3%	10%	40%	42%
Cheese and yogurt	0%	4%	3%	23%	68%
Average	7%	6%	8%	29%	48%

Asking the same question but regarding to different situations food waste could be generated, we have found another pattern of responses asking the same question but regarding to different situations in which food waste could be generated (Table 3). In this case, three of the five food groups have the mean answer in a *Small amount*. Differently, 53.2% claim not to throw anything that *you cooked/prepared too much of but didn't serve up*; and 62.3% give the same answer regarding *to products you bought but didn't open*.

Finally, Table 4 shows the amount of food discarded without specifying food groups. Overall, 70% of the sample claims they throw *a small amount* and 16.8% do not throw anything. When we analysed the responses of the survey sample in the metropolitan area of Barcelona, we observed that the distribution was similar to the results of the Eurobarometer Flash EB Series #316 (The Gallup Organization, 2011).

Table 3. Food waste situations indicator-answers.

How much of the following have you thrown away into the bin?	Quite a lot	A reasonable amount	Some	A small amount	None
Food left on the plate after the meal	3%	5%	10%	50%	30%
Food you cooked/prepared too much of but didn't serve up	0%	1%	8%	35%	53%
Food from previous meals that you initially saved but didn't get around to eating/using up	1%	3%	16%	54%	24%
Products you opened and used but didn't finish	1%	5%	9%	45%	39%
Products you bought but didn't open	0%	0%	3%	32%	62%
Average	1%	3%	9%	43%	42%

Table 4. Overall food waste indicator-answers.

	Quite a lot	A reasonable amount	Some	A small amount	None
Thinking about the different types of food waste we have just discussed, how much uneaten food would you say you throw away in general?	0.0%	0.9%	10.9%	70.0%	16.8%

Given a list of reasons to ending up throwing food, food purchasing stands as the main reason of the fact of throwing food, regardless of the person concerned are *'Buying too much when shopping'* (41%); *'Trying to buy more fresh food (that doesn't last as long)'* (40%); *'Not eating the foods that need to be eaten first'* (32%); *'Food visibly gone bad or smells bad'* (31%).

Environmental concern

People participating in the study have revealed a high environmental concern by answering four questions with a 5-point Likert scale. They are aware of the necessity to protect the environment.

Food habits
- *Purchasing.* Half of the respondents use to do their main shopping trip once a week (49%) in malls (64%). On the other hand the small purchases are made every two, three days (51%) in small stores (73%). Regarding to household expenditure in food, previous studies have indicated that 14.4% of Spaniards wage is spent buying food and non-alcoholic beverages (INE, 2011). The sample considered in this study follows a normal distribution with a higher percentage spending 16-20% of their salary in food purchasing (18.2%).
- *Cooking.* In order to prevent consumers' food waste it is essential to identify the culinary habits and foods that make up the diet of individuals. Almost half of the sample plan their meals (46%), more than half have revealed culinary skills (56%), namely they are capable to cook and eventually half of the sample (55%) know how to combine meals (use leftovers to make new dishes).

Waste management
- *Recycling behaviour.* Most of the participants (83%) claim to separate their household waste, which shows a clear adoption of recycling behaviour not equally distributed into the different fractions. Those who have revealed they separate into different fractions, 95% recycle plastics, 92% glass, 80% paper and 61% organic waste. Half of the sample (54%) separates the four fractions.
- *Knowledge waste prevention.* The Waste Framework Directive (2008/98/EC) defines a waste hierarchy, in which prevention is on the top. The literature has shown, that prevention and recycling behaviour are not always positively related (Barr, 2007) and we want to evaluate the participants' knowledge about the priority among three general concepts: (1) prevention; (2) recycle; and (3) landfill. In the survey they have to order the three concepts regarding to waste management priority. Almost half of the sample (46.8%) correctly ordered the concepts.

21% scores recycled as a first action and prevent as the second. It is significant in terms of lack of information that 18.2% of people have not answered the question.

Regarding the actions associated with the food waste management, there is a specific hierarchy which transposes the waste into food (UE Res. 2011/2175 (INI)). Respondents were asked to order the concepts from most to least important: source reduction, feed hungry people, feed animals, industrial uses, composting and landfill-incineration. *Feeding hungry people* is the action that has the highest rank. The less priority action is considered *landfill-incineration*. Unless a rank could be distinguished, it could be said that in the metropolitan area there is no clear opinion about food waste management hierarchy.

4. Conclusions

After all we can conclude that inhabitants of the Barcelona metropolitan area participating in the study are not aware of their food waste generation. Mainly due to a lack of knowledge on: their self-behaviour, the priority on waste prevention, the food waste concept definition and about the responsibilities along the food supply chain. Despite the lack of awareness it does exist a commitment due to ethical and monetary reasons.

Taking into account the results of our investigation we suggest that to increase the awareness about food waste generation the topic should be addressed from a social perspective at a local level. It is important to put emphasis on the waste hierarchy and to treat the food waste equally as a food and waste behaviour.

Acknowledgements

This study was financially supported by the Metropolitan Area of Barcelona Prevention Grants.

References

Barr, S. (2007). Factors influencing environmental attitudes and behaviors: a U.K. case study of household waste management. Environment and Behavior 39, 435-473.
Bio Intelligence Service (2010). Preparatory Study on Food Waste Across EU 27. Technical Report 2010-254. Available at: http://ec.europa.eu/environment/eussd/pdf/bio_foodwaste_report.pdf.
Cecere, G., Mancinelli, S., Mazzanti, M. (2014). Waste prevention and social preferences: the role of intrinsic and extrinsic motivations. Ecological Economics 107: 163-176.
FUSIONS. (2014). FUSIONS definitional framework for food waste full report. Available at: http://www.eu-fusions. org/index.php/publications#.
Gustavsson, J., Cederberg, C., Sonesson, U., van Otterdijk, R., Meybeck, A. (2011). global food losses and food waste – extent, causes and prevention. Rome. Available at: http://www.fao.org/docrep/014/mb060e/mb060e00.pdf.
HISPACOOP, De Cooperativas, and De Consumidores. (2012). Estudio Sobre El Desperdicio de Alimentos En Los Hogares Estudio Sobre El Desperdicio de Alimentos En Los Hogares. Available at: http://tinyurl.com/o5ptur5.
HLPE (2014). Food losses and waste in the context of sustainable food systems. A report by The High Level Panel of Experts on Food Security and Nutrition of the Committee on World Food Security. Rome, 2014. Available at: http://tinyurl.com/oau7z9f.

Idescat/IERMB, (2011). L'Enquesta de Condicions de Vida i Hàbits de la Població de Catalunya Available at: http://www.iermb.uab.es/htm/descargaBinaria.asp?idPub=197.

Kosseva, M.R. (2013). Introduction: causes and challenges of food wastage. In: Food industry wastes. assessment and recuperation of commodities. Elsevier, Amstredam, pp. xv-xxiv

Parfitt, J., Barthel, M., Macnaughton, S. (2010). Food waste within food supply chains: quantification and potential for change to 2050. Philosophical Transactions of the Royal Society of London. Series B, Biological sciences 365(1554): 3065-3081.

Programme, Retail, Food Waste, Final Report, and WRAP. (2007). Food behaviour consumer research : quantitative phase. Available at: http://tinyurl.com/ofuxkwo.

Quested, T. E., Marsh, E., Stunell, D., Parry, A.D. (2013). Spaghetti soup: the complex world of food waste behaviours. Resources, Conservation and Recycling 79: 43-51.

The Gallup Organization (2011). Flash Eurobarometer Series 316, Attitudes of Europeans towards Resource Efficiency Analytical Report Attitudes of Europeans towards Resource Efficiency. Available at: http://ec.europa.eu/public_opinion/flash/fl_316_en.pdf.

19. You have the power to stop wasting food

S. Juul

Stop Wasting Food Movement Denmark (Stop Spild Af Mad), Copenhagen 1117, Denmark; info@stopspildafmad.dk

Abstract

Consumers are the key players in reducing food waste. Stop Wasting Food movement Denmark (Stop Spild Af Mad) is Denmark's biggest NGO against food waste, counting 30,000+ members and impacting weekly 250,000+ people. Stop Wasting Food has influenced the value chain from farm to fork, as well as politicians, MP's, MEP's and the media – and generated visible results: every 2nd Dane has reduced food waste. The movement is an official Partner of FAO initiative SAVE FOOD, EU collaboration FUSIONS and FAO/UNEP campaign Think.Eat.Save. Stop Wasting Food has also launched the European United Against Food Waste event campaign in several countries – and is a co-developer of MILAN PROTOCOL, co-signer of FAO SAVE FOOD Declaration and Every Crumb Counts Declaration and a co-developer of the Joint Declaration Against Food Waste. The movement published an international award winning Leftovers Cookbook 'Stop spild af mad – en kogebog med mere' by Selina Juul. Among the many results, Stop Wasting Food inspired the retail chain Rema 1000 to drop quantity discounts in all Danish stores, introduced doggybags in Danish restaurants in collaboration with Unilever and put food waste on the agenda of numerous Danish and international media. Stop Wasting Food also delivers tons of good free surplus food to homeless people of Denmark.

Keywords: food waste, citizen movement, public participation

> Remember what our grandmas always used to tell us: do not waste food – think of all the hungry children in Africa.

In my TEDx talk, I mentioned that global food waste could feed every starving child, man and woman on this planet – three times over in fact! Here is some food for thought.

A global shame

Globally, human beings produce enough food waste to feed 3 billion people: over 30% of the world's food supply is wasted. The annual food waste in Italy could feed 44 million people – all of Ethiopia's undernourished population. The annual food waste in France is enough to feed the entire population of the Democratic Republic of Congo. Just 5% of USA's food waste could feed 4 million people for one day.

United Nations Secretary-General Ban Ki-Moon noted that there is enough food in the world, yet millions are still starving – and unless we take action, it will devastate our planet. Who could possibly disagree: food waste is a global shame, especially in a world in which over a billion people are starving. And yet: everybody is waiting for somebody else to take action.

Leire Escajedo San-Epifanio and Mertxe De Renobales Scheifler (eds.) *Envisioning a future without food waste and food poverty*   **165**

DOI 10.3920/978-90-8686-820-9_19, © Wageningen Academic Publishers 2015

Can we send our leftovers to starving children in Africa? No, that is clearly not a permanent or sustainable solution. The problem in Africa is food loss. The amount of food lost per year in sub-Saharan Africa could feed 48 million people. Due to poor harvesting facilities, storage, packaging, distribution and the lack of a stable infrastructure, good food is lost in the fields before it even has a chance to reach peoples' bellies.

Food loss and food waste

In the West we waste approximately 40% of our food. This 40% happens at the end of the food value chain – by retailers and consumers. The same percentage of food, 40%, is lost in developing countries, though here the food losses happen at the beginning of the value chain. If we look at global food wasters, according to the Food and Agriculture Organization of the United Nations (FAO), we find that Western populations, such as EU member states, top the list. Here approximately 179 kg is wasted per capita per year. In developing countries, only 6-11 kg is wasted per capita per year.

This global imbalance must be corrected. But how does the food we waste in our homes in the Western world actually affect developing countries and hungry children in Africa? Does it actually matter? Indirectly, it does.

I participated in a panel debate during the People's Meeting (Folkemødet) in Bornholm at which the Secretary-General of the Danish Red Cross, Anders Ladekarl, said the following:

> The Western world's overconsumption of food is affecting global food prices: The more we in the West consume (and the more we throw out), the greater global demand for food becomes – and the higher food prices rise globally.

Let's imagine a pile of bananas, grown and produced in a developing country, transported all the way across the globe to a Western country just to be wasted because of some silly cosmetic reason. People in the very same developing country lack food. Imagine looking those hungry people in the eyes and telling them that the good bananas grown in their very own country are being thrown away just as fast they arrive in the Western world.

Food is the new gold

At my panel debate at the Barilla Center for Food and Nutrition 4th Annual Forum on Food and Nutrition, I addressed the food challenges of future generations. One of the speakers at the Forum, globally respected author and founder of the Worldwatch Institute, Lester R. Brown, mentioned that food was the new oil. I would say, however, that food is the new gold.

Why? Because fighting food scarcity will be one of the central geopolitical issues of the future.
- 1st fact: Population growth. By 2050, the earth's population will reach 9 billion people. By then, food production must be increased by 70% to meet demand. Today we already produce enough food waste to feed 3 billion human beings. Reducing food waste should number among our key focal areas. The UN estimates that in just 20 years, the earth's population will need at least 50% more food, 45% more energy and 30% more water.

- 2nd fact: Climate change. The increasing changes to our climate affect the world's agriculture and thus, the production of food. Floods, droughts and other increasingly irregular climate patterns will only worsen in future. More and more farmers are being forced to use GMOs and pesticides to ensure the survival of their harvest due to a changing climate, which in turn affects the loss of biodiversity.
- 3rd fact: Increasing food prices. This third fact has its roots in the 1st and 2nd facts, but additional factors include: the financial crisis, land grabbing (and the resulting desertification and deforestation), the world trade market structure, the global imbalance in food distribution, global and local food policy making, a lack of infrastructure, and a general lack of transparency in the food production value chain from farm to fork.

We must remember that there is enough food in the world, more than enough. Yet billions are still starving.

Who is to blame?

So, who should we point the finger at? Who is to blame? The industry? The politicians? The farmers? The retailers? Ourselves? At a conference in Bonn where I was a panel speaker, I learned from a fellow panel speaker from sub-Saharan Africa that African countries' agriculture often needs aid from the Western world. Unfortunately, this agricultural aid is often not used to improve agriculture. Instead, the money is appropriated by local politicians due to the lack of local infrastructure. In this particular case, a local politician bought 4 limousines from the agricultural aid money. This demonstrates that people on the ground are needed in Africa too in order to ensure that the money is indeed used to improve agriculture.

Sometimes I wonder if the global food waste scandal is a self-perpetuating system. Why have we, the consumers, become accustomed to such high standards that we cannot accept wonky fruit and vegetables in our supermarkets? Our choices affect the entire food production value chain and force farmers to toss out perfectly good fruits and vegetables because of the way they look.

Would there be a paradigm shift if Western countries were be able to cut their food waste? Would it affect developing countries? Would there be enough food for everyone? Is it even possible? Yes, I think it is. Imagine if every child, man and woman on this planet had enough food. Imagine what it would do to our human civilization. If every single human being on this planet had enough food, it would change our societies. It would stop wars, put an end to suffering and even change the course of human history. It could create a paradigm shift, a new era of peace on this planet.

And I strongly believe that we can achieve that paradigm shift. That is why I have been working – for over 7 years and putting in over 40 volunteer hours a week – on the Stop Wasting Food movement. Because I strongly believe that humanity can and will come up with a solution. And I think about it every day. Our work achieved great results: food wasted in Denmark was increased by 25% since the year 2010.

We must remember that food is the most powerful basic necessity for human beings. It is what keeps us going. It is what is keeping us alive. Food waste is a clear indication that there is something fundamentally wrong with our civilization.

Look at nature: There is no food waste in nature whatsoever. Everything is used and recycled. Every resource is used intelligently. The only species on this planet unable to cut down on food waste is us humans.

You are in control

So, what are the solutions to the global food waste scandal? Are we still waiting for everybody else to do something about it? Consumers have the power to change the entire system. And it would take just one simple personal step: stop wasting food. Will you continue to waste your food – and your money – after reading my article? Don't you think it's time for action?

Buy only what you actually need. Cook leftovers. Share food with your neighbours. Use it up. It is the simple wisdom of our grandmas, the very same grandmas who admonished us as children not to waste food and to think of all the hungry children in Africa.

Do the industry and the retailers dictate your shopping habits – or do you? Who is actually in control of this situation? You are of course. You are in control. Demand wonky fruit and vegetable in the stores. Don't fall for quantity discounts if you don't need that amount of food. Don't overstuff your plate at the cafeteria if you already know that you can only eat half. Ask for a doggy bag at a restaurant.

Speak your mind: Encourage positive action everywhere! Encourage the food industry to donate edible surplus food to charities. Join consumer movements. Encourage politicians to act.

The European alliance against food waste

You see, I am neither a celebrity chef nor a politician. I am just a simple consumer. In 2008, I got very tired of food waste. I created a group on Facebook: 'Stop Wasting Food'. Today, over four years later, the Stop Wasting Food movement Denmark (*Stop Spild Af Mad*) has become Denmark's largest non-profit consumer movement against food waste.

Stop Wasting Food movement Denmark counts today over 35,000+ members reaching weekly 250,000+ people – and enjoys the support of high-ranking politicians who even include our former Prime Minister. We have published an award-winning leftovers cookbook, we have brought the topic of food waste to numerous Danish and international print media, radio and TV – counting 18,000 media exposures all around the world.

We distribute good surplus food to homeless people, we have convinced a large retail chain, Rema 1000, to drop all quantity discounts, and we have helped put the topic of food waste on the UN and EU agendas.

We are all just ordinary consumers, ordinary people. But ordinary people can do extraordinary things. Consumers cannot fight food waste alone though. All the stakeholders in the food production value chain must be involved: farmers, industry, retailers, canteens, restaurants and food services.

Right now, we are establishing the Food Waste Think Tank: the world's first joint and transnational Think Tank to end food waste. The Think Tank will activate, engage, and support the stakeholders in food waste value chain in order to make a global reduction of food waste. But don't wait for the Think Tank, the industry, the EU, politicians or someone else to act. Take action yourself.

No matter who we are, we are all consumers, we all eat, we all waste food – and we are all a part of the problem. And thus, we are also part of the solution. The next time you are considering feeding good food to your rubbish bin, ask yourself: how many starving African families would approve of your actions? Not one. There, you have your answer. You have the power. You have the knowledge. Don't wait for someone else to take action. Do it yourself.

Stop wasting food.

20. Potential of market based instruments and economic incentives in food waste prevention and reduction

L.H. Aramyan[1]*, N.I. Valeeva[1], M. Vittuari[2], A. Politano[2], P. Mahon[3], C. Balazs[4], K. Ujhelyi[4], S. Scherhaufer[5], M. Gheouldus[6] and D. Paschali[7]

[1]LEI, Wageningen UR. Hollandseweg 1, 6701 KN, Wageningen, the Netherlands; [2]Faculty of Agriculture, University of Bologna, Via Fanin 50, Bologna, Italy; [3]WRAP Blenheim Court, 19 George Street, Banbury, Oxfordshire, OX16 5BH, United Kingdom; [4]Hungarian Food Bank Association, HFA, 3 Lokátor ut, 1172 Budapest, Hungary; [5]University of Natural Resources and Life Sciences BOKU Vienna, Muthgasse 107, 1190 Vienna, Austria; [6]BIO by Deloyte, 185 Avenue Charles de Gaulle, 92200 Neuilly-sur-Seine, France; [7]ANATOLIKI SA, Development Agency of Eastern Thessaloniki's Local Authorities, 1st Km Thermi-Triadi, Golden Centre (No 9), 57001 Thermi, Greece; lusine.aramyan@wur.nl

Abstract

Preventing and reducing food waste along supply chains is a high priority in many countries. Market based instruments and economic incentives are important tools for stimulating food supply-chain operators and households to food waste prevention and reduction. This study analyses the potential of market based instruments and economic incentives as specific policy measures to prevent and reduce food waste along the whole supply chain. The results show that a successful implementation of a tool or incentives requires accurate advance planning and should include a detailed analysis of the potential impacts and barriers. A mix of complementary policy measures increases the possibility of successful implementation.

Keywords: policy measures, environment, health, household waste

1. Introduction

Roughly one third of food produced for human consumption is lost or wasted globally, while millions of people in the world do not have enough food to eat. Food waste however translates not only into human hunger, but also into a tremendous waste of resources and financial losses to food producers, resulting in less efficient food production. Preventing and reducing food waste along supply chains is a high priority in many countries. Market based instruments and economic incentives are important tools for stimulating food supply-chain operators and households to deliver food waste prevention and reduction. They are tools that encourage behavioural change through market signals rather than through traditional regulations (Driesen, 2006; Gupta *et al.*, 2013). Market based instruments and economic incentives can be stimulated, developed and implemented by governments, by private organizations and can also be stimulated by means of interaction of government and private mechanisms.

Government can use a full range of policy tools and market based instruments to provide incentives that might stimulate prevention and reduction of food waste. Some examples are public payment schemes (e.g. tax measures, subsidies, etc.). Market based instruments and economic incentives

Leire Escajedo San-Epifanio and Mertxe De Renobales Scheifler (eds.) *Envisioning a future without food waste and food poverty*   **171**

DOI 10.3920/978-90-8686-820-9_20, © Wageningen Academic Publishers 2015

can also be used by private parties in prevention and reduction of food waste. Market forces along the food supply chain on their own can promote voluntary improvements. Some examples of such voluntary improvements are self-organized private market instruments and incentives, marketing labels and certification schemes, bonus/penalty systems, long-term supplier contracts, preferred supplier agreements, applications offering discounted prices and other purchasing incentives for perishables approaching their expiration dates. These voluntary incentives appeal to private sector actors because of improved corporate reputation and positive brand impact, strong corporate governance, competitive advantage, etc. Besides this, interactions between governmental policies and voluntary improvements can also stimulate supply chain stakeholders to prevent and reduce food waste. In this case private incentives may be designed in combination with complementary policy initiatives.

Although such instruments can be very promising, there is a lack of knowledge on the potential of such tools for food waste reduction. This study identifies and analyses the potential of market based instruments and economic incentives as specific policy measures to prevent and reduce food waste along the whole chain. Policy is defined as a course or principle of action, proposed or adopted by a government, party, business or individual, intended to influence and determine coherent decisions, actions, and other matters; usually with a common long-term purpose.

This study is a part of the EU project FUSIONS (Food Use for Social Innovation by Optimising Waste Prevention Strategies) funded by the European Commission Framework Programme 7 and is a work in progress. For more information regarding FUSIONS project please see http://www. eu-fusions.org/.

2. Methodology

The study is based on a literature scan on market based instruments and economic incentives for stimulating food waste prevention and reduction. Since the topic of food waste is rather new, a literature review of market based instruments and economic incentives, as applied not only to food waste but to other sectors and areas, was carried out. This was done in order to identify previous applications of incentives and tools from other sectors/areas that might be applicable to the food waste issue also. The identified areas were: (1) environment; (2) sustainability; (3) product quality; (4) food safety; (5) animal health (incl. prevention of antibiotic use); (6) human health and obesity; (7) waste (industrial); (8) household waste recycling; and (9) biofuels.

The selection of these areas was conditioned by the fact that these issues have been on the agenda of governmental and business policies for a much longer period, compared to the food waste issue; thus they may offer workable tools/incentives that can potentially be applied to food waste. Given the limitation on the length of the paper for this conference we have selected a limited number of areas to discuss further in detail in the paper. The focus areas, including household solid waste management, environmental pollution and nutritional health and obesity, have been selected as having the potential to offer lessons that could be learned in the implementation of such tools in the area of food waste reduction and prevention.

The selection of market based instruments and economic incentives from the literature was based on one or more of the following criteria:

- general principle of action (e.g. taxation, subsidy, certification schemes);
- tool/incentive is advised and/or applied by: government-EU, national governments/countries, local governments, supply chain operators (producers, retailers, food services, etc.);
- tool/incentives is advised, but not implemented yet or implemented;
- efficiency/positive impact;
- potential applicability to food waste.

3. Results

The discussion of results starts from a short review of instruments and incentives from the literature on food waste generation. Food waste causes are numerous. At an early stage of the FUSIONS project a literature review on food waste drivers was carried out which identified about 200 drivers that contribute to food waste generation along the entire food supply chain (http://www.eu-fusions. org/index.php/publications). Due to the limitation of the length of this paper, in this study we have decided to present the most cited and interrelated food waste drivers that can be used in developing market based tools and incentives to prevent and reduce food waste.

Donations and redistribution

Redistribution and/or donation of unsold food can be a powerful strategy to prevent and reduce food waste. However there are several problems related to the realization of these incentives. One of them is EU VAT legislation, which often hinders cooperation between retailers and food banks for the following reason. According to the directive VAT has to be paid on food intended for donation. The basis for the VAT on the products is the purchase price at the moment of the donation adjusted to the state of those goods at the time when the donation takes place. The problem is the legal uncertainty on whether the food which is close to its 'best before/use by' date has a countable – taxable value (therefore a VAT-able base) or a small or zero value (no VAT to be paid). In some Member States no VAT is paid when food is donated to food banks since in these countries the value of the donated food close to its 'best before/use by' date is interpreted as being small or zero. The EU VAT Committee agreed on new guidelines to harmonise the application of the Directive across EU Member States. However, it does not address the grey area of the value of donated food close to its 'best before/use by' date. Hereby arises also the liability problems with regards to donated food. However, Italy has addressed this problem by adjusting food donation liability legislation (taking away liability from donating parties) and providing an incentive to food produces in food donation.

Italy is a frontrunner in food donation liability legislation, being the first EU country to have passed Good Samaritan legislation (similar to the USA Good Samaritan Food Donation Act), which exempts food donors from legal liability for donated food. Other measures to drive food donation, including for example fiscal incentives such as tax breaks, can also be taken at a national level. Another option is to pass the liability for donated food (after the 'best before' and 'use by' dates) to consumers, by using existing models such as the Good Samaritan legislation. Expansion of non-profit organizations by encouraging proper networks and structures for the redistribution of food for human consumption where donors (farmers, retailers, etc.) and communities (food banks and others) will be easily linked is also an incentive that can contribute to address the redistribution problems and waste generation. However, it should also be noted that rules making it easier to redistribute food to charitable organizations might also lead to negative effects on food waste generation because they might favour overproduction, challenging the capacity of charitable organizations to redistribute the food they receive.

Labelling

EU legislation establishes the rights of consumers to safe food and to accurate and reliable information. Labelling helps consumers to make informed choices while purchasing; however, label information is sometimes unclear. Confusion amongst consumers about the different meanings of 'best before', 'use by', and 'sell by' dates is still a cause of food waste. The adjustment of food label dates related to the usage and shelf-life of foods used on packaged food and drink is another key factor for reducing food waste (Schneider, 2008). The confusion caused by product dates is as a result of different terms used on packages, i.e. 'sell-by', 'best if used by', 'used-by', etc. It can be easily overcome by education of consumers, by means of awareness campaign and by adjustments of requirements on labelling (e.g. clear information using one common term). Private actors/retailers can also provide incentives to consumers in buying products close to expiration dates by providing discounts, developing apps for consumers with latest information on product discounts or apps/websites that may stimulate the more rapid redistribution of close to expiry date products.

Nutritional health and obesity

For nutritional health and obesity, the policy measures identified in this study are underpinned mainly by voluntary agreements, rather than regulations. Some of the measures (e.g. awareness campaigns) are not specifically market-based instruments or economic incentives, but are included in brief in this analysis as they are essential components of the overall policy directions on nutritional health and obesity. The policy measures that have been identified for nutritional health and obesity are fairly standard, and confirm that some policy approaches could potentially be used for stimulating food waste prevention and reduction. They are described below.

Food pricing and subsidies is a policy measure with significant potential to contribute to improving nutritional health. A study by Powell *et al.* (2013) provided an over-arching review of research in the USA on food pricing and subsidies in order to collate existing knowledge for informing policy. The study concluded that taxes and subsidies are increasingly being considered as policy instruments to provide incentives for consumers to improve their food and drink consumption patterns. The practice of putting taxes on certain types of foods is quite well-established in some countries, and often it is the type of food and drink considered to be particularly unhealthy (e.g. soft drinks, sweets, snack foods) that are taxed more heavily. Taxes might be applicable to food waste prevention (in that higher food prices would encourage less food waste). However, this policy could have negative social implications. Minimum pricing for alcohol is another identified pricing incentive, which may, for example, reduce alcohol consumption. This policy measure was adopted into law in Scotland in 2012. It sets a floor price for a unit of alcohol, making sure that alcoholic drinks are not sold below cost price. This economic incentive would, however, have limited applicability for policies to prevent food waste.

A number of measures could be implemented to influence purchasing and potentially improve nutritional health. Some retailers in the UK have been introducing a voluntary policy to remove confectionery from displays at the point of purchase (i.e. near the check-out counters in stores). This has been proposed by the Scottish Government in a draft policy framework, with a focus on voluntary participation by retailers. Limited information is available on the effects of such a policy, but some studies indicate a positive impact is likely (Hawkes *et al.*, 2015). However, the concept of selecting the types of product displayed at the point of purchase would probably have limited applicability to prevent food waste.

There is limited research and data available on the effects of food portion and pack sizes, but there has been a trend in increased portion sizes, which is likely to be contributing to the increased levels in obesity (Steenhuis *et al.*, 2009). In terms of policies and incentives related to portion and pack sizes in restaurants and retailers, the focus tends to be on voluntary measures, linked to product strategies, labelling, responsible advertising, offering a wider range of portion sizes, etc. (EIRIS, 2006; Steenhuis and Vermeer, 2009). A recent study by McKinsey (2014) identified portion control as the most effective measure to tackle obesity. Measures to decrease portion and pack sizes would have high applicability to the prevention of food waste (WRAP 2014a,b), particularly for those people living alone.

Supermarket loyalty cards provide detailed information on consumer behaviour, and there have been discussions in the sector about whether the data could be used to target education to shoppers that buy more unhealthy products. However, there are concerns, including those over data protection issues. Web searches on news items indicate that there has been only limited discussion on the potential use of loyalty award schemes to provide incentives to shoppers that make healthier purchases. There is a patent application in the USA related to such a component of loyalty card schemes (Doak, 2013). Loyalty card schemes could be applicable for the prevention of food waste, for example by encouraging people to purchase tools to help manage their food better, by providing information and advice based on actual purchases and by helping to mitigate against price barriers (for example smaller packs may be relatively more expensive than larger ones), although the cards would not provide data on food waste generation.

Labelling is another potential instrument, which has been mentioned earlier. The EU Regulation 1169/2011 on the provision of food information to consumers (FIC) includes mandatory nutrition information on processed foods. The obligation to provide nutrition information will apply from December 2016. As an example, the EU FIC is being implemented in the UK through the UK Food Information Regulations (SI No. 1855) (2014), which includes requirements on nutrition and ingredient information. Traffic light labels on food sector products are used in some European countries on a voluntary basis, particularly in the UK. The labelling scheme shows how much fat, saturated fats, sugar and salt are in the product by using the traffic light signals for high (red), medium (amber) and low (green) percentages for each of these ingredients. The concept is that the traffic light label on the packaging is easier to identify and interpret than Guideline Daily Amount labelling. Several studies have researched the potential effects of the traffic-light nutrition labelling on foods, and most of the ones reviewed have concluded that the measure is likely to have significant positive impacts in nutritional health (Lobstein and Davies, 2008; Sacks *et al.*, 2011; Sonnenberg *et al.* 2013). Information on packaging labels to encourage food waste prevention is highly applicable.

Furthermore, there have been numerous awareness campaigns about nutritional health and obesity. Awareness campaigns to encourage food waste prevention are highly applicable, for example Love Food Hate Waste in the UK, which has proved to be very successful in raising awareness, enabling behaviour change and reducing food waste (WRAP, 2013).

Household solid waste management

A range of market-based instruments and economic incentives for household waste recycling were found in the literature. Most of them have a monetary character. Monetary incentives for household waste management can generally be divided into three main categories (Jones *et al.*, 2010). The first category includes negative incentives such as revenue taxes or 'pay as you throw' policies where the public pays depending on the volume or weight of the waste. Pay-by-volume and pay-by-weight policies appeared to be one of the most favourable measures for household waste reduction. Such policies provide benefits in environmental terms (increase in recycling and reduction of waste to landfill) and in economic terms (collection and treatment costs are adjusted to the weight treated) and can be argued to be about fairness, as people are billed according to the waste they produce. At their best, such systems can lead to an increase in waste separation, an increase in home composting, an increase in second-hand sales and most importantly greater prevention of waste. Critics argue that more waste may end up being disposed of illegally through burning or illegal dumping (e.g. disposing of waste at the workplace or at other municipalities with no such fee system). Some studies (Dahlen and Lagerkvist, 2010; Dunne *et al.*, 2008) have suggested that no correlation could be found between increased sorting for recycling and decreased amounts of residual waste in pay-by-weight or pay-by-volume schemes. This can be explained that residents might have adapted a lifestyle producing less waste, or they may have disposed of some waste outside the ordinary waste collection system. The implementation of such a measure for the purpose of food waste reduction could lead to the same negative effects (e.g. disposal via the sewer). Waste charges on food waste or charge for food waste above a predetermined quantity are measures which may be suitable for food waste reduction, but may not be popular among the public and therefore may be difficult to implement. This may be addressed through additional monetary incentives for the separate collection of waste. Most waste taxes (such as landfill tax, incineration tax) encourage the recycling rather than the prevention of food waste, but in combination with other waste prevention policies they could also enhance food waste reduction.

The second category of monetary incentives involves positive incentives, where funding opportunities or tax reduction are made available to those who engage in waste minimization or participate in recycling projects (Jones *et al.*, 2010). Examples include the setting of recycling rates, recycling certificate schemes or subsidising the use of biogas use, and some other measures which ensure that waste is going to an adequate disposal pathway, such as producer responsibility schemes, provision of bins and bags to households by municipalities. Such instruments do not appear to be directly relevant to food waste reduction or prevention.

The third category includes mixed incentive instruments, which have long been established in the field of solid waste management. Deposit-refund systems are the best known example (Jones *et al.*, 2010). Though this measure has proved to be effective in increasing recycling rates, it does not seem to be applicable to the reduction of food waste. In addition to monetary incentives, non-monetary incentives were also identified. These aim at an increase in the participation rate for recycling or an increase in the amounts collected for recycling. Incentives which are based on the quantities of household waste collected for recycling could be similarly applied to quantities of food waste which are saved (e.g. shop vouchers, community rewards, charitable donations, school rewards, leisure vouchers). These incentives might be adapted to food waste prevention and reduction. Other non-monetary incentives include participation in pilot projects which may result in a learning-by-doing effect and as a consequence enhance food waste prevention. Such pilot projects need to be promoted by local authorities and may be provided in combination with incentives, such as leisure or shop vouchers. Many such non-monetary incentives could potentially result in a positive attitude from householders to food waste prevention and reduction.

Environmental pollution

A range of market-based instruments identified in the literature for environmental pollution mainly focus on the 'Polluter Pays Principle' (PPP) as an economic principle for environmental policy. The basic tenet of PPP is that the price of a good or service should fully reflect its total cost of production, including the cost of all resources used. Thus, the use of water, air or land for the emission, discharge or storage of wastes is as much a use of resources as other labour and material inputs. Market-based systems of incentives and disincentives are instruments adopted to put the Polluter Pays Principle into practice. They aim to modify economic behaviour by promoting the internalization of environmental costs. Among the categories of market-based instruments able to stimulate environmental positive behaviour or 'green' investments, there are three types of effects that economic incentives can cause: direct alteration of price or costs levels; indirect alteration of price or costs levels via financial or fiscal means; market creation and market support. Direct alteration of price and cost level occurs when charges are levied on products (product charges) or on the processes that generate these products (emissions charges, input charges, feedstock charges), or when deposit-refund systems are put into operation (Turner *et al.*, 1994). Among this category, the most common market-based instruments are:

- Emission charges: charges (e.g. taxes or tariffs) on the discharge of pollutants into air, water and soil and related to the quantity and the quality of the pollutant and the damage costs inflicted to the environment.
- User charges: related to treatment cost, collection and disposal cost, recovery of administrative costs depending on the context of application. However, they are not directly related to environmental damage cost.
- Product charges: they are levied on products harmful to the environment and related to the environmental damage costs that they cause.
- Deposit-refund systems: they involve a deposit paid on potentially polluting products. If products are returned to some authorized collectors a refund is paid.

Indirect alteration takes place when incentives are provided to induce environmentally clean technologies (Turner *et al.*, 1994). This category includes direct subsidies, soft loans, fiscal incentives, enforcement incentives, such as non-compliance fees and performance bonds.

Some connections with the food waste issue are evident. First, some of the market-based incentives and disincentives are addressed to waste that results from production and consumption activities. Food waste is a type of waste and it also impacts the environment. If the environmental cost of food waste was estimated, a tax at the socially optimal level could be levied in order to transfer this cost to the responsible subject. Furthermore, it is possible to imagine a set of market-based instruments to facilitate the reduction of food waste. For example, taxes on waste or fees on food that is disposed of as waste, By contrast, a firm that reduces or prevents food waste generation would, in fact, be creating a positive externality, and in such cases subsidies (e.g. fiscal incentives) could be provided to the firm in direct proportion to the value of this external benefit. For example, incentives for retailers who decide to donate unsold – but safe – products. This kind of instruments could also stimulate social innovation, in terms of networking, bottom-up initiatives or inclusive strategies to solve a social problem like food waste.

4. Conclusions

Based on a brief literature review this study identifies a number of market-based instruments and incentives that could potentially be applied to design policies for food waste reduction and prevention. Analysing the literature we can conclude that, the successful implementation of a measure requires accurate advance planning and should include a detailed analysis of the potential impacts and barriers. A mix of complementary policy measures increases the possibility of successful implementation. The study provides insights into the potential of market-based instruments and incentive systems in the food chain aimed at preventing food waste. The results may support public and private decision-makers in selecting specific solutions to prevent and reduce food waste across the whole supply chain.

References

Dahlen, L., Lagerkvist, A. (2010). Pay as you throw: strengths and weaknesses of weight-based billing in household waste collection systems in Sweden. Waste Management, 30: 23-31.

Doak, J., Bridi, G. (2013). Systems and Methods for Incentive-Based Nutrition Enhancement (2013); US Patent App. 13/792,584, 2013.

Dunne, L., Convery, F.J., Gallagher, L. (2008). An investigation into waste charges in Ireland, with emphasis on public acceptability. Waste Management, 28: 2826-2834.

Ethical Investment Research Services (EIRIS) (2006). Obesity concerns in the food and beverage industry. Available at: http://tinyurl.com/oddo7ww.

Jones, N., Evangelinos, K., Halvadakis, C.P., Iosifides, T., Sophoulis, C.M. (2010). Social factors influencing perceptions and willingness to pay for a market-based policy aiming on solid waste management. Resources, Conservation and Recycling, 54: 533-540.

Hawkes, C., Smith, T.G., Jewell, J., Wardle, J., Hammond, R.A. (2015). Smart food policies for obesity prevention. The Lancet, 385: 2410-2421.

Lobstein, T. and Davies, S. (2008). Defining and labelling 'healthy' and 'unhealthy' food. Public Health Nutrition, 12(3): 331-340.

McKinsey (2014) Overcoming obesity: An initial economic analysis. http://tinyurl.com/p7fdgwe.

Powell, L.M., Chriqui, J.F., Khan, T., Wada, R. and Chaloupka, F.J. (2013). Assessing the potential effectiveness of food and beverage taxes and subsidies for improving public health: a systematic review of prices, demand and body weight outcomes. Obesity Reviews, 14: 110-128.

Schneider, F. (2008). Wasting food. urban issues & solutions, Alberta Canada, 10pp.

Sonnenberg, L., Gelsomin, E., Levy,D., Riis, J., Barraclough, S., Thorndike, A. (2013). A traffic light food labelling intervention increases consumer awareness of health and healthy choices at the point-of-purchase. Preventive Medicine, 57 (4): 253-257.

Steenhuis, I., Leeuwis, F., Vermeer, W. (2009). Small, medium, large or supersize: trends in food portion sizes in The Netherlands. Public Health Nutrition, 13(6): 852-857.

Steenhuis, I., Vermeer, W. (2009). Portion size: review and framework for interventions. International Journal of Behavioural Nutrition and Physical Activity 2009, 6: 58.

Turner R.K., Pearce D., Bateman I. (1994). Environmental economics: an elementary introduction, Harvester Wheatsheaf, London.

WRAP (2013). Household food and drink waste in the UK 2012. Available at: http://tinyurl.com/na8cbo6.

WRAP (2014a). Household food & drink waste – a product focus. Available at: http://tinyurl.com/kkqh6rm.

WRAP (2014b). Household food & drink waste – a people focus. Available at: http://tinyurl.com/pyl7leo.

Part 5. Social Responsibility of universities: contributing to reduce food waste and food poverty

21. New university-society relationships for rational consumption and solidarity: actions from the Food Banks-UPM Chair

I. de los Ríos, *A. Cazorla, S. Sastre and C. Cadeddu*
Technical University of Madrid. Avenida Puerta de Hierro 2, 28040 Madrid, Spain;
ignacio.delosrios@upm.es

Abstract

In industrialized countries and in developed societies there is a contradiction on food surpluses and the existence of poverty pockets and marginalization. In these societies, though there are indications that awaken the spirit of solidarity, it is necessary to spread human and cultural values to become aware of the social, environmental and economic consequences of food consumption. Consumption is a collective behaviour and, therefore, consumer protection means defending the whole society. Universities have great potential for articulation relationships with both the civil society and the economic sectors in order to address society's major problems by creating values in young people that result in an improvement in people's lives. The aim of this paper is to show a new form of relationship between universities and civil society, to contribute jointly to some of the most serious challenges mankind has to face: hunger and malnutrition. The relationship takes the form of University Food Banks-UPM Chair as a means to establish a long-term strategic partnership, in order to carry out training, research and knowledge transfer in areas of common interest. After two years of joint activities, the performed projects highlight the importance of these new relationships, which sit as a novel way to generate a humble mind and rational consumption.

Keywords: food waste, social awareness, competences, higher education volunteering, project-based learning

1. Introduction

According to FAO, 25,000 people worldwide die every day because of hunger (FAO, 2014). The problem is not the lack of production to meet the world food needs, but the fact that a third of the food produced in the world annually is lost, representing an annual waste of 1.3 billion tons of food (FAO, 2014).

In the context of the European Union, the reality is no less alarming; in 2014, 125 million people (25%) are at risk of poverty or social exclusion and 55 million (9.6%) have no possibility to have a quality meal every two days (Eurostat Nov, 2014). This contrasts with the 89 million tons of food that are wasted each year in Europe (European Union Committee, 2014). Food waste from households contributes the most to waste (with 42% of the total), followed by the manufacturing sector of food and beverage with a 39% (European Union Committee, 2014).

Rational consumption and solidarity are therefore a need for society (FAO, 2014). The forms of irrational consumption affect others, causing great economic losses and damage to natural resources

Leire Escajedo San-Epifanio and Mertxe De Renobales Scheifler (eds.) *Envisioning a future without food waste and food poverty*
DOI 10.3920/978-90-8686-820-9_21, © Wageningen Academic Publishers 2015

on which mankind depends for food (FAO, 2014). The consequences of consumption affect future generations in both the most developed countries and southern populations of the world (European Union Committee, 2014)

New actions to solve a serious challenge are needed, creating awareness and training processes indispensable for society (Montagut and Gascón, 2014). A critical and rational education on the phenomenon of consumption, should start in the family and continue throughout the lifetime of the people (Corrales, 2006). Schools, universities and training authorities, should be responsible for sensitizing people developing rational, civic and supportive attitudes to avoid food waste.

The university volunteer, students in higher education, has demonstrated to be of great value and potential for social change (Francis, 2011; Holdswortha and Quinnb, 2010; Hustinxa *et al.*, 2005), and it is also an instrument for implementing and adapting services to society (Anthony *et al.*, 2000; Bringle and Hatcher, 1996) as well as being experiential learning for students.

This paper presents a new way to build university-company-society relationships, implementing services to society and raising awareness and acting towards rational consumption. This work takes the form of projects and activities organized by the Food Banks-UPM Chair which began in July 2013. The idea arises at the Spanish Federation of Food Banks (FESBAL) who looked for Research Group GESPLAN of the Technical University of Madrid (UPM).

The context of action is Spain, the sixth EU country in food waste with 7.7 million tons, equivalent to 63 kilograms per person per year (FAO, 2014). There are over three million people living in severe poverty in Spain (FESBAL, 2012). Faced with this situation, Spain ranks first in Europe solidarity with 55 food banks out of a total of 254 banks in Europe (FAO, 2014). FESBAL activity has grown 20% in 2014 and 70% in the last five years (FESBAL, 2015), both in the amount of tons distributed and in the number of associates and beneficiaries the food reaches.

FESBAL's activity is integrated into the FEBA (Fédération des Banques Alimentaires Europenne) which brings together 257 banks in 22 countries in Europe. FEBA was founded in 1986 to combat hunger and waste. There are 13,000 volunteers who collaborate in their daily operations; through its 31,000 charitable organizations 402,000 tons of food are distributed, which is equivalent to 804 million meals to 5.7 million people in Europe (eurofoodbank.eu). The university-society model and its actions undertaken through the framework 'working with people' is shown below.

2. The working with people model

The methodological framework of the University Food Banks-UPM Chair (FBC) is the model working with people (WWP) (Cazorla *et al.*, 2012), which is the result of 25 years of experience in the field of management of rural development projects, in different European contexts and emerging countries. This framework has also been applied in educational innovation projects from project-based learning and experiential education, with emphasis on the development of personal and contextual skills for future professionals (De los Rios *et al.*, 2010, 2015). The actions of the FBC are applied this way in the educational strategy, which aims to promote, in addition to certain technical and contextual skills, social responsibility and commitment from solid ethical standards. According to the rules of the UPM, the FBC is configured as a stable and dynamic structure that favours the interaction between university and society within the framework of joint objectives for UPM and FESBAL (De los Rios *et al.*, 2015).The three components of this WWP framework are shown in Figure 1.

Ethical-social component

This component covers the area of behaviour, attitudes and values of people that interact to promote, participate, manage and benefit from the activities and projects. This component is identified with one of the priority objectives of the FBC: promote corporate social responsibility from a human dimension. As it includes the conducts and moral behaviours of people, this component that surrounds all actions of FBC, lays the 'foundation' for people – of different public and private spheres – to work together, with commitment, trust and personal freedom. Relations between university and society from the FBC do not have a 'neutral' character, but are based on an ideal of service that is guided from values. Thus it seeks to improve the skills and behaviours of people and one of the principles on which the activities of the Food Banks are based, awakening the spirit of solidarity and spreading human values.

Technical-entrepreneurial component

The FBC actions are aimed at improving the effectiveness and efficiency of the 'volunteer projects' in their various forms (programs and projects for training, diffusion, collection campaigns, knowledge

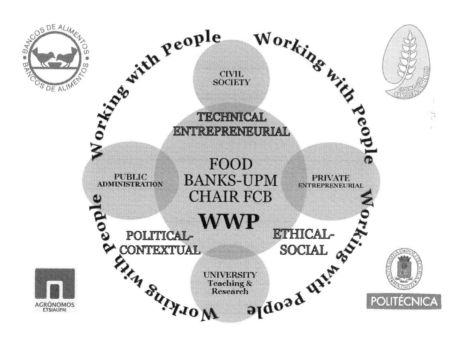

Figure 1. Working with people model applied to the Food Banks-UPM Chair.

transfer). These volunteer projects are the 'technical' instruments capable of generating services to society, meeting strategic objectives, which must also meet certain requirements and quality standards. This component promotes continuous innovation through new formulas that generate new volunteer projects, both from young volunteers from the university and from the closest collaborators of FESBAL and the food banks. The experience from GESPLAN group focuses on teaching and research in the area of human development and projects (planning, evaluation and project management) encouraging the development of skills under the international standards of the International Project Management Association (IPMA, 2010).

Political-contextual component

The idea is to provide the FBC with strategic elements to improve relations with society, with FESBAL and food banks. This area covers the ability of the university and its various research groups to relate to the public administrations of the different (international, national, regional and local) levels and other actors of civil society. This ability to interact depends on the University's ability to adopt an internal organization that facilitates participation and social activism. The FBC has an instrumental character, of service, and therefore its actions are changing according to learning and the new information generated.

3. First actions undertaken and results

Below are the actions and projects conducted by the FBC and the first results. There are three main areas of action: schools (CORAL program), awareness campaigns and social networks.

CORAL program

A group of university partners, coordinated by professors and researchers of the FBC has been commissioned to design the CORAL program (rational food consumption), creating a school awareness campaign on 'zero waste' in different Spanish schools with schoolchildren. After two years of activities, the results have been very positive and have reached more than 1,300 pupils (5th and 6th year of primary and 4th year in secondary schools) in 20 schools in different Spanish cities. The dynamic is being done through various activities (games, workshops, dynamic presentation, awareness videos) and in primary schools it concludes with a drawing competition on 'zero waste' at regional and national level. For secondary school students, the FBC has developed a methodology of 'project-based learning' simulating a real business environment to increase teamwork and the development of skills such as communication, negotiation, listening, planning and leadership. The team of university students that goes to schools is highly interdisciplinary, with students from different degrees (industrial, agricultural, civil, aviation, forestry, naval, mines engineering and architecture). Currently, after a conference held at the university to present the CORAL program, there is a database of students, who have expressed interest in getting involved as volunteers in the activities taking place for the academic year 2015-2016. Besides continuing with the 15 pilot schools, currently a schedule for 150 students of 7 new schools in Madrid, Badajoz and Ciudad Real is planned.

Social awareness program

Another axis of the actions taken consists of a series of activities around university, including in the first place awareness campaigns against food waste, aimed at university students to spread the message of rational consumption. A new action has been to bring together the final matches of football leagues in a 'sports and solidarity day', with chair information, tips and games to raise waste awareness, reaching a significant number of university students. Moreover the students working for the chair designed three food pick up campaigns in the different schools, sending products to the food bank of Madrid. The most important was the 'Great Tin Food Collection' on San Isidro, as a charity event held for two days in University Campus Metro stop. In parallel, they are working actively in the 'Great Pick up campaign' of FESBAL, which gets organized each year in Spain in the last week of November. A Food Banks-UPM Award for best undergraduate and graduate thesis has also been designed that helps *spread the culture of rational consumption*. The jury, consisting of two representatives of FESBAL and two representatives of the UPM, has awarded two students from different Spanish universities, one prize of € 1000 and one second prize of € 500. Moreover, researchers and collaborators of the chair try to raise awareness and influence through scientific conferences. On July 16, 2014 at the XVIII International Congress on Management and Project Engineering, a scientific paper on the activity of the 'Food Banks-UPM Chair as a training tool for the dissemination of culture of rational consumption' was presented, and was awarded with a prize. Furthermore 6 students participated in 2015 in the 'VII Congress of College Students in Science, Technology and Agricultural Engineering' presenting their research and bringing students to scientific and experimental world. One goal of this congress is to acquire the transversal competence ability to communicate in Spanish and English in oral and written form.

Finally, the 'studentship program' is supporting many of the above activities. So far they have offered to coordinate a total of 18 studentships in FBC, FESBAL, the food banks and collaborating companies (Carrefour, Lidl, Danone, etc.). These practices try to get UPM students to apply project-based learning, team competences, involving young people and guiding them to apply skills valued in the business world (teamwork, leadership, creativity, communication, values of solidarity ...).

Social networks program

Another large block of action consists of disseminating the FBC activities and other educational contents through social networks in order to achieve a broader spectrum of society. After designing the logo of the FBC a Twitter account was opened, as it was considered the most 'simple' social network which was then followed by a Facebook page, and a website dedicated within GESPLAN's (www.ruraldevelopment.es) with over 800 visits/day. The publications are organized according to a thematic order throughout the week, with recipes, tips, content for kids, documentaries, and always giving priority to the latest news and breaking news. From September 2014 until today it has gone from 145 to 425 followers on Facebook and from 127-244 on Twitter. Facebook is also a useful tool to keep contact with the food banks and share useful information around food waste reduction.

All these activities coordinated from a single WWP conceptual framework, are proving to be very useful and of great interest to young people, and for the organizations involved, opening new projects from university-civil society relations.

4. Conclusions

The success of the FBC is overcoming the merely 'tangible' aspects of the results achieved through the volunteer projects, as it is achieving intangible benefits in the form of values expansion to society from the academic world. After the first two years of joint activities, projects from the FBC, show the right approach from the WWP model, to build new relations which aim to overcome the current 'culture of waste', and replace it with supportive and respectful mentality towards a more equitable and humane society. The balance in the university-society actions from the three components of the model WWP (ethical-social, technical-entrepreneurial and political-contextual) is confirmed as a necessity to advance together towards one of the most serious challenges facing humanity, hunger and malnutrition.

We conclude this presentation with the testimony of one of the young volunteers, a student of aeronautical engineering working at the FBC after his volunteer experience:

> If everyone in the world jumped at the same time, it would generate an earthquake. But how difficult it is to get everyone to come together to do anything, even if it is to make things better. I understand the need to contribute and it is true that an objective can be achieved through the work of wonderful people. We just need more people willing to jump with us.

References

Bringle, R.G., Hatcher, J.A. (1996). implementing service learning in higher education. The Journal of Higher Education 67(2): 221-239.

Cazorla, A., De los Rios, I., Salvo, M. (2013). Working with people (WWP) in rural development project: a proposal from social learning. Cuadernos de Desarrollo Rural 10 (70): 131-157.

Corrales, L.M.T. (2006). Educar en el consumo crítico y responsable. Revista digital 'Investigación y Educación' 26, Agosto de 2006 – Vol. III. Sevilla.

De los Ríos-Carmenado, I., Cazorla, A., Díaz-Puente J.M., Yagüe, J.L. (2010). Project-based learning in engineering higher education: two decades of teaching competences in real environments. Procedia-Social and Behavioral Sciences 2 (2):1368-1378.

De los Ríos-Carmenado, I., Rodríguez, F., Sánchez, C. (2015). Promoting professional project management skills in engineering higher education: Project-based learning (PBL) strategy. International Journal of Engineering Education 31(1B): 1-15.

European Union Committee (2014). Counting the cost of food waste: EU Food Waste Prevention. 10th Report of Session 2013-14. House of Lords, London, Available at: http://tinyurl.com/q3b4oa3.

FAO (2014). Food losses and waste in Latin America and the Caribbean. Food and Agriculture Organization of the United Nations. Available at: http://www.fao.org/3/a-i3942e.pdf.

Federación Española de Bancos de Alimentos (2014). Memoria anual 2014. Available at: http://issuu.com/fesbal/docs/pdf_memoria_2014.

Francis, J.E. (2011). The functions and norms that drive university student volunteering. International Journal of Nonprofit and Voluntary Sector Marketing 16(1): 1-12.

Friedmann, J. (1987). Planning in the public domain: from knowledge to action. New Jersey: Princeton University Press, Princeton, 229 pp.

Holdswortha, C., Quinnb, J. (2010). Student volunteering in English higher education. Studies in Higher Education 35(1): 113-12.

Hustinxa, L., Vanhovea, T., Declercqa, A., Hermansa, K., Lammertyna F. (2005). Bifurcated commitment, priorities, and social contagion: the dynamics and correlates of volunteering within a university student population. British Journal of Sociology of Education 26(4): 523-538.

International Project Management Association (IPMA). (2010). NCB – Bases para la competencia en dirección de proyectos. Versión 3.1. AEIPRO (Spanish Project Engineering Association), Valencia (Spain), 236 pp.

Montagut, X., Gascón, J. (2014). Alimentos desperdiciados. Un análisis del derroche alimentario desde la soberanía alimentaria. Icaria editorial, Barcelona; Quito, 160 pp.

22. Science and technology serving the human right to food: corporate responsibility of universities in Kenya

J.L. Omukaga
University of Kabianga, P.O. Box 2030, 20200 Kericho, Kenya; jlikori@yahoo.co.uk

Abstract

The United Nations global community tasks all member States with the duty to serve every citizen's human right to food as law. While this statement presumes the world's potential in food security, the reality is the opposite in some countries. In Kenya, hunger has increased even as the Millennium Development Goals target year, 2015, comes to its close. The prevalence of such a situation is unacceptable in a society that respects the rule of law and embraces modern science and technology. But what sustains such a situation? Lack of education plays a major role. This paper explores the potential of the Universities in the fight against hunger in Kenya. With a case study of the University of Kabianga, it will assess the possibility of a cooperate approach generated by the University academic programmes and facilitated by its community outreach arrangements. It will argue that the Universities provide suitable institutional milieu in which society can create and implement relevant innovations to aid war against hunger. By a critical analysis of the traditional 'idea of the University' education, the paper will propose a tripartite administrative matrix tailored to facilitate integrated outreach programmes against hunger. In its conclusion the paper will plead for the entrenchment of analytic structures, creative innovations and character formation within an integrated administrative system as a lasting contribution of the universities to the fight against hunger.

Keywords: universities against hunger, corporate responsibility, right to food

1. The current situation of hunger in Kenya and its impact

At the close of 2014, the number of the world's undernourished stood at 805 million (FAO, 2014), down from 946 and 840 million in the periods 2005-2007, and 2008-2010 respectively. This represented 'a decline of more than 100 million people over the last decade and of 209 million since 1990-1992', and 42 present reduction in the prevalence of undernourished between 1990-1992 and 2012-2014. The sub-Saharan region however, markedly registered insufficient progress. Further within the region, the countries in the horn of Africa saddle the biggest burden. In Kenya, progress against hunger is in the reverse direction. With regard to the WFS target of 1996, Kenya's situation is classified as 'no progress or deterioration'. With regard to the Millennium Development Goal 1c, the report considered Kenya's progress as 'insufficient'. Generally, Food and Agriculture Organization of the United Nations (FAO) report considered these targets to be within reach, on condition that considerable efforts are invested, 'particularly in countries where progress has stalled'. The question is whether the country can even just reverse the trend from the current situation.

Hunger deprives people of their energy and skill to work productively. The effects of poor nutrition reflect in the low birth weight, wasting or stunting of the children below the age of five, exposing them to higher mortality in Kenya now standing at 80 infants per 1000 live births, (FAO, 2014).

Those who escape into adulthood risk impaired physical and cognitive growth, adult men risk reduced working capacity while women risk continued births of babies with low birth weight. The prevailing situation of under nutrition in Kenya suggests a big gap between the standard food requirement for normal operations and the actual food accessed by each individual reducing general life expectancy at birth to 60 year, (UNICEF, 2015). Hunger keeps low the enrolment of infants of school going age, its effects either disable the young to face the challenges of school life, or push out many into child labour to struggle for survival. But due to its quality, the labour force available does not meet the demands of sufficient food production due to lack of skills or physical inability. Dependency is high and exposes the masses to manipulation either by the political class or the unscrupulous business class. This situation triggers endless social inequalities.

2. The right to food law

The normative content of the right to adequate food was established by General Comment 12 (3). Thus:

> The right to adequate food is realized when every man, woman and child alone or in a community with others, have physical and economic access at all times to adequate food or means for its procurement. The core content of the Right to Adequate Food implies the availability of food in quantity and quality sufficient to satisfy the dietary needs of individuals free from adverse substances, and acceptable within a given culture, (and) the accessibility of such food in ways that are sustainable and that do not interfere with the enjoyment of other human rights (Eide and Kracht, 2005).

This legislation provides a clear definition of the right to food structured from the existing United Nations (UN) norms, especially the economic, social and cultural rights. Experts site Arts. 3, 22, 25 and 28 of the Universal Declaration on Human Rights to be those closely related with the right to food. The Geneva Conventions III and IV of 1949 and their two Protocols of 1977 also provide relevant norms on the right to food, but perhaps Article 11 of the International Covenant on Economic, Social, and Cultural Rights, remains the most operational as a solution to this problem, (Alston, 1992). In 1974, the General Assembly further endorsed the Universal Declaration on the Eradication of Hunger and Malnutrition, (see Article I). Locally, Article 20, 21, 22, 23, 43 and 53 of Kenya's new constitution can be sited as closely related to the right to food. Article 43 and 53 provide the core content of this right almost as elaborated on in the international law. Thus: 'Every person has the right ... to be free from hunger and to have adequate food of acceptable quality', and 'Every child has the right ... to basic nutrition ...'. Other articles corroborate this basic provision in an attempt to fully domesticate the international provisions. Their content includes empowering individuals to claim their right to food when denied, securing courts' jurisdiction to prosecute cases of abuse or denial of this right to take recourse to a court of law claiming that a right has been denied, (Article 22), and giving a high court jurisdiction to hear and determine such applications (Article 23).

The right to food is structured to reflect priority of the right to life and its decent upkeep, the place of community, local and international, and its inclusivity whereof it is enjoyed by people of all ages everywhere. Most notable however, is the discretion of the local government to cause the realization of the individual right to food, the state organizes the resources that ensure the enjoyment of the right to food. All of these provisions enjoy the same force; each applies to all individuals on the basis of equality and without discrimination. Despite their details however, this legislation does not enforce implementation on the national level. Each country member of the UN must therefore adapt and operationalize these provisions independently.

3. Exploring new opportunities of combating hunger

Hunger, food insecurity and malnutrition are complex problems that cannot be solved by a single stakeholder or sector. Considering the demands posed by major global trends in agricultural development and world's under nutrition, FAO, in particular, set Strategic objectives representing main areas of work on which FAO will concentrate its efforts in order to achieve its vision on global hunger. The first of these strategies outlined its renewed effort 'to eliminate hunger, food insecurity and malnutrition'. On this strategy, FAO resolves to forge partnerships where it identifies its mandate being to support members in their efforts to ensure that people have regular access to enough high-quality food. In its vision, strategic partnerships mobilize the best available knowledge and capacities and provide the most effective services in working toward common goals. In this spirit FAO established linkages with a number of universities worldwide with the intention: '... to align higher education to further achievement of food security for all and ensure people have regular access to enough high-quality food to lead active, healthy lives'. Their reasoning largely motivates this paper, thus: 'Academia and Research Institutions nurture critical thinking and generate knowledge and state-of-the-art innovations that are essential for eliminating hunger and food insecurity'. Consequently, FAO recently signed strategic agreements with Mississippi State University, the John Cabot University, and the International University of Languages and Media in Milan. Through partnership with academia, FAO is ultimately bridging the gap between the development and adoption of research outcomes and innovations. In their view, applying knowledge strategically can lead to increased agricultural productivity, raised income and food security and ultimately bring about real improvement for smallholder farmers and their communities.

The principle behind strengthening FAO-academia partnerships is to use each partner's comparative advantage to make greater impact on the ground. The philosophy behind this undertaking resonates with the motivation of this paper and is summarized in five points: Firstly, to strategize university education as source of common understanding of current and future needs of food and agriculture systems. Secondly, to stimulate capacity development, inter-disciplinary research, knowledge exchange and help shape new policies. Thirdly, to mobilize diverse range of stakeholders to support fight against hunger particularly with the involvement of the youth, professional teachers and researchers. Fourthly, to increase outreach, dialogue and visibility to mobilize agriculture's next generation of thinkers and doers. Lastly, to cultivate efficiency in food security policies through shared resources and innovative home-grown solutions facilitated by academic programmes.

4. The relevance of a university reviewed

The history of university education has oscillated between two visible poles. On one hand, university education sponsors critical mind development endeavour that emphasizes the development of the individual to full all-round personality with ability to solve his/her own or society's problems as they come. This perspective of university education sees a university as an *alma mater* commissioned to nurture individual potential to full maturity in a formational exercise. Envisioning this image in his work, *The Idea of the University*, Henry Newman saw a University as 'a seat of truth', a place where one went to seek for the knowledge as a refinement of the self that leads to the uncompromising truth. The grasping of truth was to him the entire end of a university education thus, the essence of a university lies in teaching, not research. Teaching is needed to promote intellectual culture and the training of the mind.

However, times have since changed. In every age there are aspects of the university which flourish. The postmodern view of University education triumphs in specialization and vocational training. The level of knowledge and technical expertise on offer nowadays and the level of organization in the university are rather overwhelming, economic considerations have become the dominant driving force and much has been sacrificed in the relentless drive for efficiency. University systems are more driven by targets, focus is directed more on the measurable impact. Instead of entering higher education to become a 'general scholar' immersed in becoming proficient in the entire curriculum, we now meet a type of scholar whose concern is relevance of university education to the needs of society. However, as the universities expand to admit an ever wider number of students the issue of formation and competition should find a balance that defines a university of our time. Interestingly, the recent development in the structuring of university education attests to the desire for this balance. Comparing 'the *Magna Carta* of the European Universities' in the Bologna declaration, 1988, and the priorities of the Commission of University Education in Kenya, the striking feature is three fold, namely: the interest in quality teaching, the central place of research and innovation and the imperative of extensions or community service. These three aspects today define the potential of a university institution with emphasis on the extension programmes.

5. Capacity of the university institution: case of Kabianga University in Kenya

The University of Kabianga is situated in the south-western end of the Rift Valley province of Kenya. Although the region boasts the unique and richest climatic conditions found in the country, its attention over the years has been commercial tea growing and small scale dairy farming. The university's insertion into such a background could be looked at as providential in the light of critical problem of hunger. The University of Kabianga was recently established with a view of creating more opportunities of higher learning and productive research. Currently the university has enrolled around 8,000 students and offers 68 programs for graduate and undergraduate degrees, and Diploma and certificate awards. The University programme is dominated by courses in Agricultural sciences which occupy more than 62% of the entire course offering (Figure 1). Conspicuous in the science offering are the course directly affecting agricultural concerns that can immediately be related with the issues affecting food security and nutrition. They include agriculture and agriculture economics, agricultural education extension, agro forestry and rural development, biochemistry and human resource management.

Figure 2 captures the pattern of enrolment in this agriculturally related course in the last four years. From this comparison, the relevant science course does not only occupy the central concerns of the university training offered in this period, most importantly, they have attracted a consistently growing

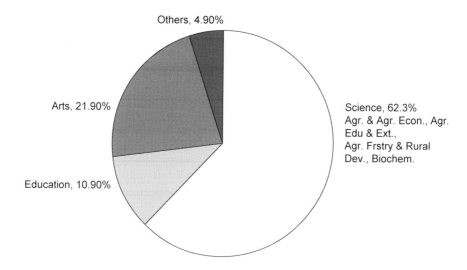

Figure 1. Pie chart showing student enrolment in the five available areas of study (UoK Statistical Data On-line; http://tinyurl.com/nuqxzc4).

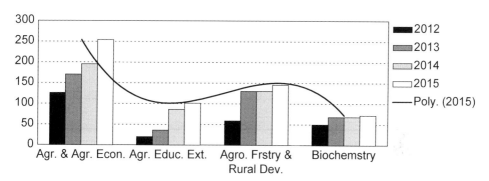

Figure 2. Student distribution in agriculturally-oriented science programs since 2012 (UoK Statistical Data On-line; http://tinyurl.com/nuqxzc4).

enrolment. Students headed to this university (by choice and government placement) seem to be set for a conscious option for agriculture related courses.

From the above graphics, it is visible that enrolment into the courses in agriculture and agriculture economics remained highest throughout the four-year period. Courses in agro forestry and rural development enrolled the next highest number of students, while programs in agriculture education and extension, and biochemistry are picking up rather consistently. To accompany this trend is the performance in human resource management which exhibits very similar behaviour in history. This potential grants the university the position of full command of all the instruments necessary in the war against hunger that is, the capability of creativity, agricultural technology and extension and above all, personnel management and consequent outreach into the rural community environment.

6. Tri-partite extension model: towards innovation for relevance

This paper proposes a tri-partite administrative structure to anchor and synergize outreach operations within the management. It comprises of the social, moral and academic pillars. The social pillar constitutes the first administrative level in which the students organize for their welfare as a community. Student administration individual bio data, interest in club and societies, religion, hobbies and sports. These constitute mandatory extracurricular undertaking intended to project the student's attention to community service in outreach. The target in this arrangement is to develop students' sensitivity to the needs of community life. The students will ensure thorough compliance to group membership and strict adherence to group obligations. Student leadership and organization superintends smooth club and society activities, sports, religious participation, scouting and all group activities registered in the university.

The moral pillar comprises of the University chaplaincy, club and society directors, the university deans and counsellors and the support staff director. It anchors the first level administrative structures into the mainstream university structure superintended by appropriate experts. The core role of this unit is to draft, discuss and implement outreach programmes aligned to food security based on the students' abilities. At the draft level, the unit clusters course offerings according to their relevance to a particular outreach undertaking regardless of the faculty or department from which they are drawn. At the discussion level, the unit builds consensus among every member as a crucial participant. At implementation the unit sets the programme, allocates duties and supervises the outreach to completion. It will marshal the entire potential of the university, moral and academic, deemed related to food security and channel it towards society through the student groups established from the students' initial free choice of an extracurricular undertaking. The academic pillar shall provide the first level of administration concerned purely with the academic and academic support operations of the university life handled by qualified experts of different academic areas. It structures courses, teaching material, academic staff, support staff, physical facility, academic resources and sets academic standards. It originates and runs the approved academic curricula and the research undertakings and guides them to a successful completion. Administration champions academic excellence, innovation and advances research to an ever higher level in competition with other universities.

The arrangement of these administrative units underscores autonomy of each as much as it fosters creative bond between these parts. Autonomous in terms of operations and obligations but tightly bound together to one goal in terms of service to society demands, namely: bringing into effect the practical relevance of university education to society. Bearing in mind that university operations are a familiar phenomenon, the three pillars propose a 'personalized' matrix in administrative procedures tailored to specifically pilot an integral university outreach mechanism that engages the entire campus lifestyle.

In conclusion, this paper has reviewed the relevance of university institutions and underlined their potential in the fight against hunger. The universities provide the best ground from which to grow practical solutions to the problem of hunger. Their potential for research and innovation disposes them favourably in estimating the value of important ingredients of food security: agricultural extension, food technologies, food wastage, environmental implications and others. It is only important that universities adjust and integrate their administrative structures to accommodate the implied demands of this undertaking.

References

Alston, P. (1984). International Law and the right to food. In: Eide, A. (ed.) Food as a human right. United Nations University, Tokyo, Japan, pp. 162-174.

Commission for University Education (2014). University standards and guidelines. Available at: http://tinyurl.com/nu7lnzn.

Eide, A. (1989). Right to adequate food as a human right. Human Rights Study Series. UN Center for Human Rights, Geneva.

Eide, W.B., Kracht, U. (2005). The right to adequate food in human rights instruments: legal norms and interpretation. In: Eide, W.B., Kracht, U. (eds.) Food and human rights in development, Vol. I, Antwerpen: Intersentia, pp. 9-118.

FAO (2015). The state of food insecurity in the World. Available at: http://tinyurl.com/qa5u6kg.

Humphreys, J. (2015). Has Cardinal John Henry Newman's vision for the Universities died? The Irish Times, April 1, 2015.

23. Improving efficiency of food distribution in the city of Seville: a proposal

H. Barco Cobalea

Environmental adviser, Sevilla, Spain; hectorbarco21@gmail.com

Abstract

According to the current food waste figures available along with the problems of access to food supply of a large part of the planet's population, the need to tackle this issue is apparent, through a wide range of initiatives. This paper is intended to contribute to a possible solution at the local level, thanks to scientific research and the collaboration with those who are part of the agro-food chain.

Keywords: food waste, agro-food chain, food bank, food management, geographic information systems, decision-making models

1. Background

One of the latest reports published by the Food and Agriculture Organization of the United Nations (FAO) titled 'Global Food Losses and Food Waste' (FAO, 2011) states that 'roughly one-third of food produced for human consumption is lost or wasted globally, which amounts to about 1.3 billion tons per year'. This situation has been also stressed by the European Parliament thanks to the report of 30 November 2011, where national authorities are requested to do their utmost to reduce the current food profligacy because the European citizens waste about 95 to 115 kilograms of food per person per year.

In this respect, the Ministry of Agriculture, Food and the Environment of Spain has published the strategy 'Más Alimento y Menos Desperdicio' – More food and less waste – in 2013 (Ministerio, 2013). This is a programme for minimizing waste and losses of food and the recovery of discarded food.

The business sector is also exploring this new way of opportunities available to add value to the product that currently treats as garbage. The Manufacturers and Distributors Association of Spain (AECOC), about 25,000 associates, is heading the programme 'Food Waste' (AECOC, 2012) which seeks to cover every phase of the agro-food chain, both for decreasing food wastage and for the use of surpluses generated.

All these initiatives should be based on experiences at the local level. This scale is the most appropriate in order to carry out these pilot projects through which good practices could be learned and adapted to other municipalities.

'Improving efficiency of food distribution in the city of Seville' is a research proposal which aims to move in that direction of searching for appropriate solutions to the current squandering of food along the food supply chain, using a well-defined geographical framework as in the case of the

Leire Escajedo San-Epifanio and Mertxe De Renobales Scheifler (eds.) *Envisioning a future without food waste and food poverty*   **199**

DOI 10.3920/978-90-8686-820-9_23, © Wageningen Academic Publishers 2015

metropolitan area of Seville. This local scale is the most suitable to serve as a pilot experience and it could be possible to adapt different conclusions and good practices to other municipalities at the regional, national and international level.

Thus, this could help to solve the problem of excessive demand for food that charities such as food banks, 'Caritas' (a Catholic NGO) and Red Cross are experiencing.

The project would open a new line of research with high potential in order to comply with the applicable requirements from the European Commission for the current research grants called 'Horizon 2020'.

2. Core objectives

The objectives of the research to pursue a doctorate are:
- An overall estimate of supplies of and demand for the populations in need over the next few years as well as the establishment, analyses and validation of distribution strategies for these products.
- The application of digital tools such as decision-making models in the industry and optimization techniques in order to improve the management and the distributions of foodstuffs, the identification of responsibilities and the traceability of food products.
- Such tools should complement with the use of ICTs (Information and Communication Technologies), highlighting the geographic information systems and the development of common databases for the network of actors. The possibility of updating data thanks to the use of internet would be studied. This should lead to the effective contribution of solutions to increase both the quantity and diversity of products that are available due to a current lack of fresh food. The management of this type of food requires an exhaustive study to prevent any health problem.

In performing these tasks, the supervision and management of the doctoral dissertation would be headed by the Department of Industrial Organisation and Business Management at the School of Engineering of the University of Seville, a department specializing in using the most appropriate techniques for modelling and analysis of management and organization problems in the industrial field and the services sector (http://www.aicia.es/).

In order to allow identification of the different participants involved in the system of distribution in Seville, a preliminary investigation during one-and-a-half years has been completed. The final result of this phase has been the obtaining of many letters of support as well as the establishment of a network of contacts including the Food Bank of Seville, charities, consumers associations, public bodies at the local, regional and national levels and food distribution companies.

Furthermore, we have collaborated with the University of Seville and the Food Bank of Seville in a report called 'Report about the food impact distributed by contributing entities with the food bank of Seville' (Banco de Alimentos *et al.*, 2013) and we have also produced one of the chapters to put forward the research proposal. This report has been the first publication by a food bank in Spain (Banco de Alimentos *et al.*, 2013).

The research proposal received the special mention of the jury in the 1st call for the award 'City of Seville for Sustainable Development' organized by the Seville City Council.

This line of pioneer research at the national and international scale is in line with the new European Union's criteria for the EU research and innovation programme called Horizon 2020. In this way, the research proposal is in tune with the topics and projects which are encouraged by the European Commission since 2014.

Therefore, the budget for this research proposal is considered an investment towards the creation a new R&D pathway where:

- The most disadvantaged citizens effectively reap the benefits thanks to a higher quantity and variety of foodstuffs available.
- Charities and the Food Bank of Seville will also benefit from this because they would obtain more food to share and it would help to solve the problem of their possible dependence on financial support at the local, regional and national level.
- Enterprises would increase their profitmaking capability because of the use of surpluses which currently are wasted, thus reducing the total amount of residues generated. Moreover, there are some possibilities of taking an advantage from the tax credit in corporation tax rates and it is possible to create a positive impact thanks to the enhanced corporate image of the company. This may be an opportunity for customer loyalty, especially for the consumer profile called 'socially-conscious consumer', which is increasing around the world according to the report on daily consumption habits by the corporation Nielsen titled 'The global socially conscious consumer' (The Nielsen Company, 2012).

References

Banco de Alimentos de Sevilla, Fundación Cajasol y Universidad de Sevilla. (2013). Resultados de la Encuesta del Programa de Distribución del Banco de Alimentos de Sevilla. Seville. Available at: http://goo.gl/oJ4tH.

FAO. (2011). Global food losses and food waste – Extent, causes and prevention. Rome. Available at: http://www.fao.org/docrep/014/mb060e/mb060e.pdf.

La Asociación de Fabricantes y Distribuidores (AECOC). (2012). Comunidad AECOC contra el despilfarro alimentario. Madrid. Available at: http://tinyurl.com/nc29zgy.

Ministerio de Medio Agricultura, Alimentación y Medio Ambiente. (2013). Estrategia 'Más alimento, menos desperdicio'. Madrid. Available at: http://tinyurl.com/ov7qf9y.

The Nielsen Company. (2012). The global, socially-conscious consumer. At: http://tinyurl.com/p93b6wr.

24. Connecting learning processes and strategies towards the reduction of food poverty: a collaborative experience between the University of the Basque Country and social inclusion NGOs

M. Merino Maestre[1]* and L. Escajedo San-Epifanio[2]

[1]Department of Applied Mathematics, Statistics and Operations Research, Faculty of Science and Technology, University of the Basque Country, Spain; [2]Department of Constitutional Law and History of Political Thought, Faculty of Social and Communication Sciences, University of the Basque Country, Spain; maria.merino@ehu.eus

Abstract

Food poverty is one of the social scourges which has most taken hold in recent years. In the so-called Global North (or developed world), to which the Autonomous Community of the Basque Country (CAPV in Spanish) belongs, a growing number of people have no physical and economic access to healthy, safe food, appropriate for their nutritional needs. Thus, diseases caused by malnutrition as well as pathologies such as obesity or cardiovascular diseases associated with an inappropriate diet are growing. According to the Ombudsman of the CAPV, hunger and malnutrition affect 6.5% of the population of the Autonomous Community of the Basque Country and 73 million people in the EU. Given that hunger and malnutrition occur in an environment in which we waste more than 50% of the food produced (official data of the EU), it is clear that deficient accessibility to food does not have much to do with a lack of food. As the EU recognizes in its Horizon 2020 program, socio-economic, legal, structural and logistical obstacles have to be removed for these people to access the food they need. Backed up by our experience as teachers, researchers and/or volunteers we have had with organizations which attend people in situations of social exclusion, we believe it is possible to implement Undergraduate Projects that cover various aspects of Social Responsibility and will provide students with the opportunity to learn through service.

Keywords: social responsibility, educational innovation, undergraduate project, civic competences, ethic competences

1. Food poverty in EU urban societies and social responsibility

According to Caritas-Europe, 79 million people are food insecure in the EU (Caritas Report, 2014). This happens while the current food wastage in EU27 is 179 kg per capita (89 million ton per annum), as several EU institutional publications, reports and statistics have stated in the last two years. URBAN ELIKA is a collaborative and participatory forum funded by the University of the Basque Country and aims to contribute to a better use of food resources in EU urban environments. Among other action areas, the group offers professional counselling to NGOs that provide hunger relief while respecting and dignifying each individual.

The goal to help people in need sounds straightforward, but professionals and volunteers who are immersed in everyday practice find many obstacles. An outsider who has never been actively involved

Leire Escajedo San-Epifanio and Mertxe De Renobales Scheifler (eds.) *Envisioning a future without food waste and food poverty*

203

DOI 10.3920/978-90-8686-820-9_24, © Wageningen Academic Publishers 2015

in these kinds of activities can easily forget the economic, political and social context in which social problems are defined. Shortcomings in human and financial resources should also be taken into account. In the last five years the amount of formal or informal charities that have emerged in Spain to attend food insecure people is impressive. URBAN ELIKA research group strives to identify good and not-so-good practices in order to improve emergency hunger relief programs, and to help those immersed in its daily practice. Taking food surpluses and giving them away is a complex process. The characteristics of each type of food (i.e. packaged or unpackaged, perishable or not, refrigerated or not, cooked or raw) need to be carefully considered to manage them in a way which guarantees their hygienic safety.

Wishing to give students the opportunity to join this initiative, teachers and researchers involved in URBAN ELIKA have designed an *Educational Innovation Project* related to the undergraduate project (UP), required to complete undergraduate degrees at Spanish universities. The name of the innovation project is *Transversal competences in undergraduate projects for food inclusion*. These projects carry a work load of between 6 and 30 credit hours and should assess the mastering of the competencies related to the degree.

A number of projects will be proposed to the students to collaborate directly with social organizations and NGOs which work in the Autonomous Community of the Basque Country (CAPV in Spanish) to improve people's access to healthy safe and nutritious food and, thus, their health.

UPs with an emphasis on social responsibility will involve the application of technical and scientific knowledge about NGOs fighting against food poverty in the CAPV. The common aspect for UPs in such diverse disciplines as applied mathematics, law, human nutrition or microbiology is social responsibility and transversal competences with an ethical and civic-social commitment.

2. Assessing the learning outcomes on social commitment and ethics skills

Nowadays social commitment and civic responsibility are not seen as peripheral in higher education. Many phrases and terms signal our new academic programs towards engaging community-based models of teaching, learning and research. But the set of core competences called *social and/or ethical commitment*, as well as *civic responsibility* or similar, still need development and implementation. Diversity in its different dimensions is a characteristic of democratic societies, as is also the fact that in this plurality, respected formulae of coexistence are obtained among the differences on values and ways of seeing life in society. In other words, in this type of society plurality and civic ethics (or minimum ethics) coexist: an ethics agreed by consensus which contributes to peaceful coexistence, *'within the respect and tolerance for the different concepts of the world'* (Cortina Orts, 1986). From this awareness arises the conviction that education, at the various levels, has a lot to contribute so that new generations are able to participate actively. The question now is: How can we develop these competences in university degrees? That is the question which has informed the present research project to which we have tried to reply within the framework of the IKD model (Ikaskuntza Kooperatibo eta Dinamikoa; Dynamic and cooperative learning; explained below).

Learning via active methodologies is indicated within the ambitious IKD model of the University of the Basque Country (UPV/EHU) as one of the tools which *'contributes to placing students in charge of their learning'* (vice-rectorate, 2012). It is a cooperative and dynamic model, centred on the student in which teachers guide the students' learning as a way to develop the teaching. It is framed within the Bologna Agreements (1999) and those which followed with the aim of creating the Higher Education European Area and making Europe one of the most competitive knowledge economies in the world.

This process of academic convergence between the Spanish and the European systems of higher education has involved important changes in the design of the degree programs and has been one more in a series of factors which have opened, to a greater or lesser extent, depending on the university and/ or the academic discipline, a reflection about *what and how we teach* (Ramsden, 2010), together with *what it is that should be evaluated* in the learning process and *how this evaluation must be.* In addition, it is important to point out that improving the quality of all these teaching dimensions considers that university teaching should be results-based *teaching* (Biggs/Tang, 2007), an understanding which has had important projections. Thus, among other things, we re-designed the contents of each subject, the teaching and learning methodology and the evaluation tools, to address competences in addition to contents competences are defined as: '*a dynamic combination of attributes combining knowledge, skills, attitudes and responsibilities which describe the results of learning of an educational program or what the students are able to show at the end of an educational program*'. These competences, as they are called, should be understood as academic and professional abilities and skills, and their acquisition must be subject to some type of *evaluation.*

The ethics and social and civic commitment competences are transversal competences: that is to say, they are competences related to personal development which do not depend on a thematic dimension of a specific discipline. By definition, they are useful for all dimensions of professional and academic action (Mir Acebron, 2010). Currently, whether social commitment or ethics can and/or should *be taught* is intensely debated. The main idea is that students should use their understanding of these competences to develop new ways of observing society and new ways of being and acting as professionals serving society. In other words, faced with competences specifically related to a given discipline, transversal competences which imply social commitment, ethics or sustainability force students *to engage with the world* (Barnet, 2007). Civic imagination, civic values, and civic habits are cultivated in part by exposure to the complex struggle for social justice. Understanding that the self is always embedded in relationships, a social location, and a specific historic moment and understanding how communities can both include and exclude individuals are to be learned by experience (McTighe Musil, 2009).

3. Transversal key competences and the promotion of access to food

As we have already said, we promote undergraduate projects which apply technical and scientific knowledge from different professional areas such as applied mathematics, nutrition and dietetics, legal consultancy or social work, among others. Due to their characteristics, however, these projects imply, moreover, the implementation of cross skills or transversal competences such as team work, critical thinking and social commitment.

Each Faculty and degree program has the coordination systems established for the technical and scientific aspects of the undergraduate project. The coordination promoted by this cross-disciplinary project will not overlap that system. We have designed a coordination system to make explicit how the learning outcomes regarding ethic and civic competences are derived and how they are assessed in different Undergraduate Programs and Faculties of our universities. Regarding the said element, the coordination of the Undergraduate Project will propitiate a scenario where it is possible to reflect about the way in which the said competences have been incorporated into degrees of the UPV/EHU as well as regarding the ways in which they are interpreted, applied and evaluated at present.

These are the tasks to be performed. In the first place, a joint identification and reflection regarding the way in which, currently, ethical and civic-social commitment competences are interpreted and implemented in the degrees involved, together with a comparison with other experiences in the UPV/EHU and other European universities. The aim of the second task is to consolidate the network of contacts with organizations in the CAPV and other autonomous communities dealing with food poverty.

As a third task we will proceed to a coordinated design of the UPs, in collaboration with the social entities and the students participating in this project. This design will be discussed with external experts (of other disciplines and/or universities) in a seminar format, before offering and carrying out the UPs. And, finally, we will tackle in a joint manner the question of experience and then draft a document with our conclusions. As one aspect of the evaluation, we will measure the results of the Project by conducting pre- and post-tests of the students' and the teachers' perceptions and levels of understanding of the skills involved in this project.

Acknowledgements

This work has been funded by the project '*Transversal competences (TCs) in degree final projects for food inclusion*', an educational innovation project (PIE/HBP) supported by the Educational Advisory Service (SAE/HELAZ) of the University of the Basque Country, 2014-2016 (PI, M. Merino Maestre). URBAN ELIKA is funded by the University of the Basque Country (University & Society Research Grant Call for 2014-2016, US 14/19 Project; PI = L. Escajedo San-Epifanio).

References

Antonio, A.K., Astin, H.S., Cress, C.M. (2002). Community service in higher education: a look at the nation's faculty, Review of Higher Education, 23 (3), 373-398.

Banks, J.A. (2008). Teaching for social justice, diversity and citizenship in a global world. The Educational Forum, 68, 4, 289-298.

Bansal, A. (2014). An empirical study of teaching ethics on accountancy students. International Journal of Business and Administration Research Review 1, 3, 152-157.

Beaumont, E., Colby, A., Ehrlich, T., Stevens, J. (2003). Educating citizens: Preparing America's undergraduates for lives of moral and civic responsibility. Jossey-Bass, San Francisco.

Berger, J.B., Milem, J.F. (2002). The impact of community service involvement on three measures of undergraduate self-concept. NASPA Journal, 40 (1), 85-103.

Burton, B.K., Hegarty, W.H. (1999). Some determinants of students corporate social responsibility orientation. Business & Society, 38:188-205.

Caritas Report (2014). The European crisis and its human costs – a call for fair alternatives and solutions. Available at: http://tinyurl.com/puz3r55.

Cortina Orts, A. (1986). Etica minima: introducción a la filosofía práctica. Tecnos, Madrid, Spain.

Ehrlich, Y. (2002). Civic responsibility and higher education. Oryx Press, Phoenix.

Gray, R., Bebbington, J., McOhall, K. (1994). Teaching ethics in accounting and the ethics of accounting teaching: educating for immorality and a possible case for social and environmental accounting education. Accounting Education3, 1, 51-75.

Hurtado, S., Ruiz, A., Whang, H. (2012). Assessing student's social responsibility and civil learning. Paper presented at the 2012 Annual Forum fo the Association for Institutional Research, New Orleans, Louisiana.

Karp, S. (1997). Educating for a civil society: the core issue is inequality. Educational Leadership 54, 40-43.

Kuh, G.D. (2008). High-impact educational practices: what they are, who as access to them, and why they matter. Association of American Colleges and Universities, Washington DC Available at: http://tinyurl.com/odzc6jl.

López, D. (2009). Introducing sustainability and social commitment concepts in a computing degree. Published at the Global University Network for Innovation. Available at: http://tinyurl.com/ouhv5as.

Moran, E.T. (2003). Civic responsibility and teaching macroethics. Teaching Ethics 3 (2): 27-39.

Morrison, T. (2001). How can values be taught in the university? Michigan Quarterly Review, 40, 273-278.

Musil, C.M. (2009). Educating students for personal and social responsibility: The civic learning spiral. In: B. Jacoby (ed.) Civic engagement in higher education. San Francisco, CA: Jossey-Bass.

Nicholson, C.Y., DeMoss, M. (2009). Teaching ethics and social responsibility: an evaluation of undergraduate business education at the discipline level. Journal or Education for Business, 84, 4, 213-218.

Parkin, S., Johnston, A., Buckland, H., Brookes, F., White, E. (2004). Learning and skills for sustainable development. Developing a sustainability literate society. Guidance for Higher Education Institutions. Edited by Forum for the Future, London.

Shor, I. (1992). Empowering education. Critical teaching for social change. The University of Chicago Press.

Siegel, B.L., Watson, S.C. (2013). Terms of the contract: the role of ethics in higher education. Journal of Executive Education, 2, 1: 7.

25. Social commitment competence in the University of the Basque Country degree programmes

M. Lledó Sáinz de Rozas and *A. Inza-Bartolomé*

University School of Social Work, University of the Basque Country (UPV-EHU), Los Apráiz 2, 01006 Vitoria-Gasteiz, Spain; mariadelmar.lledo@ehu.eus

Abstract

The aim and mission of any educational agency are built amongst all the people who make up that entity and the society it serves. Thus, the University of the Basque Country (UPV/EHU) in its statutes, Article 4.3 thereof, acquires a strong social commitment by taking necessary measures for its achievement. The starting point is the responsibility for creating an organizational structure based on ethical criteria that, in turn, can provide answers to the needs of the environment and the society. Within its educational project, IKD, the UPV/EHU wants to ensure that their students may face concepts (knowing), procedures (know how) and attitudes (being), to new situations and conflict related to the satisfaction of human needs. This challenge will require people to take into account the principles of equality, respecting diversity and difference, and carry out their technical, ethical and social commitment according to sustainability and balanced representation. This paper draws the broad scope of the 'social and ethical commitment' in the UPV/EHU and highlights how different degrees refer to different meanings in the arrangement of their transversal and generic skills, generally taking into account those aspects related to their subject area with the concept itself.

Keywords: competences, university, social commitment, ethical commitment

1. University aim and mission

The aim and mission of any educational agency are constructed amongst all the people who make up that entity, or in other words the educational community and the society for and with which it works. The university therefore constitutes a system resulting from the functional or dysfunctional relations established between the members of the university community and society. Hence, the people who make up a university have the responsibility to create structures which provide responses to the needs of the surrounding environment, through a sense of ethics and commitment. Therefore the teaching/learning process must be built in such a way that students, male and female alike, are able to deal with new situations of conflict by using creativity and cooperative work, as well as critical, analytical sensitivity, with a commitment to equality and ethical development.

The university is the main role-player in this relationship between academia and society, based on the ethical commitment that defines it and allows it to identify itself as a 'healthy' university (Gallardo and Martínez, 2013). If we bear in mind the concept of 'health' proclaimed at the first International Conference for Health Promotion in Ottawa (WHO, 1986), in which a positive concept thereof was determined, accentuating social and personal means as sources of physical, psychological and mental welfare, it seems logical that universities would promote this concept as part of their strategic guidelines. At present, there is a global movement among universities which are using this concept of

Leire Escajedo San-Epifanio and Mertxe De Renobales Scheifler (eds.) *Envisioning a future without food waste and food poverty*

DOI 10.3920/978-90-8686-820-9_25, © Wageningen Academic Publishers 2015

209

public health as a point of reference, joining the effort to achieve the goal of 'Health for all in the year 2000', in which it was established that peace, education, housing, food, income, a stable ecosystem, social justice and equality were the essential conditions and prerequisites for achieving health. As a result, playing an active role in health meant creating favourable environments and reinforcing action by the community, as well as other activities. The concept of health thereby went beyond a relationship with just diseases and disorders, also including welfare in every aspect (Taylor, 2007).

Therefore, the Spanish Network of Healthy Universities (REUS, 2015) promotes a vision of university and society, defining it through a document that establishes the premises regarding the role which the university is to play, as well as the methods for pursuing its goal of promoting the health and social welfare of the student body, faculty, service providers and society for which it works.

Reflection on how to build a community proves to be the horizon which we must look towards. This means including human rights as a prospect and thinking about what social factors are related with health, how to empower people and how to turn society into the core which integrates them into healthy environments. This requires the implementation of public policies aimed at spurring community action.

A sense of coherence as a guideline in life is what gives meaning to the relationship between university and society, making it possible to understand and manage the gears in this machine. It is a psychological construct that allows us to understand life as that which is truly worthwhile, and to foresee how to handle new and/or difficult situations through the integration of environments, groups and people on the basis of three foundations: manageability, comprehensibility and meaningfulness (Palacios-Espinosa, 2008). Life is what is most important, and therefore dealing with its challenges in a sustainable way is, as well, relating the socioeconomic system with the ecosystem (UN, 1987). The challenges of both sustainability and social commitment lie in determining what demands truly arise from a need and what needs have not yet been identified as demands (Sánchez *et al.*, 2010). In terms of the social factors to be considered within the concept of sustainability, reference may be made to the social impact of the profession's solutions and ethics, legal aspects, development aid, minority cultures and languages, democracy and equal opportunities.

In this sense, the CRUE Work Group for Environmental Quality and Sustainable Development approved a document in 2005, in which it established the 'Guidelines for introducing sustainability into the curriculum' (CRUE, 2005). This document defines sustainability as a concept which includes the pursuit of environmental quality, social justice and a more equitable, viable economy in the long term, promoting a more equitable and just distribution of resources. For instance, it deals with the concept of sustainability as an ethical imperative which is related with the social commitment to reduce poverty, promote health and human rights, a culture of peace, responsible production and consumerism, and equality in access to resources, as laid down in the Declaration of the Decade of Education for Sustainable Development, 2005-2014 (UN, 2002), the purpose of which is based on the universal right to access the satisfaction of basic needs and an education through which individuals can learn the values, behaviours and lifestyles coherent with a sustainable future. This objective is conditioned by the need for equality, respect for human rights and promotion of participative democracy.

2. Ethical and social competence at the university

One can speak of professional ethics or professional morals, ethical and moral values, etc., though today these concepts are distinguished from one another. Morals are regarded as that which is experienced and practiced by a collective group, whereas ethics consists of theoretical reflection on morals experienced and practiced (Rojas, 2011). Therefore, morals would be a subject of study by ethics. The most accurate way to discuss this would be to use the terms 'moral values' and 'codes of professional morals', in the same way that professional ethics would go above and beyond deontology and just those standards themselves, and therefore beyond just adherence to the rules which limit an entity's behaviours.

The Tunning Project highlights 'ethical commitments' amongst all forms of personal competence. There is no single definition which describes this, because each author takes into account different cognitive, affective or behavioural variables (López *et al.* 2004). This competence has to do with the free will to find fulfilment in a specific way as a person, citizen and professional (Villa and Poblete, 2007) and is related with values and attitudes such as independence, responsibility, cooperation, mutual respect, integrity, social awareness, etc.

It comes in different names: consciousness of ethical value and an ethical sense (ethics and commitment), ethical commitment, ethics and values (ability to understand ethical responsibility) and professional ethics (García, 2009).Villa and Poblete (2007) define it as the ability to respond independently and responsibly, showing mutual respect and social sensitivity towards problems. It means overcoming the challenge of coherently integrating professional life and a project for a personal life, which involves being able to identify and value the ethical aspects which make up each professional identity, or in other words, to acknowledge the ethical foundations that must lie within each professional behaviour, so that we may all understand each other and approach the ultimate goal of attaining social justice. It means creating a sturdy framework of reference to guide activity. People become professional when, in addition to possessing knowledge, they follow ethical principles that guide them as free individuals responsible for their actions (Rojas, 2011). If, moreover, we view a profession as a social practice, it must be aimed towards the common good (Bolivar, 2005; Cobo, 1993; Traver and Garcia, 2006). The potential for social development and the advancement towards that goal will depend upon each professional, through the impact on employment, environment and society which his or her work produces in environments supported by academic training and the completion of research (Aguirre *et al.*, 2012). The student body becomes a social role-player that produces prosperity and human coexistence, ensuring sustainable development in an ongoing dialogue with society to meet its needs. The university's social responsibility commitment is measured by the impact it has on the surrounding environment in terms of achieving collective welfare, as a part of its institutional policy beyond just charitable actions. All in all, it means committing to the furtherance of people and professionals with a sense of ethics (Esteban, 2004), prepared to act towards the common good and committed to social improvement.

Some authors also include the concept of solidarity as an attitude and value which forms part of competence in 'ethical commitment', which requires training on cooperative and collaborative work, using different techniques (Traver and García, 2006).

3. Developing educators at the University of the Basque Country

In today's society, students must become prepared by receiving broad training that does not only place an emphasis on acquiring technical knowledge, but also teaches the essentials for interpreting and dealing with a complex social reality rife with uncertainty (Commission of the European Communities, 2010), which they must handle in an ethical, responsible and competent manner by understanding the surrounding environment in a systemic, sustainable manner, more through a prism of inclusion than one of exclusion (De Miguel, 2006).

From this starting point, degree programmes are managed by designing teaching/learning processes in which students take responsibility for their learning process, entering into a more conscious personal management model, more directly related with their personal and social needs and with the implementation of change and transformation processes. The objective of encouraging the student body to play an active role in dealing with social realities is highlighted, putting the spotlight back on their ability to achieve changes which they can experience as satisfactory, and the belief that they are capable of doing so.

The teaching objectives are articulated on the basis of the way in which the university views itself. For instance, in order to organise these objectives, the University of the Basque Country (UPV/EHU) is developing the IKD, or *Ikaskuntza Kooperatibo eta Dinamikoa (dinamic and cooperative learning)*, which includes the foundations for curricular development of the official degree programmes designed by the UPV/EHU, approved by the governing board in April 2010 and developed as part of its strategic plan for 2012-2017.

The IKD model points out the general guidelines for teaching and learning throughout the entire UPV/EHU and proposes interaction between educators, student body and surrounding environment, undertaking to solve the problems and surmount the challenges faced by society. It is based on the principles of equality, sustainability and social responsibility. To do so, this model has designed an active, dynamic, cooperative teaching and learning process, bearing in mind the need to develop educational and institutional policies which include it. It is on the basis of ethics that the need arises to reflect upon the current university structure, as well as adapting and improving the relationship between university and society.

Despite the efforts to theorise a teaching/learning process that enables professors and student body to pursue the goal of responding to the needs and problems which arise in different social contexts, it is costly to achieve the acceptance and application of new methodologies which question old, deeply-rooted teaching approaches. If, up to a few years ago, concerns were focused on the contents of each subject matter, and the students were mere recipients of the knowledge conveyed by an expert individual, new social realities mean we must also be concerned about how students learn, interpret and manage information using reliable, critical, thoughtful and ethical criteria.

Thus, teaching based on ethics (Norma and Ramljak, 2002) requires us to reach a consensus-based definition of the rules to be followed by those participating, because, by establishing and upholding a series of clear principles upon which we can build the teaching/learning process, it may become possible to move forward in terms of the construction of the university and society dichotomy

(Luengo and Elichiguerra, 1997; Palacios *et al.* 2006). Teaching/learning processes must identify what competence-based objectives will work favourably towards defining a systemic community relationship amongst the members of the university community (educators, students, external agents, employers, etc.) (Pirla and Traveria, 2014). Competence as an integrated, coordinated set of strategies to solve a problem or meet a specific, regular demand within some context consists of the tools which connect the university with the profession, in which acquiring knowledge becomes a means for developing competence (López *et al.* 2004). One must keep a creative, innovative attitude in the event of social changes, adapting them to academic and professional profiles, as a way to respond to social needs and any changes in social reality.

The concept of competence is broad and diverse. Poblete (2006) defines it as good practice in diverse, real, authentic contexts, based on the integration and activation knowledge, rules, techniques, procedures, abilities, skills, attitudes and values. It has to do with knowledge, knowing how to be and *savoir-faire*, or in other words it brings into play knowledge, logic, reflection, critical thinking, controlling feelings, creativity, and more. Therefore, if competence has to do with the ability to respond successfully when given different situations in wide-ranging fields of personal and social life (Climent, 2010), education based on competence must gradually be adapted to new academic, social and labour-related changes in a flexible way (Masseilot, 2000). Working based on competence means placing students at the centre of learning and making them responsible for it, surpassing just the definition of being able or an expert, so as to be understood as being responsible in social life, using all of the knowledge, abilities, attitudes and resources necessary (Sarramona *et al.*, 2005). It does not simply mean adding dexterity to knowledge, but also combining attributes involving different levels within the whole individual that have to do with knowledge, *savoir-faire* and knowing how to be (Jover *et al.*, 2005). Therefore, they require ongoing redefinition and comparison.

4. Ethical and social competence in the degree programmes at the UPV/EHU

Most of the degrees offered at UPV/EHU are committed to developing ethical and social competence in general throughout the entire study plan, or in other words, through all of the courses and subjects. As a result, this competence is deemed necessary and advisable in order to perform the professional and academic work of all students who graduate. Therefore, each degree must include a clear definition of this competence, establish different levels of mastery to be acquired throughout the degree program and set indicators or benchmarks that make it possible to evaluate each student's level of development.

In a preliminary analysis of the expression and form of this commitment in the different UPV/EHU degree programmes, we find differences in terms of the importance assigned to their dissemination, as well as the degree of specificity and rigor with which they are defined. Certain degrees mention the need to be able to provide a critical, valued, reasoned response, as well as making reasoned value judgments on relevant topics in different areas, including the social realm (degree in business administration and management and degree in fundamentals of architecture). Others include the need to enable members of the student body to take on their ethical and social responsibilities, bearing in mind the social and environmental impact of their activities (degrees in: electrical engineering/ industrial and automated electronics, information technology engineering for management and information systems, mechanical engineer/industrial chemical engineering, biology, maritime engineering and food science and technology).

Other degree programmes highlight the need for cooperation amongst different sectors of the educational community and social environment, promoting democratic values for an active citizenry (degrees in children's education, primary education, nursery, philosophy, biology and social anthropology). Likewise, they mention the need to be able to regulate contexts of diversity, with respect for gender equality, equity and human rights as values for education in being a citizen. They include the concepts of individual and collective responsibility as requirements for achieving a sustainable future (renewable energy engineering, degree in pre-school education, degree in technical architecture).

There are also those which add the need to be able to promote these values of social responsibility when participating in organisations and institutions, in such way that the principles of professional ethics are upheld, above and beyond just criteria of quality (degree in social work). Similarly, some make reference to the importance of knowing about and acting in accordance with the ethical principles of the profession as regards the legislative, regulatory and administrative provisions of law governing the exercise of the profession in question (degrees in: pharmacology, physical activity and sports sciences, nursing, political science, medicine and psychology), taking into account that the social environment to which they must respond is undergoing transformation.

Some degrees relate the acquisition of basic knowledge with the ability to use the results thereof while bearing in mind social and ethical principles, with a view to identifying environmental problems (degree in environmental sciences) and territorial development. More specifically, some mention social justice as a principle in professional practice, assisting people and taking into account their habits, beliefs and cultures, as well as their responsibility to participate in social furtherance activities (degree in human nutrition and dietetics). Others include the concept of the culture of peace and democracy (degrees in: creation and design, maritime engineering, criminology and law). However, one must highlight the introduction of the concept of social attitude, which allows individuals to understand and show sensitivity towards different forms of exclusion and promote a commitment towards improvement, as a way to define the competition (degree in labour relations and human resources) permitted by ethics in labour relations.

This competence is dealt with in a less specific way in certain degree programmes, which mention the need to promote attitudes that favour integration and tolerance towards multilingual and multicultural diversity (degrees in English studies, Basque studies, philology and teaching), and the relationship between the physical environment and the social and human realms (degree in geography and territorial organisation).

Ethical and social commitments govern the relationship between the university and society in a way that builds the foundations for a more fair and equitable future, as well as one with respect for human rights and diversity. As demonstrated by the evidence collected, competence in these areas is viewed from a wide range of focuses, with characteristics determined by the fields of knowledge involved in the different degrees, which requires a commitment to continue studying and analysing the specific implications that this 'ethical and social commitment' translates into within each university degree programme. This will allow us to promote their development to the greatest extent possible.

Acknowledgements

This work has been funded by the project 'CTs en TFGs de Inclusion Alimentaria', an Educational Innovation Project (PIE/HBP) supported by the Educational Advisory Service (SAE/HELAZ) of the University of the Basque Country, 2015-2016 and the US 14/19 Project, URBAN ELIKA, supported by the University of the Basque Country (University & Society Research Grant Call for 2014-2016). M. Lledó Sainz de Rozas and A. Inza Bartolomé are members of the Parte Hartuz Research Group (IT300-10).

References

Aguirre, R., de Pelekais, C., Paz, A. (2012). Responsabilidad social: compromiso u obligación universitaria. TELOS. Revista de Estudios Interdisciplinarios en Ciencias Sociales 14, 1: 11-20.

Bolívar, A. (2005). El lugar de la ética profesional en la formación universitaria. Revista Mejicana de Investigación Educativa 10, 24: 93-123.

Climent, J.B. (2010). Reflexiones sobre la Educación Basada en Competencias. Revista Complutense de Educación. 21, 1: 97-106.

Cobo, J.M. (1993). Educación ética para un mundo en cambio y una sociedad plural. Endymion, Madrid.

Commission of the European Communities (2010). European Commission Report on Progress towards the common European objectives in education and training. Available at: http://tinyurl.com/q8v5veu.

CRUE (Conferencia de Rectores de las Universidades Españolas) (2005). Directrices para la introducción de la Sostenibilidad en el Curriculum. CRUE-CADEP, Valladolid, Spain.

De Miguel, M. (2006). Modalidades de enseñanza centradas en el desarrollo de competencias. Ediciones Universidad de Oviedo, Oviedo, Spain.

Esteban, F. (2004). Excelentes profesionales y comprometidos ciudadanos. Un cambio de mirada desde la universidad. Desclée de Brouwer, Bilbao, Spain.

Gallardo, C., Martínez, A. (2013). Universidad saludable: alfabetización en salud en la educación superior. In: Botello, E.; Moreno, J.; Sánchez, J., Zamora, F. (eds.) Ética en la Universidad. Transversalizar las competencias éticas en el Espacio Europeo de Educación Superior, ENTIMEMA, Madrid, Spain, pp. 149-164.

García, M.J. (2009). Evaluación de competencias Transversales. FIB-UPC, Madrid, Spain.

Jover, G., Fernández, C., Ruiz, M. (2005). El diseño de titulaciones y programas ante la convergencia europea. In: V. Esteban (ed.) El Espacio Europeo de Educación Superior. Universidad Politécnica de Valencia, Valencia, Spain, pp. 27-93.

López, N., Iriarte, C., González, M.C. (2004). Aproximación y revisión del concepto de competencia social. Revista Española de Pedagogía, 62, 227: 143-158.

Luengo, M. J. and Elechiguerra, C. (1997). ¿Es posible la calidad en la docencia sin ética?. In: C. Olalde, C. Lobato, P.M. Apodaca, F. Arbizu (eds.). Orientación Universitaria y Evaluación de la Calidad, Congress Papers, pp. 117-128.

Masseilot, H. (2000). Competencias laborales y procesos de certificación ocupacional, Competencias laborales en la formación profesional. Boletín Técnico Interamericano de Formación Profesional, 149: 73-94.

Norma, B., Ramljak, L. (2002). Ética y docencia en trabajo social. Margen 25. Available at: http://www.margen.org/suscri/margen25/ramljak.html.

Palacios, J.M., Cordero, N., Fernández, M.I., Fernández, A.M. (2006). Valores profesionales en los alumnos/as de Trabajo Social en la Universidad Pablo de Olavide de Sevilla. Documentos de trabajo social: Revista de Trabajo y Acción Social 38: 33-68.

Palacios-Espinosa, X., Restrepo-Espinosa, M. (2008). Aspectos conceptuales e históricos del sentido de coherencia propuesto por Antonovsky: ¿Una alternativa para abordar el tema de la salud mental? Informes Psicológicos 10, 11: 275-300.

Pirla, A.J., Traveria, R. (2014). Comunicar lo comunitario. Cuadernos de Trabajo Social 27, 1: 139-152.

Poblete, M. (2006). Investigación en educación matemática. In: Actas del X Simposio de la Sociedad. Española de Investigación en Educación Matemática, Huesca, 6-9 Sep, pp. 83-106.

REUS (Red Española de Universidades Saludables) (2015). Principios, objetivos y estructura de la red. Available at: http://tinyurl.com/pd6uo3c.

Rojas, C.E. (2011). Ética profesional docente: Un compromiso pedagógico humanístico. Revista Humanidades 1: 1-22.

Sánchez, F., López, D.,García, J. (2010). El desarrollo de la competencia Sostenibilidad y Compromiso Social en la Facultat d'Informàtica de Barcelona. In: XVI Jornadas de Enseñanza Universitaria de la Informática, JENUI, Santiago de Compostela, July 2010, pp. 249-256.

Sarramona, J., Domínguez, E., Noguera, J., Vázquez, G. (2005). Las competencias en la Secundaria y su incidencia en el acceso a la universidad. In: V. Esteban (ed.). El Espacio Europeo de Educación Superior. Unive. Politécnica de Valencia, Valencia, pp. 199-251.

Taylor, S. (2007). Psicología de la salud. McGrawHill/ Interamericana Editores, México.

Traver, J.A., García, R. (2006). La técnica puzzle de Aronson como herramienta para desarrollar la competencia 'compromiso ético' y la solidaridad en la enseñanza universitaria. Revista Iberoamericana de Educación, 40, 4.

UN (1987). Report of the World Commission on Environment and Development: Our Common Future. UN. Available at: http://www.un-documents.net/our-common-future.pdf.

UN (2002). Resolución 57/254 de la Asamblea General de Naciones Unidas. Available at: http://www.oei.es/decada/resonu.htm.

Villa, A., Poblete, M. (2007). Aprendizaje basado en competencias: Una propuesta para la evaluación de las competencias genéricas. Bilbao, Spain: Universidad de Deusto.

World Health Organization (WHO) (1986). Ottawa Charter for Health Promotion, First International Conference on Health Promotion, Ottawa, 21 November 1986. Available at: http://www.who.int/hpr/docs/ottawa.html.

Part 6. Sociology and anthropology of food waste and food poverty

26. Eating in a time of 'crisis': new health and dietary contexts in Spain

M. Gracia-Arnaiz

Universitat Rovira i Virgili, Av. Catalunya 35, 43002 Tarragona, Spain; mabel.gracia@urv.cat

Abstract

Although questions relating to food insecurity have a high priority in the political, economic and health agendas of industrialized societies all over the world, the focus has been on precautions for keeping the food chain free of contaminants in order to minimize the danger of adulteration and food poisoning. Without underestimating the importance of these policies, this papers poses the question of why, in Spain, food insecurity must increasingly be understood as lack of access to sufficient food resources to guarantee the survival, reproduction, and well-being of part of the population. In order to determine why this is the case, we investigate whether certain positive developments that seemed firmly established, such as the leveling out of social differences in food consumption and the democratization of dietary patterns, are now being reversed. Despite relative food abundance, we cannot safely assume that a standard of living that includes sufficient food to maintain health, recognized as a basic human right by the United Nations, is guaranteed in 21st century Spain. The economic crisis and austerity policies have significantly increased both unemployment and underemployment, social conditions that affect both the quantity and the quality of the diet, with negative consequences for health. Using an ethnographic approach, we examine to what extent the 'crisis' marks a shift in ways of eating and thinking about food among persons in precarious economic circumstances, analizing what kinds of resources and strategies they deploy in order to obtain daily sustenance, and how have both government and civil society attempted to ensure food for all.

Keywords: food system, economic crisis, precariousness, Spain

1. Introduction

After periods of recurring malnutrition, there is today a widespread feeling of food affluence in industrialized societies, and apart from a few exceptions, it can be said that everyone has access to food[1]. For over half a century, and since the major consequences of the Second World War came to an end, 'eating' has ceased to be a primary objective of social organization and has at least in theory become an internationally enshrined right: according to Article 25 (1) of the Universal Declaration of Human Rights (1948), 'Everyone has the right to a standard of living adequate for the health and well-being of himself and of his family, including food'. While this right has never been achieved for millions of people in the world's poorest countries (Gracia-Arnaiz, 2012), international political and economic events in the last decade have also called it into question in European societies. In view of the global economic crisis, we should consider whether some of the positive trends engendered

---

[1] This paper summarizes the theoretical assumptions underpinning the ongoing investigation *Eating in a time of 'crisis': New health and dietary contexts in Spain* (CSO2012-31323, 2013-2015). The ethnographically-based study includes three levels of analysis based on a literature review, treatment of documentary sources and statistics, and empirically supported fieldwork in the regions of Catalonia and Murcia. A longer version is available in Gracia-Arnaiz (2014).

by the industrial food system are changing, such as the progressive democratization of food, and reduced social differences in food consumption and security (Mennell, 1985; Pynson, 1987). Spain is a very good example.

2. Food insecurity in societies of abundance

Food insecurity has undoubtedly been exacerbated by the global economic crisis that began between 2006 and 2008, and has some very unusual characteristics. First, because despite the definition of a crisis – an acute and temporary problematic period – it has become prolonged over time, calling into question whether the crisis is a cyclical episode of difficulties and therefore the term is appropriate in this case. Second, because of its global scope, it is known as the 'crisis of the developed countries': it originated in those countries and has had negative effects on their economies and populations. The situation has been aggravated by the austerity policies that states have adopted in order to reduce their public deficit, involving reduced spending on basic services and increased taxes.

Numerous theories have been suggested as to its origins and expansion, such as the high price of raw materials (oil and food), the overvaluation of some products, the real estate bubble, poor harvests in various parts of the world, poor investments, increased demand from emerging economies and hedge funds. In any event, it is the result of a combination of interconnected difficulties caused by practices in the real estate, credit and labour markets (among others) that were not always legal or legitimate. While the impact has been relatively less dramatic in industrialized societies than in countries that have few resources and are dependent on food imports, it has led to a significant impoverishment of the population. According to the indicator AROPE[2], 115.3 million people in Europe live below the poverty line – a figure that has risen steadily in recent years (Eurostat, 2011).

The problem has become particularly acute in the countries of southern Europe. In its report on social protection and integration (Eurostat, 2013), the European Commission said that in Spain in 2008, the overall proportion of the population at risk of poverty was very high – 24% before and 20.7% after receiving social benefits, according to the figures for 2007, despite the international economic progress and the advances in the labour market in those years. This shows that economic growth is not necessarily accompanied by a reduction in social inequalities. The country has been hard hit by the global crisis, and has faced unique challenges since 2008, such as the fragility of the banking system, the bursting of the housing bubble and job losses, which have further weakened its economy and challenged government bodies that are unable to provide responses apart from bank bailouts, labour market reforms and cuts to social rights. With an economy that is stagnant – or even in recession – and with an unemployment rate of 26.02% in 2012 (INE-EPA, 2012) – the threshold of severe poverty rises to three million people, double the figure for 2008 (Cáritas, 2013).

We also know that the growing social inequality takes several different forms, including increasing precariousness and difficulties with meeting basic needs (Dávila and Quintana, 2009). The new forms of poverty are experienced by various groups: families with unemployed adults, divorced couples with children who reduce the amount of money available, older people with reduced pensions,

---

[2] Arope (At Risk Of Poverty and/or Exclusion) is an indicator provided by the European Union, which refers to the percentage of the population that is at risk of poverty and/or social exclusion.

illegal immigrants, etc. A new profile of applicants for aid has appeared, consisting of middle-class families whose daily life has been undermined by the constant lack of resources for subsistence. This is consistent with the concept of precariousness described by Paugam (2000), which reflects a dynamic idea about what is considered a process in terms of socio-economic instability. It is not necessary to be in extreme poverty to experience precariousness. A worker or a small trader may lose their job or their customer base and experience a period of economic and social instability. Precariousness does not only refer to a financial indicator, but also to changes in terms of restrictions on consumption, difficulties paying housing costs, the mortgage, electricity and gas bills, or purchasing food. Ultimately, it involves the idea of social atomization, and there is a *continuum* between integration and exclusion which even affects the health of those concerned. According to Seveso and Vergara (2001), it is also possible to speak in terms of 'precarious bodies', meaning those people with living conditions with disadvantages and barriers that have accumulated due to a process of precariousness which prevents them from obtaining regular access to formal employment, healthcare and food, and whose physical manifestations can range from extreme thinness to the most severe obesity.

3. The effects of a crisis foretold on the daily diet

Impoverishment has many effects the ways people eat, and some of these affect health. A great deal of food is still produced in Spain, and major distribution channels continue to supply food markets and shops with plenty of goods, and indeed, many new businesses that open in the neighbourhoods of cities sell food. This is another unusual feature of this crisis. The food supply remains plentiful, but at the same time there has been a steady increase in precariousness among citizens that leads to different food choices. Some changes are related to changes in the amounts purchased and consumed in order to minimize expense and waste, other changes are related to more extensive purchases of the store brand products, and others to the product's price, being price the priority as the main criterion for choice. At the same time as the worsening of the recession, Oxfam (2012) noted that 46% of the Spanish population had changed its eating habits, and specified these changes as involving buying cheaper food. The annual survey by the Food Consumption Panel in 2012 showed an increase in the consumption of staple foods such as fresh fruit, eggs, bread and dairy products (MAGRAMA, 2012). Depending on the social class, these differences in consumption are apparent in the type of products consumed and quantities purchased. Upper- and upper-middle class households have a higher per capita meat consumption (5.6 kg above the average) than working-class households, where consumption is falling (6.6 kg less than average; Martín, 2010). This difference in consumption is readily apparent in the volume of food purchased: most food groups are consumed in greater amounts among people in the upper- and upper-middle classes, except for cereals and cereal products, eggs and vegetables that are consumed most by the middle and working classes.

Meanwhile, low income levels prevent more than two million people in Spain from having regular access to food, and make them dependent on public and/or private social resources (Food Aid Plan, 2013). The number of people who as a result of their increasing impoverishment ask institutions for food assistance and eat largely what those institutions offer is increasing. In the middle of an economic crisis, soup kitchens, food banks and in general, initiatives by institutions distributing food to those most in need have been overwhelmed by the exponential growth of pockets of poverty and marginalized groups. According to the Spanish charity Cáritas, the number of families and people it provided food for tripled during the period 2007-2011, exceeding one million in 2011, 300,000 of whom came to the charity for the first time, and in Barcelona alone, the food bank has doubled its aid to charities in recent years, and distributed over ten thousand tons of food in 2012 (FBAB, 2012).

Among those people with the fewest opportunities, efforts to adapt in order to live on very limited resources have led to the adoption of new and old strategies – subsistence from food thrown away in rubbish skips, the use of allotments, recycling of leftovers, begging, theft, etc. In many cases this occurs at the same time as the emergence or expansion of the social support networks which are essential for an understanding why hunger has not affected many people. Many initiatives seek to ensure the right to food: Food kits, *Muévete contra el Hambre y la Pobreza* (Move Against Hunger and Poverty), *Restaurantes contra el hambre* (Restaurants against hunger), *Alianza Contra el Hambre y la Malnutrición* (The Alliance Against Hunger and Malnutrition), etc. Some of these projects have been backed by the authorities after the consequences of their drastic cuts became apparent, and they are often supported by charitable institutions. However, others are the result of the growing mobilization of a civil society that is assuming more and more of the responsibilities that should have been shouldered by those causing the recession.

There is as yet little empirical evidence for the effects of the crisis on health (Cortès-Franch and González, 2014), and the effects on nutritional status are subject to various hypotheses. The Spanish Association of Primary Care Paediatricians (AEPAP, 2013) rules out the existence of malnutrition due to economic reasons among children although it warns that it may appear in the near future. Some authors link the economic difficulties with the increase in diabetes mellitus type II (Escolar, 2009) and although there are no studies indicating a direct relationship between increased poverty and increased obesity in Spain (Antentas and Vivas, 2014), the epidemiology indicates that overweight is more prevalent among people with lower incomes and lower levels of education, and especially among women (Quiles, 2008).

It is true that the current situation is not comparable to the food crises that ravaged Spain in the past. The country's ports are not witnessing the arrival of cargo ships with tons of powdered milk to reduce child malnutrition, as they did in the 1950s. However, subsistence strategies among people in precarious situations have involved reducing the number of daily intakes and the amount of food consumed. At present, we do not know the consequences of this for their health. Most clinicians believe that malnutrition is not a real problem. In fact, the white paper on nutrition in Spain (FEN, 2013), published at the peak of this crisis, barely mentions the nutritional status of the population in terms of its income level. When it discusses malnutrition or undernourishment it describes them as a mere 'common phenomenon in hospitals' associated with other diseases; and not therefore one of the potential problems caused by deprivation. For example, it neglects to mention that Spain is the country that receives the second largest amount of funds from the Food Aid Plan for the most deprived people in the European Union, or that food insecurity has continued to increase over the last five years. Although at present the food donated by the institutions is more varied, they still offer thousands of litres of pasteurized milk and follow-on milk that bring to mind the emergencies of yesteryear. Although 'being hungry' in Spain may indeed be a hardship that is different to the problems described by international organizations in Africa or Asia, it covers experiences of shortages that people describe as involving 'eating too little during the day', 'skipping meals' or 'drinking plenty of water to keep the stomach quiet'.

4. Conclusions

The peculiar characteristics of this 'crisis' suggest that we are not in a period of cyclical instability caused by predictable socioeconomic changes, but we are instead witnessing a structural change in trends due to the cuts that have been implemented in some European societies and their consequences for the deterioration of the material living conditions of millions of people. In Spain, it is a turning point that highlights not only the paradoxes of insufficient policies, but also the limits of a precarious welfare state that has failed to guarantee fundamental rights that were previously considered non-negotiable, including food.

We therefore must not and cannot continue to talk about societies with food abundance with the same ease as a couple of decades ago, or make the claim that differences in consumption have diminished in countries where productive systems increasingly engender an ever widening disparity between rich and poor, highlighting the failure of the neoliberal economic model. The impact of the cuts in Spain shows the contradictions and limits of a food system that is as abundant as it is irrational, and emphasize the duality mentioned by Warde (1997), according to which although it is true that production is more flexible and particularized than ever, social class, which now has more fluid boundaries, continues to be the main explanatory variable for a highly varied and unequal diet. Today, the food consumption patterns of people with the fewest socioeconomic resources remain the same in terms of some historically defined issues: they suffer greater exclusion from variety, quality and frequency. In view of the above, the main problems of late modernity are not those caused by abundant food or the various characteristics resulting from the agroindustrial system, but instead those of guaranteeing access to healthy, culturally acceptable and economically sustainable foods for all citizens.

Whether all the actions undertaken in Spain by the civil organizations and government bodies will be sufficient to cope with the increasing difficulties in living conditions – unemployment, evictions, the loss of benefits ... – and prevent health being affected is open to question. Social cuts entail risks, and not only those arising from the gradual breakdown of the support networks that soften the impacts of hardship for thousands of people; but also those arising from designing ineffective policies which due to ignorance, vested interests or urgency force short-term action be taken, and affect individuals rather than the agents and factors responsible for the crisis. Programmes promoting nutritional health in Spain have also failed to place inequality and deprivation on the political agenda, and continue to present diseases as a simple deterioration in habits linked to unhealthy lifestyles or poor decisions by individuals, while disregarding the micro- and macro-factors that explain differences in health status.

References

AEPap Stament (2013). Asociación Española de Pediatria and Societat Catalana de Pediatria letter. Avalaible at: http://tinyurl.com/nueuoow.

Cáritas Española (2013). VIII Informe del Observatorio de la Realidad Social. Avalaible at: http://tinyurl.com/nchdz7n.

Cortès-Franch, I., González, B. (eds.) (2014). Crisis económica y salud. Informe SESPAS 2014, 28, Monográfico 1, pp. 1-146.

Dávila, C., González, B. (2009). Efectos económicos de la crisis. GacSanit 23(4): 261-265.

Escolar, A. (2009). Determinantes sociales frente a estilos de vida en la diabetes mellitus de tipo 2 en Andalucía. GacSanit. 23 (5): 427-432.

Eurostat (2011). Income and living conditions in Europe. Avalaible at: http://tinyurl.com/obdfunt.

Eurostat (2013). People at risk of poverty or social exclusion. Avalaible at: http://tinyurl.com/oefuzva.

Food Aid Plan (2013). Fondo Español de Garantía [sitio de internet] Plan 2013 de Ayuda Alimentaria a las personas más necesitadas. Avalaible at: http://tinyurl.com/namo6k8.

FBAB (2012). Memoria Annual. Barcelona: Fundacio Banc dels Aliments Barcelona.

FEN (2013). Libro Blanco de la Nutrición en España. Madrid: Fundación Española de la Nutrición.

Gracia-Arnaiz, M. (2012). La faim dans le monde, in Poulain JP, (2012)(dir). Dictionnaire des cultures alimentaires. París: P.U.F, pp: 541-555.

Gracia-Arnaiz, M. (2014). Comer en tiempos de 'crisis': nuevos contextos alimentarios y de salud en España'. Salud Pública México, 56(6): 648-653.

INE-EPA, (2012). Instituto Nacional de Estadística. Encuesta Población activa 2012. Avalaible at: http://www.ine.es/daco/daco42/daco4211/epa0412.pdf.

Martín, V. (2010). Consumo de carne y productos cárnicos. Distribución y Consumo, 5: 5-23.

MAGRAMA (Ministerio de Agricultura, Alimentación y Medio Ambiente) (2012) Consumo alimentario en el hogar y fuera del hogar en España 2012. Avalaible at: http://tinyurl.com/oml5tyb.

Mennell, S. (1985). All manners of food. Oxford: Basil Blackwell Ltd.

Oxfam (2012). El aumento de los precios de los alimentos está cambiando en todo el mundo. Avalaible at: http://www.oxfam.org/es/pressroom.

Paugam, S. (2000). La disqualification sociale, essai sur la nouvelle pauvret. Paris: Presses Universitaires Françaises.

Quiles Izquierdo, J., Rodrigo, C.P., Majem, L.S., Román, B., Aranceta, J. (2008). Situación de la obesidad en España y estrategias de intervención. Rev Esp Nutr Comunitaria 14(3): 142-149.

Seveso, E., Vergara, J. (2012). En el cerco. Los cuerpos precarios en la ciudad de Córdoba tras la crisis argentina de 2001. Papeles del CEIC, 79: 1-38.

Warde, A. (1997). Consumption, food and taste. London, Sage Publications.

27. Food altruism in human beings: facts and factors

E. Rebato

Department of Genetics, Physical Anthropology and Animal Physiology, Faculty of Science and Technology, University of the Basque Country (UPV/EHU), 48080 Bilbao, Spain; esther.rebato@ehu.eus

Abstract

Food sharing is associated with important elements of social cohesion in both humans and other primates (the best-known are the African anthropoid apes of genera *Pan*: chimpanzees and bonobos). A growing body of evidence shows altruistic behaviours not only in primates but also in various animal taxa: eusocial insects like bees, ants and termites; carnivores like lions and wolves; some species of birds (e.g. crows), and vampire bats, among others. Nevertheless, the complexity that sharing food shows in humans makes this behaviour be 'unique' with respect to any other organism. Throughout our evolutionary history the food interchange can be viewed as a survival strategy of the group in front of the risks of an unpredictable surroundings, more than an act of altruism in biological sense, that benefits the receiver and has a cost in terms of biological efficacy (fitness) for the actor. As we will see, the main mechanisms proposed by the theory of evolution (kin selection, group selection and reciprocal altruism) lead us to consider that prosocial tendencies exist in humans because we are genetically predisposed to act prosocially. But apart from a biological predisposition, various investigations also point to an important socialization in the development of food exchange. This paper examines some of the behaviours that society defines as beneficial to others, with special reference to food altruism (food sharing). Besides, the recent debates about the nature and the origin of cooperative behaviour in our species are reviewed from an ecological and evolutionary perspective.

Keywords: altruistic behaviours, sharing food, social life, primates, Hominin evolution

1. Introduction

Cooperation has been decisive in the human evolution and essential in the social life of all human populations, both past and present. As altruistic primates, the human beings often act out of self-interest but also feel concern for the welfare of others (Silk and House, 2011). Among the altruistic behaviours, food sharing is one of the most widely cited characteristics of human social groups, and it is considered an universal feature of human forager societies and an important subject of research in the domain of several disciplines as anthropology, evolutionary biology, human ecology, neurosciences and social psychology, among others. Sharing food has been linked to the evolution of sociality, the sexual division of labour, the transition from hominoids to hominids and morality, and even referred to as a prime mover in the transition between protohominids and modern humans (Gurven *et al.*, 2000).

Food sharing can be defined as 'the unresisted transfer of food from one food-motivated individual, the possessor, to another, the receiver or recipient' (Jaeggi and Van Schaik, 2011). This prosocial behaviour has important implications for the evolution of cooperation in our species, enabling the interpretation of altruism, an evolutionary paradox in Darwinian terms, whereby a recipient gains fitness benefits – that is, in its ability to survive and reproduce in a certain environment – at the

Leire Escajedo San-Epifanio and Mertxe De Renobales Scheifler (eds.) *Envisioning a future without food waste and food poverty*

DOI 10.3920/978-90-8686-820-9_27, © Wageningen Academic Publishers 2015

225

expense of a donor (Hockings *et al.*, 2007). Food sharing is associated with important elements of social cohesion in both humans and other primates like the anthropoid African apes of genera *Pan* (chimpanzees, *P. troglodytes*, and bonobos, *P. paniscus*) and the brown capuchin monkey from Central and South America (de Waal, 2000). An excellent revision about the occurrence of sharing food in primates is the paper of Jaeggi and Van Schaik (2011).

2. Facts

A growing body of evidence shows altruistic behaviours not only in primates but also in various animal taxa: eusocial insects like bees, ants and termites; carnivores like lions and wolves; some species of birds (e.g. crows), and vampire bats, among others. Nevertheless, the complexity that food sharing shows in humans, the varied forms that it takes, the diverse characteristics and individual variation in both social preferences and motivations, and mainly its impact in social life, make this behaviour be 'unique' comparing to any other organism. For example, in humans, food sharing between mothers and offspring is only not limited to lactation during infancy, in fact, human parents feed their children until adulthood in our societies. There is also an intergenerational cooperation, when adults provision children and when middle age women forgo reproduction to invest instead in their younger children and their grandchildren (Sterenly, 2007). In *Homo sapiens* food sharing has a higher adaptive value than in other species, which has probably led to a more active and prosocial sharing psychology in humans (Jaeggi *et al.*, 2010).

According to Kaplan *et al.* (2009), human life histories, as compared to those of other mammals and non-human primates, have several remarkable characteristics: (1) an exceptionally long lifespan; (2) an extended period of juvenile dependence; (3) support of reproduction by older post-reproductive individuals; (4) male support of reproduction through the provisioning of females and their offspring; (5) substantial cooperation in production and food-sharing between kin and non-kin; and (6) a large brain relative to body size that allows great cognitive abilities (e.g. perception, attention, memory, reasoning, comprehension and production of language, among others). The extension of human lifespan and the investment in a long childhood depended on controlling extrinsic causes of mortality: reducing threats of death through predation or accident and reducing the risk of starvation when ill, injured or unlucky (Hill and Kaplan, 1999); in addition, adults (mostly males) generate more resources than they consume. Some of these features could be linked to the particular human ecological niche, a variable physical and biological environment – foraging niche included – but also cultural. Humans adapt themselves to their environment through a unique combination of culture and biology; both are inherent to our existence and reflect the dual aspect of human nature. In every environment humans have a very flexible foraging strategy, consuming different foods in different ecosystems, and this flexibility has allowed our species to survive successfully in all of the world's terrestrial environments (Kaplan *et al.*, 2000).

Sterelny (2007) stated: 'Hominin evolution is hominin response to selective environments that earlier hominins have made'. This author argues that 'in contrast to social intelligence models, that cannot explain the unique cognition and cooperation explosion in our evolutive lineage, hominins have created and responded to a unique foraging mode that is social in itself and which has further effects on hominin social environments'. On the other hand, and unlike the non-human primates, humans are very dependent on tools. Tool use fundamentally altered the way our earliest ancestors interacted with nature, allowing them to eat new types of food and exploit new territories. According to Yamamoto and Tanaka (2009), humans might have uniquely expanded and developed altruism in the form of food sharing according to the increase of our dependence on meat; in other words, the human dependence on animal foods might also have played a role as a selection favouring altruism and reciprocity. Besides the studies in other animal species, particularly in Neotropical primates (de Waal, 2000; Wolowich *et al.*, 2006), and great African and Asian apes (Liebal *et al.*, 2014), models of food sharing have been usually studied in small groups as hunter-gatherers and forager-horticulturalists (see Gurven, 2004, and also Kaplan *et al.*, 2009 for an overview about the evolutionary and ecological basis of human social organization). These researches have yielded qualitative and quantitative data on this altruistic behaviour and allowed to propose some models of exchange from an evolutionary ecological perspective. Also, psychologists and economists have tried to understand the motivations for altruistic, 'other-regarding' behaviour in western societies with market economies (Gurven, 2004).

This paper examines some of the behaviours that society defines as beneficial to others, with special reference to food altruism (food sharing). Besides, the recent debates about the nature and the origin of cooperative behaviour in our species are reviewed from an ecological and evolutionary perspective.

3. Factors

There is a general agreement that social behaviours such as 'food sharing' are central to human social life, but the origin of these behaviours is still debated. Helping, sharing, and comforting are the main domains investigated regarding the ontogenetic and phylogenetic roots of human altruism. It has been observed that young children have a biological predisposition to help the others achieve their goals, to share resources with the others and to inform them of things they need; these behaviours are not originally inculcated by culture, but it is obvious that social experience and cultural transmission become increasingly more influential over ontogeny (Warneken and Tomasello, 2009a,b). Because they emerge early in ontogeny, it has been stated that humans are prosocial by nature and that empathy and sympathy motivate such behaviours from the beginning of their lives (Hoffman, 2001, in Liebal *et al.*, 2014). According to this, both positive emotions (in particular the empathy) are considered essential aptitudes to maintain the complex social life of our species. Hamburg *et al.* (2014) have proposed the potential use of food sharing for empathic emotion regulation (EER), suggesting an origin in the positive emotions engendered by food provisioning early in life. This proposal has been examined by Alley (2014) in an evolutionary perspective, indicating a 'limited influence of EER on food sharing'. Petersen *et al.* (2014) have explored the 'psychology of food sharing', with special attention to the psychology of hunger, concluding that short-term hunger could increase support for social welfare nowadays. According to the explanations of these authors 'The hungry individuals have developed strategies to take food from others and/or persuade others to give them food peacefully'. It is important to point out that our Palaeolithic ancestors probably experienced periods of starvation/scarcity (along with other abundance stages) in which they were not reliable to feed themselves and their families (Kaplan *et al.*, 2000); so they had to be able to develop strategies to survive as a species until today by taking resources from others (see Gurven *et al.*, 2000 about the theory of 'Tolerated theft' of Blurton Jones (1987), and the model of 'Tolerated scrounging'), and sharing food peacefully, as observed in non-modernized societies (Gurven, 2004). In both cases, individuals possessing more food allow others to steal from them, owed to the threat of violence from hungry individuals. An alternative hypothesis is that extensive exchange evolved to buffer the risks associated with hominid dietary specialization on calorie dense, large packages, especially from hunting (Kaplan *et al.*, 2012).

A reiterated question is whether humans share the aforementioned abilities to feel with and for someone with our closest relatives, the great apes. Although several studies demonstrated that great apes help others, little is known about their underlying motivations (Liebal *et al.*, 2014; Silk and House, 2011). Yamamoto and Tanaka (2009) have reviewed several recent experimental studies on chimpanzees' altruism and reciprocity and they propose a possible scenario for the explosion of cooperation in humans (including voluntary altruism and frequent food donation), since the common ancestor between humans and chimpanzees, that we summarized it as follows: (1) uniquely human evolution: elaborated social cognition + language and social norm + necessity of food sharing + enriched material culture; and (2) shared characteristics: recipient-initiated altruism + non-food altruism. These authors consider that 'altruism exploded in humans after divergence from the common ancestor' and that 'the common ancestor seldom showed voluntary altruism and voluntary altruism is practically absent in non-human animals'. There are different studies analysing the origins and forms of intragroup exchange in humans and other primates, as the classic work of Winterhalder (1997), where a critical overhaul of evolutionary models of mechanism and ecological models of circumstance is made: 'evolutionary mechanisms encompass individual, sexual, reciprocal, kin, group, and cultural selection; models of circumstance include tolerated theft, scrounging, marginal value, trade, show-offs, and risk reduction'. This author concludes that exchange behaviours have multicausal origins and they likely will be diverse due to differing combinations of mechanism and circumstance.

Penner *et al.* (2005) have examined prosocial behaviour in humans from a multilevel perspective, providing an excellent overview of common theories. They identify three related levels of analysis of prosocial behaviour: (1) meso: the study of helper-recipient dyads in the context of a specific situation; (2) micro: the study of the origins of prosocial tendencies and the sources of variation in these tendencies; and (3) macro: the study of prosocial actions that occur within the context of groups and large organizations. According to the micro level analysis there are three main evolutionary explanations for the origins of prosocial behaviour (final mechanisms): (a) 'kin selection' (Hamilton, 1964); (b) 'reciprocal altruism' (Trivers, 1971); and (c) 'group selection' (Wilson and Sober, 1994).

Although altruistic co-operators may mean small costs in the short term, they may gain greater selfish benefits either by increasing their 'inclusive fitness' helping their kin (a), by recouping their losses in future interactions (b), or if there is a benefit at group level (c). Kin selection is based on the premise that what matters in evolution is not individual fitness, but 'inclusive fitness', which is the successful transmission of one's genes from all sources to the next generation (Hamilton, 1964). The kin selection model requires no complex cognitive abilities, and has been reported in diverse taxa – as it had already been mentioned throughout the text – but explanations for sharing among unrelated adults, that is Reciprocal altruism, are more controversial and less common among animals (Taborsky, 2013), in fact, there are relatively few examples of reciprocal altruism and these observations are often difficult to replicate. Nevertheless, some authors (e.g. De Waal, 2000) have pointed out that there are some non-human primates that regularly share food outside the parent-offspring context: in the field, bonobos and chimpanzees share large fruits and meat and in captivity they share attractive plant foods. A proposed explanation is that the food sharing among non-kin adults has coevolved with conditions for partner choice and thus the opportunity for reciprocal exchange seems to be a main explanation accounting for the presence of sharing among unrelated adults across the non-human primates (Jaeggi and Van Schaik, 2011). In humans, some observations suggest that people are more likely to help those who offer help, and offering help increases one's status and reputation among members of one's community (Penner *et al.* 2005). A recent phylogenetic meta-analysis performed by Jaeggi and Gurven (2013a), from 23 studies on 32 independent populations (monkeys, apes and humans), demonstrates that reciprocity explains food sharing in humans and primates independent of kin selection and tolerated scrounging.

There is no doubt that humans are more inclined to help relatives than unrelated individuals, but altruism can only evolve if altruists confer benefits selectively on others who carry the same altruistic alleles (Silk and House, 2011). The 'group-level selection' position argues that 'if two groups are in direct competition with one another, the group with a broader number of altruists will have an advantage over a group comprised mainly of selfish individuals'. Thus, the altruistic group would dominate the selfish group and derive a reproductive advantage over them. At a population level, the number of phenotypic altruists would therefore increase relative to selfish individuals (Penner *et al.*, 2005). The basic requirement for this model is spatial heterogeneity in the productivity of local groups, which has an indirect effect on the group member's productivity. This requires deme structure, that is, the existence of partially differentiated subgroups (Wintherlander, 1997).

Although some of the human cooperation is sociocultural we are also products of evolution, with bodies, brains, and behaviours modelled by natural selection (Strassmann *et al.*, 2011). However, evolution by natural selection is by definition a competitive and discriminatory process, so biological altruism seems inconsistent with this evolutionary mechanism (Rachlin and Jones, 2008); natural selection acts fundamentally on genes and genes are never altruistic. Contemporary neo-Darwinian models of evolution, which define evolutionary success as the survival of one's genes in subsequent generations, generally agree that prosocial tendencies exist in humans because of: (1) genetically based predispositions to act prosocially; and (2) the evolutionary success of people who displayed such predispositions (Penner *et al.*, 2005). There must be some physiological and/or neurological processes that facilitate prosocial behaviours (biological basis) and some of these must be inherited (genetic basis). Nevertheless, although we can be genetically predisposed to act prosocially we cannot forget the important role of socialization in the development of food exchange, and that cooperation in human groups is more complex than in the animal world because it has a strong cultural component and it is reinforced by internalized moral standards and values (Toro, 2012).

In conclusion, the characteristics of altruism and the cooperation often assume that they are characteristic basic and distinctive of the humanity. Food sharing is a fundamental form of cooperation and it is particularly noteworthy because of its central role in shaping evolved human life history, social organization, and cooperative psychology (Jaeggi and Gurven, 2013b). It is possible to consider that altruism and reciprocity in humans are somewhat different in quality from those reported in nonhuman animals, since the environment, social systems, and cognitive abilities are quite different (Fehr and Fischbacher, 2003). Humans might have acquired the ability for complex forms of reciprocity and might have achieved societies based on cooperation with altruism after divergence from the common ancestor of humans and other living relatives. Although the study of cooperation is rich with theoretical models and laboratory experiments that have greatly advanced our knowledge of human 'uniqueness', the development of multidisciplinary collaborations is needed for an integrated comprehension of the origins and evolution of altruistic behaviour among human beings.

References

Alley, T.R. (2014). Food sharing and empathic emotion regulation: an evolutionary perspective. Frontiers in Psychology 5: 1-2.

Blurton Jones, N.G. (1987). Tolerated theft, suggestions about the ecology and evolution of sharing, hoarding and scrounging. Social Science Information 26: 31-54.

De Waal, F.B.M. (2000). Attitudinal reciprocity in food sharing among Brown Capuchin monkeys. Animal Behaviour 60: 253-261.

Fehr, E., Fischbacher, U. (2003). The nature of human altruism. Nature 425: 785-791.

Gurven, M. (2004). To give and to give not: The behavioral ecology of human food transfers. Behavioral and Brain Sciences 27: 543-583.

Gurven, M., Hill, K., Kaplan, H., Hurtado, A., Lyles, R. (2000). Food transfers among Hiwi Foragers of Venezuela: tests of reciprocity. Human Ecology 28: 171-218.

Hamburg, M.E., Finkenauer, C., Schuengel, C. (2014). Food for love: the role of food offering in empathic emotion regulation. Frontiers in Psychology 5: 1-9.

Hamilton, W.D. (1964). The genetical evolution of social behaviour. I, II. Journal of Theoretical Biology 7: 1-52.

Hill, K., Kaplan, H. (1999). Life history traits in humans: Theory and empirical studies. Annual Review of Anthropology 28: 397-430.

Hockings, K.J., Humke, T., Anderson, J.R., Biro, D., Sousa, C., Ohasi, G., Matsuzawa, T. (2007). Chimpanzees share forbidden fruit. PloS ONE 2: e886.

Hoffman, M.L. (2001). Empathy and moral development. Implications for caring and justice: Cambridge University Press, Cambridge, UK. 333 pp.

Jaeggi, A.V., Gurven, M. (2013a). Reciprocity explains food sharing in humans and other primates independent of kin selection and tolerated scrounging: a phylogenetic meta-analysis. Proceedings of the Royal Society B: Biological Sciences 280: 20131615.

Jaeggi, A.V., Gurven, M. (2013b). Natural cooperators: Food sharing in humans and other primates. Evolutionary Anthropology 22: 186-195.

Jaeggi, A.V., Van Schaik, C.P. (2011). The evolution of food sharing in primates. Behavioral Ecology and Sociobiology 65: 2125-2140.

Jaeggi, A.V., Burkart, J.M., Van Schaik, C.P. (2010). On the psychology of cooperation in humans and other primates: combining the natural history and experimental evidence of prosociality. Philosophical Transactions of the Royal Society B: Biological Sciences 365: 2723-2735.

Kaplan, H., Hill, K., Lancaster, J.B., Hurtado, A.M. (2000). A theory of human life history evolution: diet, intelligence, and longevity. Evolutionary Anthropology 9, 156-185.

Kaplan, H.S., Hooper, P.L., Gurven, M. (2009). The evolutionary and ecological roots of human social organization. Philosophical Transactions of the Royal Society B: Biological Sciences 364: 3289-3299.

Kaplan, H.S., Schniter, E., Smith, V.L., Wilson, B.J. (2012). Risk and the evolution of human exchange. Proceedings of the Royal Society B: Biological Sciences 279: 2930-2935.

Liebal, K., Vaish, A., Haun, D., Tomasello, M. (2014). Does sympathy motivate prosocial behaviour in great apes? PloS ONE 9: e84299.

Penner, L.A., Dovidio, J.F., Piliavin, J.A., Schroeder, D.A. (2005). Prosocial Behavior: Multilevel Perspectives. Annual Review of Psychology 56: 14.1-14.28.

Petersen, M.B., Aarøe, L., Jensen, N.H., Curry, O. (2014). Social welfare and the psychology of food sharing: short-term hunger increases support for social selfare. Political Psychology 35: 757-773.

Rachlin, H., Jones, B.A. (2008). Altruism among relatives and non-relatives. Behavioural Processes 79: 120-123.

Silk, J.B., House, B.R. (2011). Evolutionary foundations of human prosocial sentiments. Proceedings of the National Academy of Sciences of the United States of America (PNAS) 108: 10910-10917.

Sterenly, K. (2007). Social intelligence, human intelligence and niche construction. Philosophical Transanctions of the Royal Society B: Biological Sciences 362: 719-730.

Strassmann, J.E., Queller, D.C., Avise, J.C., Ayala, F.J. (2011). In the light of evolution V: Cooperation and conflicto. Proceedings of the National Academy of Sciences of the United States of America 108: 10787-10791.

Taborsky, M. (2013). Social evolution: reciprocity there is. Current Biology 23: R486-R488.

Toro, M.A. (2012). Altruismo y evolución en los grupos humanos. Revista de la Sociedad Española de Biología Evolutiva 7: 33-42.

Trivers, R.L. (1971). The evolution of reciprocal altruism. Quarterly Review of Biology 46: 35-57.

Warneken, F., Tomasello, M. (2009a). Varieties of altruism in children and chimpanzees. Trends in Cognitive Sciences 13: 397-402.

Warneken, F., Tomasello, M. (2009b). The roots of human altruism. British Journal of Psychology 100: 455-471.

Wilson, D.S., Sober, E. (1994). Reintroducing group selection to the human behavioral sciences. Behavioral and Brain Sciences 17: 585-608.

Wintherlander, B. (1997). Gifts given, Gifts taken: The behavioral ecology of nonmarket, intragroup exchange. Journal of Archaeological Research 5: 121-168.

Wolowich, C.K., Feged, A., Evans, S., Green, S.M. (2006). Social patterns of food sharing in monogamous Owl Monkeys. American Journal of Primatology 68: 663-674.

Yamamoto, S., Tanaka, M. (2009). How did altruism and reciprocity evolve in humans? Perspectives from experiments on chimpanzees (*Pan troglodytes*). Interaction Studies, 10: 150-182.

28. Whose 'everyday mundane'? The influence of class and privilege in the creation of food waste

T. Soma

University of Toronto, Department of Geography and Planning, 100 St. George Street, Sidney Smith Hall Room 5047, Toronto, ON M5S 3G3, Canada; tammara.soma@utoronto.ca

Abstract

There are a growing number of studies focusing on household food waste in the Global North. However, very little is known empirically about the impacts that class and privilege have upon food wasting practices in the Global South. According to Evans (2014), the passage of food into waste occurs more or less due to mundane household activities. This study aims to apply Evans' finding while illustrating the difficulty in defining the concept of mundane with respect to intergenerational and interclass households. Findings from studies on household food waste in developed countries are not necessarily reflective of several factors that are found within households in developing countries including but not limited to: the family structure, the kinship system, and extreme income disparity. Relatively little is known about food waste in developing countries. This paper argues that due to the corporatisation of the food provisioning landscape, along with the forces of globalisation and urbanisation in the Global South, food waste research is required in developing countries. By drawing upon an immersed field research of 21 upper (n=7), middle (n=7) and lower (n=7) income households as well as 12 key informant interviews, this paper explores the diversity of consumption and food wasting patterns occurring within households of varying incomes and sizes in urban Indonesia, raising the issue of privilege in the creation of food waste. In a developing country context, understanding the inter-class dynamic of the 'household' and the fluid notion of who belongs in a 'household' is crucial to understanding the phenomenon of food waste in Indonesia. This paper employs two theoretical frameworks namely, practice theory and a waste regime framework to analyse the social and material relations that influence food provisioning and resulting food wasting practices in Indonesia.

Keywords: developing country, Indonesia, food consumption, food waste

1. Introduction

It has been identified by several studies conducted in the Global North that households are the largest generators of food waste (Gooch *et al.*, 2010; WRAP, 2011). For example, in Canada, 47% of food waste is estimated to originate from households (Gooch and Felfel, 2014). In the UK it was calculated in 2010 that 7.2 million tons of food and drink were wasted by households with 4.4 million of the total waste categorised as avoidable food waste (WRAP, 2011). Accordingly, there are significant efforts at the international level through Food and Agriculture Organisation (FAO) reports (Gustavsson *et al.*, 2011) at the municipal level, and within the regional level (for example, through the European Parliament waste directives) to prevent, reduce, and better manage the food that is wasted. In predominantly developed countries, composting and waste-to-energy programs have been initiated in various cities around the world (Slater and Aiken, 2014). In countries such as

Leire Escajedo San-Epifanio and Mertxe De Renobales Scheifler (eds.) *Envisioning a future without food waste and food poverty*

233

DOI 10.3920/978-90-8686-820-9_28, © Wageningen Academic Publishers 2015

Sweden, households are mandated to source-separate their food waste and it is then collected through a centralised composting system (Bernstad *et al.*, 2013). A large number of initiatives to address food waste in developed countries have been focused on changing consumer attitudes and behaviours as well as increasing support for 'food rescue' social innovation (for example: Disco Boco soup kitchen in France). Food rescue or food recovery initiatives are usually run by non-profit organisations that 'rescue' surplus food that would otherwise be thrown out by supermarkets. Campaigns such as the popular 'love food hate waste' (www.lovefoodhatewaste.com) initiated in the UK offer recipe ideas for using up avoidable food waste such as potato skins and broccoli stalks, and provide tips on how to shop and better store food. Individual responses to food waste also include freeganism or dumpster diving, whereby individuals or groups of individuals collect or redistribute foods that have been thrown out by supermarkets or stores (Edwards and Mercer, 2013). Overwhelmingly, most of these studies, campaigns, and responses have explored and addressed the issue of food waste from a developed country context. However, very little is known about household food waste in developing countries and it is a challenge to find academic responses to this growing issue.

Criticism of the reports conducted by the FAO and British Institute of Mechanical Engineers (IMechE, 2013) largely cite the 'throwaway mentality' of consumers from the developed world as a major cause of food waste. Evans (2014) and Bulkeley and Gregson (2009) argue that it is important for studies to empirically 'engage more thoroughly with the household as the primary unit of consumption' (as cited in Evans, 2014). These authors argue that doing so will ensure that said policies (i.e. FAO and IMechE) and research do not put forward assumptions on how food waste is generated based upon mere 'conjecture' or 'common sensical' explanations (Bulkeley and Gregson, 2009; Evans, 2014). Therefore, it is argued that there is a need to move beyond 'blaming the consumers' or 'blaming the households'. Scholars such as Evans (2014) have argued that from a consumptive and social perspective, the generation of food waste is not 'extraordinary' nor is it due to some imaginary conspicuous consumption. Rather, according to Evans (2014), 'the passage of 'food' into 'waste' occurs *as a more or less mundane* consequence of the ways in which practices of everyday and domestic life are carried out'. In other words, 'the passage of food into waste' is considered to occur due to quite 'ordinary', 'dull' or 'banal' domestic practices (Evans, 2014). This paper will attempt to complement Evans conceptualisation by demonstrating the importance of including discussions of privilege and class in determining the so-called 'mundane' passages of 'food' into 'waste' particularly in a developing country context. This paper illustrates that the generation of food waste in Bogor, Indonesia in many cases is neither mundane nor ordinary. It is 'extraordinary' due to several factors including: the transformation of food provisioning infrastructure via modern retail outlets (e.g. hypermarkets, supermarkets), the individual's changing relationship with foods, the varieties of food available, and the changing competence required to manage 'modern' food products.

Evans (2014) has clearly expounded that his UK based ethnographic research does not seek to offer explanations along the lines of class or other social stratification. Nevertheless, the concept of mundane household practices needs to be unpacked when looking at households of varied income and class. This paper will attempt to demonstrate that the notion of 'mundane' domestic practices is socially constructed, naturally subjective, and cannot be divorced from class-based analysis. In the context of interclass dynamic, (for example, between the owner of a household and a domestic helper working or living in the household) it is useful to re-examine the notion of 'mundane consequences' through the framework of 'risk' offered by Gille (2013) global food waste regime theory. Further, it is imperative to analyse the social relations of risk through the lenses of practice theory, in order to take into account the transformations of food provisioning and food wasting practices in Indonesia. As the question of what constitutes food waste is subjective and strongly connected to cultural and class identities, it is important to consider who defines 'what is and what is not food and waste'. Hitherto, the definition of food waste and 'who gets to define what is not food and waste' has been determined primarily by Eurocentric-based studies and reports (Shilling, 2013). However, according to a recent World Bank report, in industrialised Asia, 46% of food is wasted at the consumption stage, while in South and Southeast Asia, 13% of food is wasted at the consumption stage (Lipinski *et al.*, 2013 as cited in World Bank, 2014). Taking into account the interclass dynamic within the household unit, and varied income-levels, solutions that emphasises individual agencies to practice responsible consumption and more equitable waste management policies at a municipal level is needed. Finally, this paper will also demonstrate that the fluid kinship and often 'larger' household units that exist in Indonesia makes the flow and the passage of food into waste exceedingly complex to study.

2. Global food waste regime and practice theory

Studies on food waste have significantly grown within the field of sociology (see Evans *et al.*, 2013) through ethnographic studies on household food consumption and the analysis of the passages of food into waste (Evans, 2011, 2012a, 2012b, 2014). However, food waste is still understudied in other fields of social science such as geography, planning, anthropology, and environmental studies. Within this growing literature, two studies – by Cloke (2013) and Gille (2013) – stand out by providing acknowledgment of the unequal power relations and the systemic inequality that influence the generation of food waste (i.e. the macro-level factors influencing food waste). The first study authored by Cloke (2013) raised the argument that 'wastefulness' has become a profitmaking tool in the industrial food system. As Cloke (2013) argues, in a world where food is over-produced 'of which over a third is wasted and in which waste has become a mechanism for profit, no presentation of food security that fails to take into account this systemic reality can produce valid or effective policy'.

The second study on food waste by Gille (2013) takes into account the complexity and multi-scalar nature of the industrial food system and global food consumption. Gille argues that food waste is produced through 'risks', namely, that the 'unequal organisation of uncertainty is a key structural determinant of food waste production' Accordingly, she argues for the need to consider solutions that connect geographical and scalar boundaries (Gille, 2013). This particular paper takes into account that the current paradoxical state, whereby massive amounts of food are wasted amidst global hunger, or where starvation exists amidst plenty, is premised upon an extremely unjust food system and social relationships as they relate to food consumption. This macro-level background cannot be divorced from scholarly attempts to understand daily household consumption and wasting practices.

Food waste has increasingly been studied through the lens of practice theory or every day 'ordinary' practices (Evans, 2011; Ganglbauer *et al.*, 2013). Unlike individualistic models of behavioural framework such as the Theory of Planned Behaviour (Ajzen, 1991) – which is focused on attitudes, behaviours and choices – practice theory looks at the bundle of social practices that make up everyday lives. Practice theory accounts for the contextual factors that influence behaviours (such as material infrastructure, socio-cultural norms, and social institutions) without neglecting individual agency (Lee and Soma, unpublished data). According to Shove and Pantzar (2005) there are three components to understanding practice: materials, meanings, and competence.

Understanding the processes of food procurement or the obtainment of food also entails addressing and analysing the issue of class and spatial access, which is absent from most food waste literature. Class can be a very important factor driving certain types of consumption of the 'profligate' nature, especially in industrialising and modernising countries with extreme income disparity such as Indonesia. Space is also an important component of consumption as it plays a role in where and how people shop, where people work, which further influences the types of foods that people consume and waste. Spatial considerations also influence people's access to food (supermarkets, traditional wet markets) and waste infrastructures (waste collection, dumpsite). It follows that the planning of food and waste infrastructures, particularly in developing countries, can lead to either inclusivism or exclusivism. Consumption in urban areas is more than the production and consumption of goods, it is also about the ability of various segments of the population to participate and belong (Martin, 2014).

3. Methodology

The author employed an immersed field research conducted in Bogor, Indonesia, with low (n=7), middle (n=7), and high-income (n=7) urban households for a total of 21 households. The households were recruited through snowball sampling and lived in various neighbourhoods around the city of Bogor with the exception of two households that lived in a gated elite residential housing site on the periphery of the city (within the regency of Bogor). The households were then selected and categorised based on income-level. Efforts have been made to ensure that the household composition, culture, religious beliefs, workplace and shopping locations are varied. The research includes repeat in-depth semi-structured interviews, going along (Kusenbach, 2003) with the respondents on food shopping trips, accompanying them during meal preparation, observing their management of food waste, in some cases fridge and cupboard checks, and where possible participant observation in the household as well as around the neighbourhood. Twelve key informant interviews were conducted with various actors in the food and waste sector, including (n=2) directors of a large hypermarket, a manager of a mid-size supermarket (n=1), mobile vegetable vendors (n=3), local government representatives from the waste and planning department (n=4), a neighbourhood leader (n=1), and a waste collector (n=1). All names (respondents as well as neighbourhood names) have been changed to maintain anonymity.

4. Discussion: challenging the concept of 'mundane'

Having demonstrated the issues within the dominant food waste narratives, a better understanding of the transformation of consumption patterns and class relationships in Indonesia allows for a re-framing of the narrative from a global South perspective. This will contribute to the broader academic literature on household food waste. It does so by unpacking the concept of the 'everyday mundane' which is emphasised in household food waste studies that are informed by practice theory (Evans, 2014). This problematizes the thrust of the argument that:

> the passage of 'food' into 'waste' occurs '*as a more or less mundane*' consequence of the ways in which practices of everyday and domestic life are currently carried out,

and the various factors that shape the prevailing organisation of food consumption (Evans, 2014, emphasis added)

The concept of 'mundane' is culturally constructed, it is not static, nor is it perceived uniformly across different classes, cultures and income. This is not to say that Evans (2014) argues that it is static. Rather, I engage this concept to demonstrate the importance of carefully analysing such 'norms' while giving consideration to the unequal power relations in 'risk avoidance strategies' (Gille, 2013) mobilised by individuals with different incomes, classes and knowledge in a household. In this paper, I will demonstrate how my findings illustrate that the categorisation of 'mundane practices' is subjective and this consideration involves an analysis of one's 'ability' or 'inability' to waste, access to waste collection, as well as one's ability or inability to access certain food spaces/infrastructure. As Gille (2013) argues, 'the ability to shield oneself from risk and to increase another's exposure to them is a key source and result of power'. The modern/industrial food provisioning and consumption activities that have been 'normalised' and made 'mundane by the corporate food regime have involved the purposeful (for profit) transformation of entire global food systems most often at the expense of low income communities. This transformation is occurring at a rapid pace in Indonesia and in many other developing countries as evidenced by the rise of supermarket revolutions and the accompanying industrial agri-food supply chains (Harvey, 2007; Reardon and Hopkins, 2006). Sensitivity to the issue of privilege and corresponding influences on food provisioning practices will assist in analysing how individuals determine what is defined as food or waste. Doing so will illuminate the disproportionate burden of waste (including food waste) on the poor in Indonesia.

References

Ajzen, I. (1991). The theory of planned behaviour. Organizational Behavior and Human Decision Processes 50: 179-211.

Bernstad, A., la Cour Jansen, J., Aspergen, A. (2013). Door-stepping as a strategy for improved food waste recycling behavior-evaluation of a full-scale experiment. Resources, Conservation and Recycling 73: 94-103.

Bulkeley, H., Gregson, N. (2009). Crossing the threshold: municipal waste policy and household waste generation. Environment and Planning A 41: 929-945.

Cloke, J. (2013). Empires of waste and the food security meme. Geography Compass 7: 622-636.

Edwards, F., Mercer, D. (2013). Food waste in Australia: the freegan response. In: Evans, D., Campbell, H., Murcott, A. (eds.) Waste matters: new perspectives on food and society. John Wiley & Sons Inc., London, UK.

Evans, D. (2011). Blaming the consumer-once again: the social and material contexts of everyday food waste practices in some English households. Critical Public Health 21: 429-440.

Evans, D. (2012a). Beyond the throwaway society: ordinary domestic practice and a sociological approach to household food waste. Sociology, 46(1), 41-56.

Evans, D. (2012b). Binning, gifting and recovery: the conduits of disposal in household food consumption. Environment and Planning D: Society and Space 30(6): 1123-1137.

Evans, D. (2014). Food waste: home consumption, material culture and everyday life. Bloomsbury, London, UK.

Ganglbauer, E., Fitzpatrick, G., Comber, R. (2013). Negotiating food waste: using a practice lens to inform design. ACM Transactions on Computer-Human Interaction, 20(2), 11-36.

Gille, Z. (2013). From risk to waste: global food waste regimes. In: Evans, D., Campbell, H., Murcott, A. (eds.) Waste matters: new perspectives on food and society. John Wiley & Sons Inc., London, UK.

Gooch, M., Felfel, A. (2014). $27 Billion revisited: the cost of Canada's annual food waste. Value Chain Management International Inc. Available at: http://tinyurl.com/no5ovx6.

Gooch, M., Felfel, A., Marenick, N. (2010). Food waste in Canada. Value Chain Management Centre, George Morris Centre.

Gustavsson, J., Cederberg, C., Sonesson, U., van Otterdijk, R., Meybeck, A. (2011). Global food losses and food waste: extent, causes and prevention. Study conducted for the International Congress SAVE FOOD! At Interpack 2011, Food and Agriculture Organization of the United Nations, Düsseldorf, Germany, Rome.

Harvey, M. (2007). The rise of supermarkets and asymmetries of economic power. In: Burch, D., Lawrence, G. (eds.). Supermarkets and agro-food supply chain. Edward Elgar Publishing Ltd., UK.

IMechE (British Institute of Mechanical Engineers) (2013). Global food waste not want not. Available at: http://tinyurl.com/arvrdof.

Kusenbach, M. (2003). Street phenomenology the go-along as ethnographic research tool. Ethnography, 4(3):455-485.

Lipinski, B., Hanson, C., Lomax, K., Kitinoja, L., Waite, R., Searchinger, T. (2013). Reducing food loss and waste. Available at: http://tinyurl.com/notoewv.

Martin, N. (2014). Food fight! Immigrant street vendors, gourmet food trucks and the differential valuation of creative producers in Chicago. International Journal of Urban and Regional Research 38: 1867-1883.

Reardon, T., Hopkins, R. (2006). The supermarket revolution in developing countries: policies to address emerging tensions among supermarkets, suppliers and traditional retailers. The European Journal of Development Research 18: 522-545.

Shilling, C. (2013). Series editor's introduction. In: Evans, D., Campbell, H., Murcott, A. (eds.), Waste matters: new perspectives on food and society. The Sociological Review, John Wiley & Sons, Malden: MA, USA.

Shove, E., Pantzar, M. (2005). Consumers, producers and practices understanding the invention and reinvention of Nordic walking. Journal of Consumer Culture 5: 43-64.

Slater, R., Aiken, M. (2014). Can't you count? Public service delivery and standardised measurement challenges-the case of community composting. Public Management Review. DOI: 10.1080/14719037.2014.881532.

World Bank (2014). Food price watch, 4(16). Available at: http://tinyurl.com/q9gqw4w.

WRAP (2011). New estimates for household food and drink waste in the UK. Available at: http://tinyurl.com/p5jdeu7.

29. Food and interculturality: a question of freedom and of models of protection

A. Castro Jover

Faculty of Law, University of the Basque Country, P.O. Box 644, 48080 Bilbao, Spain; adoracion.castro@ehu.eus

Abstract

At a European level, interculturality refers to the diversity management model which, on the basis of common values, recognises differences, its main tool being intercultural dialogue. The dialogue should be between people or groups who enjoy the exercise of their rights in terms of material equality. In the construction of the social services system and in the initiatives to be implemented by public and private agents, the intercultural perspective has, in some models, been included within equality amongst the principles governing the system. Food may be subject to binding rules for the individual that are based on his or her culture or religion of origin and which find constitutional protection in the right to religious freedom. The protection of this right should be taken into account in those cases in which a situation of need obliges the individual to rely upon the social services. The provision of food in situations of need provides us with an optimum vantage point from which to assess the true scope of the implementation of the cultural perspective.

Keywords: religious freedom, food, religious rules, social services

1. Introduction

Western democracies face the challenge of managing increasingly complex societies in which, on many occasions, the model of legal-political organisation upon which the great European states were built is put to the test. In particular, the characteristic homogeneity of the State-Nation model and majority rule, typical of the model of representative democracy, illustrate the difficulties involved in recognising difference and transcending majority rule in order to address the solution of conflicts posed by minorities.

Interculturality as a model proposed by the Council of Europe and the European Union focuses on these two aforementioned aspects. Firstly, by integrating it as an essential element of the model of recognition of diversity, it indicates a path that, from the legal point of view, makes it necessary to re-evaluate those laws whose application has a discriminatory impact upon minorities, and similarly, the opinion of the majority cannot legitimise decisions which infringe upon the rights of minorities.

One of the priority areas of action indicated by the Council of Europe is to be found in the informal education, which plays a very important role in educating citizens in interculturality. This education is developed fundamentally via Civil Society Organisations (CSOs) and NGOs: '*an indispensable element of pluralist democracy, promoting active participation in public affairs and responsible democratic citizenship based on human rights and equality between men and women*'. The Council of Europe states that *'migrant organisations could be enabled and funded to develop voluntary services for persons from a minority background ...'* (Council of Europe, 2008).

The projection of the model of interculturality in the sphere we might in a broad sense call the Third Sector has the virtue of, in addition to educating in an informal way in the acknowledgement of difference, fostering common values such as solidarity, the promotion of which the Council of Europe regards as a priority (Council of Europe, 2008).

Taking this starting point as a model, I will project it onto the area of social exclusion and vulnerability. These situations exist under the umbrella of social assistance and its organisation via the social services. In Spain, this area falls under the exclusive competence of the autonomous communities. Therefore, I shall confine my study to the Autonomous Community of the Basque Country. I shall select two areas of intervention: charity canteens and centres for children in situations of vulnerability, the custody of whom is assigned to the Provincial Councils. I shall pay special attention to the determination of the conditions of harmonisation of two rights: on the one hand, the subjective right, subject to compliance with the requirements established by the law, of Access to the public services of social services; and on the other hand, one of the manifestations of the right to religious freedom: namely feeding oneself according to the rules of one's religion.

2. Food and right to religious freedom

Food may be closely linked to questions of religious affiliation, its consumption being subject to rules which prohibit or oblige the consumption of certain foods. Compliance with these rules affects different phases of the food chain: preparation and production, distribution, commercialisation and consumption.

The first question to be answered is whether food constitutes part of the contents of religious freedom. The Spanish Organic Law of Religious Freedom 5/1980 does not mention food amongst the manifestations of religious freedom. The link between food and the contents of religious freedom is to be found in the agreements signed between the Spanish State and the Muslim and Jewish communities (enforced by two laws of the Spanish Parliament, 1992) which refer to the preparation, production, distribution, commercialisation and consumption of food, guaranteeing their compliance with the rules established by Islamic law with halal food and kosher in the case of the Jewish.

Thus, the European Court of Human Rights (ECHR) has declared that food laws may be considered a direct expression of beliefs and as a result come under the protection of section 9 of the European Convention on Human Rights of November 4, 1950. That was acknowledged by the ECHR in the Ruling on *Jakóbski v. Poland* (2010). The Court regards as violated the right to religious freedom of a prison inmate who, as a Buddhist, requested a vegetarian meal, a request that was denied (paragraph 45). The ECHR issued a similar decision in the case *Vartic vs Romania* (2014, paragraph 49 of the judgement). In the case of the Spanish Law, the Palencia Provincial Court declared a contrary opinion in April 20, 1999, dismissing the appeal lodged by a prison inmate requesting a strictly vegetarian diet. The judge reasoned that and the institution should ensure that inmates received a balanced diet, and that medical reason, service requirements and the security of other inmates prevented this request from being granted.

Therefore, it can be stated that feeding oneself in accordance with religious rules forms a part of the contents of religious freedom implicitly included in the constitutional block, in particular in the Spanish Constitution and Organic Law on Religious Freedom, and explicitly in the aforementioned Agreements, as well as in the jurisprudence of the ECHR (Chizzoniti, 2010). This affirmation results in positive obligations for public authorities, which take the form of the adoption of reasonable and proportional measures that, ensure the fulfilment of this right. However, no right is absolute, and in its application the public authorities may find themselves under the obligation to weigh up the rights or common good at stake; the decision resulting from this evaluation should be reasonable and proportional: in other words, the damage caused should not be greater than that one seeks to avoid.

The second question we have to ask is whether this right can be modulated and reduced in circumstances in which the individual is in a situation of dependence upon the authorities. The agreement with the Islamic Commission refers in section 14 to situations in which the individual is interned in public centres or establishments. In this regard, the article 226 of the Spanish prison regulations (Spanish Government, 1990) stated that prisons should provide food that responds amongst other criteria, '*as far as possible, to personal and religious convictions*'. In this sense, in these cases, when requested '*attempts will be made to adapt (food) to Islamic religious tenets, as well as mealtimes during the month of fasting (Ramadan)*'. Though there is no official mandate, there is a request to the Authorities to create suitable conditions for these practices.

So the specific request for a different menu should lead the administration to consider diverse aspects, from nutrition to the financial or organisational difficulties that may arise from this request. However, the decision adopted, be it positive or negative, should be proportional. In this sense, the ECHR declared in the case of *Jakóbski vs Poland* (2010), that the violation of the right to religious freedom is produced because '*the Court finds that the authorities failed to strike a fair balance between the interests of the prison authorities and those of the applicant, namely the right to manifest his religion through observance of the rules of the Buddhist religion*' (Paragraph 54). The Court followed the same line in 2014 in the case of *Vartic vs Romania* (Paragraph 54).

3. Food, intercultural perspective, social exclusion and vulnerability

The Social Services Law 12/2008 of December 5 of the Autonomous Community of the Basque Country notes in its explanatory memorandum the importance of taking into account in the construction and articulation of the system of social services that respecting the rights of individuals involves the integration of, amongst others, the intercultural perspective. Thus, in the list of principles in Article 7, Paragraph c), with regard to the principle of equality and equity, we read that '*... the Basque public authorities ... shall guarantee access to these benefits and services according to criteria of equity ... and incorporating into their actions the intercultural ... perspective*'.

I have chosen two areas of intervention in which the religious-cultural affiliations of the users of public service demand that public administrations facilitate food in keeping with their religious laws: centres for children in need and social canteens.

Centres for children in need

As we have already noted, the agreement with the Islamic Commission refers in article 14 to situations in which the individual is interned in public centres or establishments, and this event has also been referred to at the above mentioned Spanish Prison Regulations. In these cases, when requested *'attempts will be made to adapt (food) to Islamic religious tenets, as well as mealtimes during the month of fasting (Ramadan)'.* Nothing is stated in this respect in the agreement with the Jewish community.

The agreement employs a generic term, public centres or establishments in which the individual is interned, and, consequently, in a situation of dependence upon the administration. The area described by the law would also include those social care centres occupied by people in a situation of vulnerability.

When minors are in a situation of vulnerability and therefore under the guardianship of the administration, it is the responsibility of the latter to preserve and guarantee their rights. In the Autonomous Community of the Basque Country this situation was regulated by Decree 131/2008 of July 7, one of the principles of which guarantees respect for diversity and difference (Article 8.2 f), as well as for origins, favouring the conservation of their cultural and religious baggage (Article 14 a). *'In order to respect the diversity of beliefs of the children and adolescents in residence, no religious symbols of any kind shall be placed in common areas, respecting the right of residents to place such symbols in the individual space of their rooms'* (Article 47.5). The absence of religious symbols in common areas guarantees the neutrality of the public residential centre, the only means of respecting in conditions of equality the diverse cultural-religious identities of the users of these centres.

Amongst the areas of attention, special attention is made to food and socio-cultural identity. With regard to food, it establishes the need to take into account both nutritional and educational and cultural aspects, including tastes, styles, cultural or religious customs or guidelines (Article 68). With regard to socio-cultural identity, not only should this be respected, but the conservation of socio-cultural identity should be fostered, establishing the measures to be adopted and which are specified in two areas: in education and in the organisation of the residential care facility. In educational activities, the measures to be adopted focus on access to teaching or cultural materials and on the activities that permit this approach, and in the organisation of the residential care facility attention will be paid to meals, decoration of individual spaces and teaching materials (Article 76).

In particular, in the centres for unaccompanied foreign minors, most of the residents are from the Maghreb, followers of Islam and its compulsory food laws. The aforementioned legal text conserves their right to receive food in accordance with their religious rules.

Social canteens

A different analysis is required for the situation of exclusion in which some people find themselves, and as a result of which are obliged to use social canteens to obtain sufficient food to survive.

Social canteens to a large extent are run by religious organisations, partially supported by public funds. Familiar to many is the *'Cocina económica'* (which could be translated as 'affordable kitchen') managed by the Daughters of Charity of St. Vincent de Paul in several Spanish cities, or in the case of the Basque Country, and Biskay in particular, the social canteens run by religious organisations: Cáritas (Bilbao and Barakaldo) and Franciscans of Irala (Bilbao).

Today the social canteens face two types of request: one related to the adaptation of their menus to the requirements of users, 40% of whom are Muslim. Another of the requests is connected with the period of Ramadan, in which mealtimes at the social canteens may not coincide with those prescribed by Islam.

Sources consulted indicate that the attention to difference in social canteens is confined to offering adapted menus when the daily menu includes pork, but does not guarantee that the menus are halal. The non-coincidence of social canteen mealtimes during Ramadan with permitted mealtimes is resolved via the preparation of packaged food which is distributed to those who request it before the canteen closes.

Acknowledgements

I+D Project DER2013-42261-P, 'Solidarity, Participation and Co-existence in Diversity', convened by the Ministry of Economy and Competitiveness, sub-programme of knowledge generation. Research Group of the Basque University System 'Multilevel Constitutionalism and Inclusion of Diversity (Politics, Legal, Institutional and religious-cultural diversity) (IT743-13). Source of information regarding the attention of the difference at the social canteens has been Gemma Orbe, social worker of Cáritas Bizkaia in the social canteen of the apostolic centre in Bilbao.

References

Basque Government (2008). Decree 131/2008 of July 7, regulating the facilities of residential centres for children and adolescents in situations of vulnerability.
Basque Parliament (2008). Social Services Law 12/2008.
Chizzoniti, A.G. (2010). La tutela della diversità: cibo, diritto e religiones. In: Chizzoniti A.G., Tallacchini, M. (eds.) Cibo e Religione: Diritto e Diritti. Libellula Edizioni, Tricase, Italy.
Council of Europe (2008). White Paper on Intercultural Dialogue, Strasbourg.
ECHR (2010). Case of Jakóbski v. Poland of December 7, 2010. Application No. 18429/06.
ECHR (2014). Case of Vartic v. Romania of March 17, 2014. Application No. 14150/08.
Spanish Government (1990). Prision Regulations. Approved via Royal Decree 190/1996 of February 9.
Spanish Parliament (1980). Organic Law on Religious Freedom, July 5 1980.
Spanish Parliament (1992). Law 25 /1992, of November 10, bringing into force the Agreement with the Jewish Communities
Spanish Parliament (1992). Law 26/1992, of November 10, bringing into force the Agreement with the Islamic Commission.

Part 7. Food banking, food redistribution and social supermarkets

30. Food poverty and food charity in the United Kingdom

H. Lambie-Mumford

Sheffield Political Economy Research Institute (SPERI). University of Sheffield, Sheffield S10 2TN, United Kingdom; h.lambie-mumford@sheffield.ac.uk

Abstract

This chapter explores the changing nature of food poverty and the rise of food charity in the United Kingdom. Looking in particular at the effects of austerity, welfare reform, stagnating incomes and rising costs of living on people's ability to eat well, the chapter contemplates how experiences of food poverty have changed in recent years. The chapter also provides an account of the rise of food charity – on a national scale in the UK – since the turn of the century and particularly since 2010 and examines the reaction and responses, which have emerged from policy makers. The implications of food charity for the realisation of the human right to food in the UK are discussed, along with the tangible steps policy makers could take towards implementing a rights-based approach to overcoming household food poverty.

Keywords: food poverty, food charity, food banks, right to food

1. Introduction

Since the turn of the century we have seen the emergence of charitable initiatives providing food to people in need on a widespread scale in the UK (Lambie-Mumford, 2013). Whilst the provision of food to people in need by charitable organisations has a long history in the UK (McGlone *et al.*, 1999), the formalisation of this provision and its facilitation and co-ordination at a national level is unprecedented in this country. In the year 2013-2014 the largest food banking organisation, the Trussell Trust Foodbank Network, distributed 913,138 food parcels to adults and children across the country, up from 128,697 in the year 2011-2012 and 25,899 in the year 2008-2009 (Trussell Trust, 2015). In the year 2014-2015 this provision rose to over a million, to 1,084,604 parcels (*ibid.*). The last few years have been particularly formative for the emergency food movement in the UK – not just in terms of operation, as illustrated by these statistics, but also in terms of public profile and political discourse with widespread national news coverage and increasingly intense political scrutiny (see Morris, 2013, Channel 4 News, 2014 amongst many and the report from the Parliamentary Inquiry into Hunger and Food Poverty: Food Poverty Inquiry, 2014).

The most recent proliferation of emergency food provision, in the last five years (since 2010), has occurred within a context of economic crisis, recession, public finance austerity and welfare reform, all of which are impacting on people's economic security and ability to eat well. Incomes have been lagging behind rising costs of living, including housing, fuel and food prices (Davis *et al.*, 2014) which has had the overall impact of reducing the affordability of food by over 20% for the poorest income decile since the mid-2000s (DEFRA, 2014). Public sector finances have also been subject to a programme of significant cuts, some of which are yet to kick in, and there has been a high profile agenda of extensive welfare reform which has introduced caps to entitlements and increased conditionality (Beatty and Fothergill, 2013; Taylor-Gooby and Stoker, 2011). The growth

Leire Escajedo San-Epifanio and Mertxe De Renobales Scheifler (eds.) *Envisioning a future without food waste and food poverty*

DOI 10.3920/978-90-8686-820-9_30, © Wageningen Academic Publishers 2015

247

of emergency food provision has therefore occurred at a time of significant social and political-economic shifts which are impacting in real terms on the adequacy of household income and the level of state support people can access.

As the nature and extent of food poverty appears to change in the UK, the question of the role of policy becomes more urgent. As food poverty has risen in visibility there has been a considerable level of political reaction in the form of parliamentary debates, comments to the media and most significantly the launch of a Parliamentary Inquiry into Hunger and Food Poverty in the UK (Food Poverty Inquiry, 2014; Hansard, 2013). Despite this, there has been little tangible policy response in the shape of targeted interventions or strategies. This policy void occurs in the context of a general lack of policy ownership of the issue of food poverty. Traditionally in the UK, approaches to ensuring everyone has access to healthy food has been left to the operation of efficient markets in retail and employment, appropriate consumer choice and a social welfare system which is meant to enable those lacking employment to be able to purchase food (Dowler *et al.*, 2011). However, the evidence of rising food poverty and increasing numbers of people turning to charitable food sources seriously questions the efficacy of these mechanisms in protecting people from food poverty. Responsibility for household level 'food security' is currently situated within the remit of the Department for the Environment, Food and Rural Affairs (DEFRA, 2006) and whilst the issue of household access to food is located formally within DEFRA's remit, as discussed later, it nonetheless intersects with areas of responsibility in other Whitehall departments, therefore opening up the possibility of cross-Whitehall commitment and working. Despite these intersections, there has been little formal engagement with developing cross-governmental approaches to addressing the structural root causes of food poverty to date.

Questions have been raised about what is driving the growth of emergency food provision and if and how household experiences are changing in the context of economic uncertainty, rising cost of living, austerity and welfare reform (Ashton *et al.*, 2014; DEFRA, 2015). This chapter explores changing household circumstances through the lens of food poverty and charts the rise of food charity in the UK in recent years. It also explores some of the implications of rising food charity on the potential for the realisation of the human right to food in the UK. In the first instance the chapter explores changing household food poverty through the secondary sources available in the UK. It then goes on to look at the rise of food charity in particular, employing two national case studies – the Trussell Trust Foodbank Network and FareShare – which are the two largest national charities involved in the facilitation or co-ordination of emergency food provision in the UK. The chapter then discusses some of the implications of the rise of food charity from a right to food perspective, based on empirical work with these two case study organisations. For this empirical section, findings are presented which originate from a study of the nature of emergency food provision in the UK and involved an empirical investigation into how it works as a system and a critical engagement with the phenomenon, specifically from a right to food perspective (see the full research in Lambie-Mumford, 2014a). Extensive qualitative research was undertaken with the Foodbank Network and FareShare and data were collected in two stages: from local emergency food projects in several areas across the country; and at the head offices of the national organisations themselves. Fifty two interviews were conducted over the period of a year (September 2012 – October 2013). The chapter concludes by arguing that a stronger policy framework – guided by right to food principles – is required to drive forward progressive responses to growing household food poverty and increased food charity provision.

2. Changing nature of food poverty

The first challenge faced by research into constrained food experiences in the UK is the variety of, sometimes overlapping, language used to describe the experience of lack of access to food (Lambie-Mumford and Dowler, 2014). Overall, there is a lack of an agreed definition of this experience in the UK and 'hunger', 'food poverty' and 'food insecurity' are all utilised (see Food Poverty Inquiry, 2014). Whilst the phrase 'food poverty' has popular resonance (Cooper and Dumpleton, 2013; Oxfam, 2013) and a history of utilisation in academic research (Dowler *et al.*, 2001) 'food insecurity' is used by government (DEFRA, 2006) and is also commonly used by researchers from other developed country contexts (Loopstra and Tarasuk, 2013; Riches, 1997). Recently, 'food poverty' and 'food insecurity' have been used interchangeably – as synonymous – in the UK (Dowler and O'Connor, 2012). This chapter adopts this approach, but uses 'food poverty' as the chosen of the two terms given the focus of this edited collection.

The conceptualisation of food poverty employed in this chapter encapsulates a broad notion of a dynamic process, one that is experienced differently by different people who have active agency in how they manage their lives within the structural determinants constraining their food experiences (Lister, 2004; Riches, 1997). Ultimately, food poverty is understood as relative to different societies and as a construct of those societies. This conceptualisation is realised in this chapter through the definition of food insecurity offered by Anderson (1990, p. 1560):

> Access by all people at all times to enough food for an active, healthy life and includes at a minimum: (a) the ready availability of nutritionally adequate and safe foods, and (b) the assured ability to acquire acceptable foods in socially acceptable ways (e.g. without resorting to emergency food supplies, scavenging, stealing, and other coping strategies). Food insecurity exists whenever the availability of nutritionally adequate and safe foods or the ability to acquire acceptable foods in socially acceptable ways is limited or uncertain.

This is a definition which takes account of the key composite dynamics of access (broadly defined), acceptability and adequacy, and security in the longer term. It is a definition which highlights the importance of food for social participation and the value of aspirations and equity. Experiences of food poverty are seen through this interpretation as more than a symptom of poverty; they are treated as a site of analysis in their own right, as a set of experiences which both result from and contribute to social exclusion and injustice (see also Lambie-Mumford, 2015). Those experiencing food poverty can be seen as active agents within this experience; whilst their agency may be constrained by the structural determinants of their food experiences, notions of 'personal agency' and the ways in which people 'get by' in these circumstances can be accounted for (see Lister, 2004 for an analysis of agency and structural interpretations of poverty). Food poverty is an experience determined by structural forces, including the food production and retail system, the labour market, the welfare state and transport, housing and planning infrastructures (see Caraher *et al.*, 1998; Hitchman *et al.*, 2002; for international references to these dynamics, see Coleman-Jensen, 2011; De Marco and Thorburn, 2009; Kirkpatrick and Tarasuk, 2011).

Importantly, conceptualisations of food poverty/insecurity enable an appreciation of various scales and levels of severity. Whilst the focus in this chapter is on household 'food insecurity', this insecurity can be explored at the individual, community, national or global scale (FAO, 2006). Furthermore, and particularly importantly in the context of food charity, food insecurity is also described in terms of differing severity which enables analyses of both more chronic experiences (for example longer term anxiety about access to food or changes to diet in response to limited access) and more acute crisis experiences to which food charity is often posited as a response (see Loopstra and Tarasuk, 2013 for work on severity of food insecurity).

In addition to a lack of an agreed definition, in the UK there is also a lack of a systematic measurement of food poverty. Unlike in other countries including the USA and Canada where measures of household food insecurity take place, using an established methodology (Coleman-Jensen et al., 2014; Loopstra and Tarasuk, 2013), the UK does not measure it (see discussion in Lambie-Mumford and Dowler, 2015). This means that researchers, policy makers and activists in the field of food poverty have to turn to food-related measures in poverty surveys or proxy indicators.

Recently published Poverty and Social Exclusion Survey data reveal that in 2011, four million adults could not afford 'food basics', with three per cent of adults not able to afford two meals a day (up from less than 1% in 1999) and seven per cent not able to afford fresh fruit and vegetables (PSE, 2012). Other surveys and research also indicate that people have been finding it more difficult to access food since the turn of the decade (Save the Children, 2012; Shelter, 2013). The key drivers of the shifting nature of hunger lie in the fact that wages and social security incomes have been growing at a slower pace relative to costs of living, including housing, food and fuel (Davis et al., 2014; Padley and Hirsch, 2014; see also Dowler and Lambie-Mumford, 2015). This has been putting considerable pressure on household budgets and evidence tells us households have been changing their shopping and eating habits; trading down or eating less in response (DEFRA, 2014; Save the Children, 2012).

Importantly, when looking at structural determinants of food poverty and how they have changed over time (especially income levels and income security, food prices and transport costs), evidence suggests that circumstances have worsened in the last few years (see Hirsch, 2013). Relative to wages and benefit levels, prices have risen and access to services and other forms of publicly funded support have been restricted as a result of public finance austerity. This, together with evidence which suggests that when under financial pressure households will cut back on food budgets, perceived to be the most flexible element of household spending (Dowler, 1997), means that the nature of food poverty is likely to have changed for the worse over recent years.

Statistics produced by DEFRA (2014) highlight that falling income and rising costs of living, including rising food prices, have meant that food is now over 20% less affordable for those living in the lowest income decile in the UK compared to 2003. In 2012 a report by Save the Children

(2012) found that of the families surveyed with household incomes of up to £ 16,999 a year, 60.8% were cutting back on how much they spent on food, 39.1% were eating less fruit and vegetables and 25.5% were serving smaller portions.

When we look at the nature of income crises leading to food bank use it appears that there has also been an increase in vulnerability to these shocks in the last few years. Impacts of reduced or insecure incomes and problems associated with social security appear to be leaving increasing numbers of people without income or with a reduced level of income leading to food bank uptake (Lambie-Mumford, 2014b; Lambie-Mumford and Dowler, 2014). In particular, delays in benefit payments and sanctions[3], leaving people without incomes, are key triggers of food bank use (see Perry *et al.*, 2014).

This body of evidence indicates that food poverty has worsened over the so called 'age of austerity', with the effects of lagging incomes relative to other costs of living forcing households to cut back or trade down on the food they purchase and eat. Moreover, the situation appears to have worsened across the scale of severity, with acute experiences of food poverty, induced by income crises, also increasing particularly in the context of welfare reform.

3. Rise of food charity

Food charity in the UK is referred to by the government as part of a wider category of 'food aid', defined as:

> An umbrella term encompassing a range of large-scale and small local activities aiming to help people meet food needs, often on a short-term basis during crisis or immediate difficulty; more broadly they contribute to relieving symptoms of household or individual level food poverty and poverty. (Lambie-Mumford *et al.*, 2014, p. iv).

Within this broader category both state (for example welfare food vouchers) and non-state support are situated. For this chapter, non-state provided support in the form of charitable initiatives is of particular interest – especially, charitable emergency food assistance. Whilst charitable emergency food assistance would include a whole range of initiatives such as hot meal providers or soup runs, 'food banks' have come to dominate the discourse and debate in the UK in recent years. Food banks have come to be recognised as charitable initiatives, which provide emergency parcels of food for people to take away, prepare and eat (Lambie-Mumford and Dowler, 2014). This provision is usually given to help relieve some kind of (food) crisis. Whilst this food bank label is quite high profile and dominates the food charity debate, it belies significant variability amongst the projects that identify themselves as such including differences in the food provided, how they are accessed and when they are open and if other services are on offer at the project (Dowler and Lambie-Mumford, 2015). Surplus food redistribution is not part of the 'food bank' category in the UK. Whilst some food banks might obtain food for provision from these surplus food redistribution initiatives, unlike in some other countries, these redistributors are not themselves recognised as 'food banks'. Arguably,

---

[3] Sanctions are reductions in social security income for a set period of time as a result of administrating authorities deciding that a person has 'failed in the performance of some activity related to seeking or preparing for work' (Perry *et al.*, 2014).

the most important distinction in the UK for this food bank label is that the projects within it are direct providers to people in need themselves, not providers to community projects who in turn hand food out.

Examples of 'food bank' and 'surplus food redistribution' organisations are the two case study organisations for the research on which parts of this chapter is based – the Trussell Trust Foodbank Network and FareShare. These organisations form the biggest national networks of food banking initiatives and surplus food redistribution respectively in the United Kingdom. The Trussell Trust foodbank Network is a national network of locally based not-for-profit foodbank[4] franchises which individually collect and store food donations and distribute food parcels to local people in crisis. The first Trussell Trust foodbank was set up in Salisbury, where the Trust is based, in 1999 and the first not-for-profit foodbank was established in 2004 in Gloucester. As of April 2014 the network had over 400 foodbanks across the UK, including in England, Scotland, Wales and Northern Ireland. FareShare is a surplus food redistribution charity with a dual motivation of 'fighting hunger, tackling food waste' (FareShare, 2015). In 1994, the homeless charity Crisis founded Crisis FareShare with one depot in London. In 2004, FareShare became an independent charity with five depots (in London, Yorkshire, Brighton, Edinburgh and Dundee). FareShare now has nineteen depots across the country that distribute food to 1,200 charities. In 2013, FareShare reported that the food they redistributed contributed to 51,000 meals a day (FareShare, 2014).

When looking at the provision of the Trussell Trust foodbank Network in particular it is important to note that the growth of their emergency food provision, since the turn of the century but particularly in the last 5 years (when their food parcel provision rose from 40,898 to over one million), has occurred in parallel to significant changes to the welfare state. Since the economic crash in the mid-2000s the UK has seen a programme of extensive cuts to services which form part of the welfare state and widespread reforms to social security – what some have termed an 'age of welfare austerity' (Farnsworth, 2011). This fits onto a wider historical trajectory of shifts in the shape and nature of the welfare state since the 1970s and particularly since the beginning of the New Labour years in 1997 (-2010) which saw the increased and more formalised role of the voluntary sector in welfare services through programmes of diversification and a consequently more formalised and professionalised voluntary sector generally. In the context of the Conservative-Liberal Democrat coalition government (2010-2015) this process of diversification continued under the policy platform of the 'Big Society' – based on an ideology of localism and transferring power to individuals and communities (Cabinet Office, 2010). Food banks have come to be seen to represent key elements of this re-shaped welfare state. As Ellison and Fenger (2013) observe:

> For the current coalition government in Britain, food banks are viewed as a positive translation of the ideology of the 'Big Society' – a mix of libertarian paternalism and communitarian forms of social solidarity.

Research indicates that both changes to the levels of social security entitlements and problematic welfare processes are impacting on needs (Lambie-Mumford, 2014b; Loopstra *et al.*, 2015; Perry

---

[4] 'Foodbank' is the registered name of the Trussell Trust franchise; 'food bank' refers to the wider category of food initiatives.

et al., 2014). It appears that as social security entitlements are being reduced, people are becoming worse off and increasingly vulnerable to crises (Lambie-Mumford and Dowler, 2015). At the same time problems with administration of payments (particularly unexplained delays) and inappropriate sanctions (when someone's entitlement is withheld) are also reported as key reasons for food bank uptake (Perry *et al.*, 2014). A broader analysis undertaken by Loopstra *et al.* (2015) showed important links between expanding foodbank provision and austerity and cuts to welfare, highlighting that foodbanks are opening in areas which are experiencing cuts in welfare spending and benefits and with rising unemployment.

The rise of food banks has captured significant attention in the media and amongst politicians: 2012 was declared by a prominent columnist in the Guardian to be the 'year of the foodbank' (Moore, 2012); in December 2013 the first parliamentary debate on food banks was held (Hansard, 2013); and in February 2013 DEFRA commissioned a review of food aid in response to increasing numbers of parliamentary questions and widespread reporting of the rising rates of people turning to food aid (Lambie-Mumford *et al.*, 2014). In 2014 an All Party Parliamentary Group on food poverty was established and an inquiry into Hunger and Food Poverty was launched, which reported in December 2014 (Food Poverty Inquiry, 2014; Register of All Party Parliamentary Groups, 2014). The nature of the political reaction to this growth is interesting to note because on the one hand, the rise of food banks has been said to represent the success of the Conservative party's 'big society' policy and to embody calls for more responsibility to be held in local communities; on the other hand the rise in numbers visiting food banks is attributed to social-economic failures, unmanageable increases in cost of living and ultimately a wider growing 'hunger' problem (Conservative Home, 2012; Dugan, 2014).

However, these reactions have yet to translate into substantive policy responses, driven by elected members of councils, assemblies or parliament, particularly as outcomes of the Food Poverty Inquiry remain forthcoming. Similarly, despite DEFRA commissioning a Rapid Evidence Assessment on food aid in the UK (Lambie-Mumford *et al.*, 2014), at the national level there has so far been a lack of policy response from policy makers within government departments with area of relevant responsibility. Notably from DEFRA which has responsibility for household food security, the Department for Work and Pensions which oversees social security, the Department for Communities and Local Government which oversees planning regulations or the Department of Health which oversees nutrition policy. Whilst officers in devolved and local governments have worked on various responses, such as grant funding or food strategies, these have been local and often short- or medium-term responses (see Dowler and Lambie-Mumford, 2015).

4. Human right to food

The right to adequate food was originally enshrined in Article 25 of the Universal Declaration of Human Rights (ratified in 1948) as part of the right to an adequate standard of living, which incorporated adequate food (UN, 1948). As part of the range of economic, social and cultural rights, the right to food was ratified by the UK in the mid-1970s in the form of the International Covenant on Economic, Social and Cultural Rights (CESCR, 1999) (published in 1966) (Joint Committee on Human Rights, 2004; OHCHR, 1996; UN, 2015). Since then, work on the particularities of the right was published by the UN Committee on Economic, Social and Cultural Rights in 1999, specifically in the form of General Comment 12 on the Right to Adequate Food (CESCR, 1999).

There has also been the development of Voluntary Guidelines in support of the realisation of the right to food (Food and Agriculture Organization (FAO) (2005) and, since the first appointment in 2000, the right to food has had a dedicated UN Special Rapporteur on the Right to Food (SR Food, 2015).

Previous research has explored ethical and rights-based dimensions of food charity in other countries (see for example chapters in Riches and Silvasti, 2014). The research on which this chapter is based provided an analysis of the rise of food charity, in the UK specifically, in relation to two key areas of the right's content; it's normative content and the obligations and violations it sets out. These state that people should have sustainable access to adequate food and that the state – as duty bearer – has the obligation to respect, protect and fulfil the right to food (see CESCR, 1999). Each of these two elements drawn from the right to food gives rise to distinct sets of questions. Normative content surrounding the 'adequacy and sustainability of food availability and access' raises questions regarding the acceptability and sustainability of emergency food systems; and, in turn, the notion of the obligations of states raises questions around the idea of responsibility and the theoretical and practical role of charity and the state in realising the right to food.

The empirical research with the case study organisations (reported in full in Lambie-Mumford, 2014a) found that emergency food initiatives were important spaces of caring and social solidarity in local communities. They embody moral imperatives to feed the hungry and overcome social injustice. They also provide care to those in need in various ways – beyond food – including providing personal support, a safe space, and other advice or signposting. However, by right to food standards these projects are problematic. The research found that these systems are ultimately not adequate or sustainable by right to food standards which emphasise the importance of the social acceptability of food acquisition, on the one hand, and the sustainability of food access into the future, on the other. Emergency food provision forms an identifiably and experientially 'other' system to the socially accepted mode of food acquisition in the UK today – the commercial food market through shopping (Meah, 2013). Providers are not necessarily able to make food available through these systems, with their ability to do so shaped in important ways by the structure of the food industry in which they operate; in addition, people do not always have the ability to access emergency food projects and the food available from them whenever they wish, for as long as they may feel they need.

The research indicates that the state is, if anything, retreating from its duty to respect, protect and fulfil the human right to food and emergency food provision is assuming the responsibility to fulfil this right, where it can and in its own way. Whilst the rise of emergency food projects could well represent the increasing responsibility held by civil society-based social protection (within the context of a pulling back of the state's role in welfare provision), the right to food approach sets out clearly that the state is the duty-bearer. Shifts from entitlement to charity (which is not a right and accessible to all) is a particularly problematic aspect of the contemporary shift in charitable food-based social protection from a right to food perspective. The rise of food charity is therefore problematic from a right to food perspective in terms of the charities' ability and appropriateness to fulfil the right to food.

Whilst progress may not currently be being made, nonetheless rights approaches could still be suited to contemporary policy making in the UK for two key reasons. Firstly, it is well suited to current policy-making processes which incorporate multiple actors and interests. Policy network analysis highlights the ways in which policy making is conducted through informal networks which involve complex interplay between ministers, civil servants, pressure and interest groups and many others in the process of arriving at particular policies (see Hudson *et al.*, 2007 and Richards and Smith, 2002). The right to food approach fits well within this networked reality and is particularly 'actionable' within it. It is inclusive of the wide variety of actors and groups that have a stake in the agenda and takes account of the complex roles played by each and every one of them. Secondly, and relatedly, the right to food approach helps us to think about and understand the role of a whole range of stakeholders. In practice the state has – for reasons of necessity or ideology – little capacity (or political will) to respond comprehensively by itself to the problem of food poverty. Politically and practically, we are having to face up to the emergence of a much leaner welfare state and an ever-increasing reluctance to interfere with any kind of market. This more networked approach fits this reality in its focus on other actors taking responsibilities alongside the state. The notion of the state as the duty-bearer within this context, then, is particularly helpful. It places accountability with the state, but recognises the role for other stakeholders in actions towards progressive realisation of the right to food.

The right to food approach also sets out a clear strategy for the laws, institutional capacity and policy programmes which need to be put in place to enable progressive work towards the realisation of the human right to food. In the UK this would involve firstly conducting a right to food consultation (McClain-Nhlapo, 2004), followed by establishing legal frameworks and a national right to food strategy. In the whole process, the participation of affected people in decision-making will be key (DeSchutter, 2010). Legal frameworks are also essential, in order for rights to be claimed (Beuchelt and Vircow, 2012) and this will involve both constitutional and framework law. An institutional framework to support the right to food agenda must also be present in order to 'monitor and assess the right to food situation in a country' (DeSchutter, 2010). For the right to food to be put into action after these structures have been put in place, national strategies will also be required, to establish specific programmes and their implementation (DeSchutter, 2010; FAO, 2005).

5. Conclusions

The indications are that food poverty is worsening in the UK. But without agreed definitions and measurements we are very limited in what we can know and how we can compare these UK experiences with other countries (see discussion in Lambie-Mumford and Dowler, 2015). The rise of food charity has occurred in the context of a shrinking welfare state and is concerning from a human rights perspective. The UK should use the pathways this approach sets out and affinity with the policy making process it holds, to pursue more robust policy frameworks which facilitate the realisation of the right to food.

References

Anderson, S.A. (ed.) (1990). Core indicators of nutritional status for difficult-to-sample populations. Journal of Nutrition 120: 1559-1600.

Ashton, J.R., Middleton, J., Lang, T. (2014). Open letter to Prime Minister David Cameron on food poverty in the UK. The Lancet 383: 9929, 1631.

Beatty, C., Fothergill, S. (2013). Hitting the poorest places hardest: the local and regional impact of Welfare Reform. Centre for Regional Economic and Social Research Sheffield Hallam University: Sheffield Hallam University.

Beuchelt, T.D., Vircow, D. (2012). Food sovereignty or the human right to food? Which concept serves better as international development policy for global hunger? Agriculture and Human Values 29: 259-273.

Cabinet Office (2010). Building the Big Society, London. Available at: http://tinyurl.com/o5f5fb6.

Caraher, M., Dixon, P., Lang, T., Carr-Hill, R. (1998). Access to healthy foods: part I. Barriers to accessing healthy foods: differentials by gender, social class, income and mode of transport. Health Education Journal 57: 3, 191-201.

Channel 4 News (2014). The truth about food banks: dependency or welfare crisis? Available at: http://tinyurl.com/p8f2b7d.

Coleman-Jensen, A. (2011). Working for peanuts: non-standard work and food insecurity across household structure. Journal of Family and Economic Issues 32: 1, 84-97.

Coleman-Jensen, A., Gregory, C., Singh, A. (2014). Household Food Security in the United States in 2013, ERR-173, US Department of Agriculture, ERR-173: US Department of Agriculture, Economic Research Service.

Committee on Economic, Social and Cultural Rights (CESCR) (1999). Substantive issues arising in the implementation of the international covenant on economic, social and cultural rights: general comment 12 (Twentieth session, 1999) The right to adequate food (art. 11) United Nations: Geneva.

Conservative Home (2012). Food banks ARE part of the Big Society – but the problems they are tackling is not new. Available at: http://tinyurl.com/q8xbr27.

Cooper, N., Dumpleton, S. (2013). Walking the breadline: the scandal of food poverty in 21st century Britain: CAP-OXFAM. Available at: http://tinyurl.com/oqr2p63.

Davis, A., Hirsch, D., Padley, M. (2014). A minimum income standard for the UK in 2014. Joseph Rowntree Foundation: York.

De Marco, M., Thorburn, S. (2009). The relationship between income and food insecurity among oregon residents: does social support matter? Public Health Nutr 12: 11, 2104-2112.

Department for the Environment, Food and Rural Affairs (DEFRA) (2006). Food security and the UK: An Evidence and Analysis Paper, DEFRA: London.

Department for the Environment, Food and Rural Affairs (DEFRA) (2014). food statistics pocket book: in year update. London. Available at: http://tinyurl.com/o83y78u.

DeSchutter, O. (2010). Countries tackling hunger with a right to food approach. Briefing note 01, May 2010. Available at: http://tinyurl.com/pvh4euo.

Dowler, E. (1997). Budgeting for food on a low income in the UK: the case of lone-parent families. Food Policy 22: 5, 405-417.

Dowler, E., Kneafsey, M., Lambie, H., Inman, A., Collier, R. (2011). Thinking about 'food security': engaging with UK consumers. Critical Public Health 21: 4, 403-416.

Dowler, E., Lambie-Mumford, H. (2015). How can households eat in austerity? Challenges for social policy in the UK. Social Policy and Society 14, 417-428.

Dowler, E., Turner, S.A., Dobson, B. (2001). Poverty bites: food, health and poor families. Child Poverty Action Group.

Dowler, E., O'Connor, D. (2012). Rights-based approaches to addressing food poverty and food insecurity in Ireland and UK. Social Science and Medicine 74: 44-51.

Dugan, E. (2014). The real cost-of-living crisis: Five million British children 'sentenced to life of poverty thanks to welfare reforms'. The Independent. Available at: http://tinyurl.com/nz4jzck.

Ellison, M., Fenger, M. (2013). Social investment, protection and inequality in the new economy and politics of welfare in Europe. Social Policy and Society 12: 4, 611-624.

Environment, Food and Rural Affairs Committee (EFRA) (2015) Sixth Report. Food security: demand, consumption and waste, Hasard. Available at: http://tinyurl.com/onrmcqx.

FareShare (2014). Our History. Available at: http://www.fareshare.org.uk/our-history/.

FareShare (2015). FareShare Homepage. Available at: http://www.fareshare.org.uk/.

Farnsworth, K. (2011). From economic crisis to a new age of austerity: the UK. In: K. Karnsworth and Z. Irving (eds.). Social policy and challenging times: economic crisis and welfare systems, Policy Press: Bristol, UK.

Food and Agriculture Organization (FAO) (2005). Voluntary guidelines to support the progressive realisations of the right to adequate food in the context of national security. FAO: Rome. Available at: http://tinyurl.com/pynyxrt.

Food and Agriculture Organization (FAO) (2006). Food Security Policy Brief. Available at: http://tinyurl.com/q4k5rn8.

Food Poverty Inquiry (2014). Feeding Britain: a strategy for zero hunger in England, Wales, Scotland and Northern Ireland. The report of the All-Party Parliamentary Inquiry: The Children's Society.

Hansard (2013). 'Food banks' debate. Available at: http://tinyurl.com/lk64rv7.

Hirsch, D. (2013). A minimum income standard for the UK in 2013, Joesph Rowntree Foundation: Joseph Rowntree Foundation. Available at: http://tinyurl.com/plac742.

Hitchman, C., Christie, I., Harrison, M., Lang, T. (2002). Inconvenience food: the struggle to eat well on a low income. Demos: London.

Hudson, J., Lowe, S., Oscroft, N., Snell, C. (2007). Activating Policy Networks: A case study of local environmental policy-making in the United Kingdom. Policy Studies 28: 1, 55-70.

Joint Committee on Human Rights (2004). The International Covenant on Economic, Social and Cultural Rights, in H. o. L. p. 183 and H. o. C. p. 1188 (eds.): Houses of Parliament.

Kirkpatrick, S.I., Tarasuk, V. (2011). Housing circumstances are associated with household food access among low-income urban families. Journal of Urban Health 88: 2,284-296.

Lambie-Mumford, H. (2013). every town should have one: emergency food banking in the UK. Journal of Social Policy 42: 1, 73-89.

Lambie-Mumford, H. (2014a). The right to food and the rise of charitable emergency food provision in the United Kingdom. University of Sheffield. Available at: http://etheses.whiterose.ac.uk/7227/.

Lambie-Mumford, H. (2014b). Food bank provision & welfare reform in the UK. SPERI British Political Economy Brief No.4. Available at: http://tinyurl.com/ksynohh.

Lambie-Mumford, H. (2015). Britain's hunger crisis: where's the social policy? Social Policy Review 27, Policy Press: Bristol.

Lambie-Mumford, H., Crossley, D., Jensen, E., Verbeke, M., Dowler, E. (2014). Household food security: a review of food aid: Defra. Available at: http://tinyurl.com/qamk9rq.

Lambie-Mumford, H., Dowler, E. (2014). Rising use of 'food aid' in the United Kingdom, British Food Journal 116: 9, 1418-1425.

Lambie-Mumford, H., Dowler, E. (2015). Hunger, food charity and social policy – challenges faced by the emerging evidence base. Social Policy and Society 14, 497-506.

Lister, R. (2004). Poverty. Polity Press: Cambridge.

Loopstra, R., Reeves, A., Taylor-Robinson, D., Barr, B., McKee, M., Stuckler, D. (2015). Austerity, sanctions, and the rise of food banks in the UK. British Medical Journal 350: h1775.

Loopstra, R., Tarasuk, V. (2013). What does increasing severity of food insecurity indicate for food insecure families? Relationships between severity of food insecurity and indicators of material hardship and constrained food purchasing. Journal of Hunger and Environmental Nutrition 8: 337-349.

McClain-Nhlapo, C. (2004). Implementing a human rights approach to food security. International Food Policy Research Institute, Washington DC.

McGlone, P., Dobson, B., Dowler, E., Nelson, M. (1999). Food Projects and How They Work. Joseph Rowntree Foundation: York. Available at: http://www.jrf.org.uk/sites/files/jrf/1859354165.pdf.

Meah, A. (2013). Shopping. In: P. Jackson (ed.). Food words. Bloomsbury: London.

Moore, S. (2012). 2012 has been the year of the food bank. The Guardian. Available at: http://tinyurl.com/o7kfqgo.

Morris, N. (2013). Hungrier than ever: Britain's use of food banks triples. The Independent. Available at: http://tinyurl.com/pkevuok.

Office of the High Commissioner for Human Rights (OHCHR) (1996). International Covenant on Economic, Social and Cultural Rights. Available at: http://tinyurl.com/qxqfpj5.

Oxfam (2013). Food Poverty in the UK. Available at: http://tinyurl.com/nmxdyrr.

Padley, M., Hirsch, D. (2014). Households below a minimum income standard: 2008/9 to 2011/12, Joseph Rowntree Foundation: York. Available at: http://tinyurl.com/otkcjy8.

Perry, J., Sefton, T., Williams, M., Haddad, M. (2014). Emergency use only: understanding and reducing the use of food banks in the UK. Oxfam, UK. Available at: http://tinyurl.com/okdzl4m.

Poverty and Social Exclusion (PSE) (2012). Going backwards: 1983-2012. Available at: http://tinyurl.com/qez79bs.

Register of All Party Parliamentary Groups (2014). Hunger and food poverty. Available at: http://tinyurl.com/r5maqa.

Richards, D., Smith, M.J. (2002). Governance and public policy in the UK. OUP: Oxford, UK.

Riches, G. (1997). Hunger and the welfare state: comparative perspectives. In: G. Riches (ed.). First world hunger: food security and welfare politics. Macmillan Press: Basingstoke.

Riches, G., Silvasti, T. (eds.) (2014). First world hunger revisited: food charity or the right to food? Palgrave Macmillan: Basingstoke.

Save the Children (2012). Child poverty in 2012: it shouldn't happen here. Save the Children: Manchester. Available at: http://tinyurl.com/cuv763w.

Shelter (2013). 4 out of 10 families cut back on food to stay in their homes. Available at: http://tinyurl.com/pj5hlm4.

SR Food (2015). Special rapporteur on the right to food. Available at: http://tinyurl.com/oyflz8q.

Taylor-Gooby, P., Stoker, G. (2011). The coalition programme: a new vision for Britain or politics as usual? The Political Quarterly 82: 1, 4-15.

Trussell Trust (2015). Foodbank statistics. Available at: http://www.trusselltrust.org/stats.

United Nations (UN) (1948). The Universal Declaration of Human Rights.

United Nations (UN) (2015). Chapter IV Human Rights: 3 International Covenant on Economic, Social and Cultural Rights. Available at: http://tinyurl.com/ng533sw.

31. The paradox of scarcity in abundance: the contribution of food banks against poverty in Italy

G. Rovati

Università Cattolica del Sacro Cuore, Largo Gemelli 1, 20122 Milan, Italy; giancarlo.rovati@unicatt.it

Abstract

2015 is the target year set by the UN to achieve the Millennium Development Goals, but poverty persists even among the more fragile sections of populations in developed countries. Among others, Europe is also experiencing the paradox of scarcity in abundance. Food insecurity goes hand in hand not only with the problem of food waste, but also with food overproduction (surplus) for market reasons. Food surplus and waste are morally unsustainable, but a number of innovators have transformed this contradiction into a positive opportunity for the needy in the form of food banks. These are non-profit organisations specialised in recovering surplus products generated in the agri-food chain and redistributing them to persons and households in need. There are currently 264 active food banks working in Europe within the *Fédération Européene des Banques Alimentaires*. In Italy the Food Bank Foundation (FBF) has been working since 1989 and currently supports thousands of non-profit organizations that every day help poor people to extend access to food security. To fulfil its solidarity objectives, the FBF recovers food surplus produced by agri-food companies, large-scale distribution, and the food services sector. It also organises a National Food Collecting Day, when millions of people donate food while shopping for their groceries. FBF was national partner for the managing of *Programme Européen d'Aide Alimentaire aux plus Démunis* and for the new program Fund for European Aid to the most Deprived. FBF – together with five other national non-profit organizations – participates in a subsidiary public policy against food poverty. In the paper I will analyse, first of all, the incidence of food poverty in Italy during the last decade using official statistics; secondly the contribution of food supplies provided by FBF network according to empirical data collected by a recent sample research on 9,000 charitable organizations that assist the poor and needy across the country.

Keywords: food poverty, food bank, charitable organizations, non-profit organizations, third sector

1. Introduction

With the United Nations Millennium Declaration world leaders committed their nations to reduce 'extreme poverty', in particular halving the proportion of people whose income is less than $ 1.25 and halving the proportion of people who suffer from hunger by the target date of 2015. However, both relative and absolute poverty still exist even among the most fragile sections of populations in developed countries. This part of populations – which amounts to 123 million people in the EU – takes part in a 'denied feast' because they cannot have access to food in a quantity and quality sufficient to satisfy their dietary needs in a balanced way (Rovati, 2015). As the USA, also Europe is experiencing the paradox of scarcity in abundance, so definable because abundance should make it possible to defeat scarcity more easily than in developing countries, and because massive public investment is made in counter policies.

Leire Escajedo San-Epifanio and Mertxe De Renobales Scheifler (eds.) *Envisioning a future without food waste and food poverty*
DOI 10.3920/978-90-8686-820-9_31, © Wageningen Academic Publishers 2015

For this part of the population the scarcity of food goes hand in hand not only with the food *waste* generated by the majority of the population, but also with the overproduction and excess of supply generated by food companies *(surplus food)* due to many different reasons: for instance, (predicted) margins of error during the production and packaging process, and risks connected to the launch of innovative products to beat the competition and to satisfy the changing tastes of consumers (Garrone *et al.*, 2012).

The phenomenon of surplus food and food waste would appear even more absurd and morally unsustainable if some innovators did not transform this contradiction into a positive opportunity for people in need and for the entire social and economic system, establishing Food Banks, i.e. non-profit organisations specialised in the recovery of surplus food generated in the different stages of the agri-food chain (agriculture and farming, production, distribution and catering) and in their redistribution to people and households in need. Food Banks are a social invention that transforms surplus food with zero market value into goods with a positive value for people who lack food. The well-known challenge-response debate that is at the basis of all innovation, has, in practice, been transformed into a constructive solution for *all*, thanks to the sense of solidarity and justice of a *few*.

The world's first food bank (St. Mary's Food Bank) was created in 1967 in the USA where today the Feeding America network (2014), through 200 members food banks and 60,000 food pantries and meal programs, provides food to more than 46 million poor people. In Europe the same experience started in 1984, where the first food bank was established in France and today the European Federation Food Banks brings together 264 food banks in 22 European countries. The first Italian food bank (*Fondazione Banco Alimentare Onlus* – FBAO) was founded in 1989 and it daily operates at national level through a network of 21 regional non-profit organizations, creating the Food Bank Network (FBN).

2. The spread of food poverty in Italy

The annual EU-SILC survey (European Union Statistics on Income and Living Conditions), launched in 2003 and available up to the 2013 edition (Istat, 2014b) provides an estimation of food scarcity experienced by Italian populations in comparison with other European countries. In 2013, 19% of people living in Italy were at risk of poverty, 12.4% lived in a state of serious material deprivation (with at least four deprivation indicators out of a total of nine), and 11% lived in households with very low work intensity (families with people aged 18-59 that worked less than 20% of their total work potential). The synthetic indicator of the risk of poverty and social exclusion, which considers those who have at least one of these conditions as vulnerable, is 28.4%. The highest values of risk of poverty or social exclusion were recorded among residents in the south of Italy (46.2%), among large households (39.8%), single-wage families (46.1%) and those with three or more children (43.7%), especially minors (45.4%). It should be pointed out that, on a national level, in 2013 the percentage of people who declared that they could not afford a decent meal at least every two days reached 14.2% (Istat, 2014). Italy has a synthetic index value of about 4% above the European average (24.5%), below Croatia (32.3%), Hungary (33.5%), Lithuania (30.8%), Greece (34.6%), Latvia (35.1%), Romania (40.4%), Bulgaria (48%).

A more objective estimation of 'food poverty' (i.e. a lack of decent nutrition) was collected by Istat (2009) with a new measuring technique for *absolute poverty*, where the food component is accurately estimated bearing in mind the different costs of living according to the different types of households and the different areas (large, medium or small municipalities in different regional macropartitions). Even absolute poverty is thus relative to the economic and social context of developed countries and does not coincide with the physical subsistence level as opposed to the concept of 'extreme' poverty (i.e. a threshold equal to $ 1.25 per capita per day) used at global level in the context of the Millennium Development Goals (UN, 2014).

During the last few years the impact of absolute poverty has substantially increased from a national average of 4% (during the period 2005-2007) to a peak of 7.9% with a general worsening in all the macro area divisions (Figure 1, North, Centre, South; Istat, 2014a). In practice, 2,028,000 families, equal to a total of 6,020,000 individuals (10% of the entire population), lived in conditions of absolute poverty in 2013. Variations between individuals are even more marked, following the sharp increase in absolute poverty in larger families, especially those with minors. Almost half the people in absolute poverty live in the south of Italy (3,072,000), 1,058,000 are minors, 728,000 are elderly, almost 1,100, 000 are members of families where the head of the family retired from work, 1,506,000 are members of working class families, and 764,000 belong to households where the head of the family is unemployed.

Among families who live in absolute poverty, food represents a high quota of necessary purchases, but nevertheless it is often relegated to second place in order to cope with the more inflexible expenses of the home and the utilities, with serious nutritional consequences. The negative impact of this behaviour is attenuated by the possibility of obtaining food assistance both from charitable organizations and public institutions, thus satisfying both nutritional needs and the need to allocate the scarce monetary resources to other equally urgent purposes. This explains the increased demand

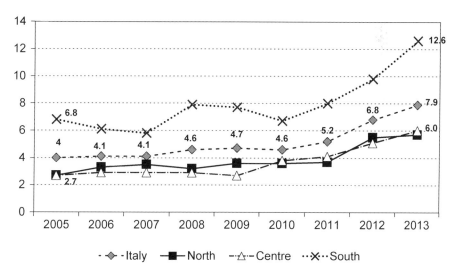

Figure 1. Incidence of absolute poverty in Italy (2005-2013).

for food assistance in the last two years from increasingly more varied categories of people, who have the common need to find a compensation for their insufficient income and supplement their deficient diet.

3. The contrast to food poverty by Italian Food Bank Network

In Italy over the years more than 15,000 non-profit organizations have been working to provide access to food security to people in need. Each one of these organizations operates independently and has established direct relations with persons in difficulty, often responding to their needs in collaboration with the public institutions responsible for social assistance policies. Since 1992 these organizations have been supported by the European Food Distribution Programme for the Most Deprived Persons of the Community and recently by the new Fund for European Aid to the most Deprived, which is managed by *Agenzia per le Erogazioni dell'Agricoltura* and of second level non-profit organizations among which there is also FBAO-FBN (AGEA, 2013).

The origins and the daily activity of the FBAO aim to answer two complementary problems: on the one hand food waste generated by the consumer society, and on the other hand food poverty of people in need, living in the same society. The operational activity of the FBAO and its national network (FBN) is focused to the recovery of surplus food and its redistribution to charitable organizations and associations that operate in Italy for the benefit of the poor and marginalized. Since its origins the FBAO has always conceived itself as a structure at the service of organizations that directly assist people in need. It operates as an intermediary between the major producers of surplus food and charitable organization that are widely present in small and large towns of our country.

Over the years the FBAO has steadily increased the amount of recovered and distributed food as well as the number of charitable organizations, which directly assist people in need, the number of people in need, and the variety of collected and donated food, including baby food products, which are as required and difficult to find. In 2000 the FBN supported 5,416 charitable organizations that provided food aid to 952,199 most deprived; in 2014 the number of charitable organized increased to 8,669 and they assisted 1,909,986 most deprived (Figure 2).

Food products recovered by the FBN are distributed to people in need following two ways: the distribution of *food parcels* and the transformation into *meals* provided by soup kitchens and daily or residential shelters with different typologies of interventions. In addition to the recovery of surplus food, the FBAO works to raise awareness about the issue of food poverty that affects many Italians and foreigners living in our country. In order to reach this goal in 1997 the FBAO started to organize the National Food Collection Day, an important event that takes place every year on the last Saturday of November. On 30 November 2014 in 11,182 supermarkets 135,000 volunteers collected 9,201 tons of donated food by 5.5 million citizens of different ages and classes.

It is important to underline that food products coming from European programmes represent only a part of the food that is recovered and distributed by the FBN. In 2013 the FBN recovered and distributed 72,654 tons of food products, of which 42,000 tons (58%) came from the EU, while the remaining came from surplus food generated by the production sector (21%), the distribution sector (7%), and from the National Food Collection Day (12%) and other initiatives (2%).

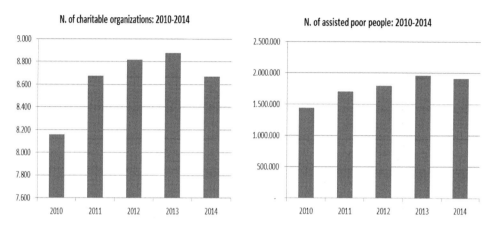

Figure 2. Italian Food Bank Network (2010-2014).

Siticibo is a further development of the subsidiary 'network' created by the FBAO, charitable organizations and public institutions. It is a programme for the recovery and redistribution of surplus food that is daily generated in the catering sector (hotels, hospitals, company and school canteens, retailers, etc.) and in the distribution sector, where large amounts of fresh food products and cooked meals remain unsold and therefore wasted when the demand for food aid from people in need continues to increase. The recovered products include first and second courses, side dishes as well as fresh foods like fruit and vegetables, bread and pastries that are delivered and consumed by charitable organizations within a few hours. The activity is possible thanks to volunteers and through a logistics network of vans equipped to transfer food surplus where the need is more urgent.

These figures demonstrate the efficiency and effectiveness of the activity of the FBN, but at the same time confirm the huge spread of food needs of the poorest part of the population in Italy.

A recent sample survey of 250 charitable agencies affiliated with FBN (Rovati and Pesenti, 2015) documented that 70% of them provides not only food aid to the needy, but also support services to job search, social counselling, medical and pharmaceutical assistance. Therefore most charities are taking charge (at least partially) of personal and familiar situations even complex. Among the poor assisted by the Food Bank Network prevail those in adulthood (18-64 years: 56%) coinciding largely with 'working poor'; however, young poor people (up to 17 years: 29%) exceed old poor people (from 65 years: 15%), and a large proportion of younger (12%) consists of children up to 5 years.

4. Conclusions

While it cannot be denied that anti-poverty policies require, *at the financial level*, a massive commitment by governments and public administrations, *at the operational level*, the most effective organizations are the non-bureaucratic ones that have been released from social control and are able to manage personalised relationships on a universalistic basis. A symbolic example of these dynamics comes from the USA experience of the Food Stamp Program and the WIC (special supplemental nutritional program for women, infants and children), which have a different degree of adhesion according to whether the agencies that provide the assistance are public or third sector organisations. The main difference lies in the centralization of the personal relations of the aid agencies, which makes it possible to evaluate the nature of the need and assistance carefully and accurately, the material aspect of which is only a part.

In practice, it is possible to prefigure a division of roles in function of the objectives pursued and the available tools. The public sector can set itself the objective of effectively reducing the risk of monetary poverty, focussing monetary grants on families, children and the disabled. The non-profit sector and social agencies can, instead, set the objective of reducing material and psycho-social deprivation on the basis of an organisation model that is different from both the public and the profit sectors.

Time, i.e. the timeliness during which the various forms of prevention, assistance and promotion come into effect, is a determinant factor for the effectiveness of social policies and individual interventions. The paths into poverty can be portrayed as an inclined plane that becomes increasingly steeper and more slippery the longer the time between the start of the poverty and the intervention of effective aid. If we slip too far down, it is easy to remain trapped and the ascent becomes more difficult and unlikely. This gives rise to the need to provide emergency services able to deal with and eliminate the difficulties that individuals and families are not able to cope with alone. The charitable organizations that distribute food aid play an active and key role in this race against time, together with the many formal and informal solidarity groups that enrich both morally and socially our society.

References

AGEA (2013). Piano di distribuzione degli alimenti agli indigenti 2013. Sintesi del consuntivo delle attività realizzate al 30-04-2013, Roma. Available at: http://tinyurl.com/oy5p4yl.

Campiglio, L., Rovati, G. (eds.) (2009). La povertà alimentare in Italia. Guerini e Associati, Milano.

Feeding America (2014). Bringing hope to 46 million people. 2014 Annual Report.

Garrone, P., Melacini, M., Perego, A. (eds.) (2012). Dar da mangiare agli affamati. Le eccedenze alimentari come opportunità. Guerini e Associati, Milano.

ISTAT (2009). La misura della povertà assoluta. Metodi e Norme n. 39, Roma.

ISTAT (2014a). La povertà in Italia. Anno 2013, Roma 14 luglio.

ISTAT (2014b). Reddito e condizioni di vita. Anno 2013, Roma 16 dicembre.

Rovati, G. (2015). The denied convivium: the paradox of scarcity in abundance. Vita e Pensiero, Milano, pp. 219-233.

Rovati, G., Pesenti, L. (eds.) (2015). Food poverty, Food bank. Aiuti alimentari e inclusione sociale. Vita e Pensiero, Milano.

UN (2014). The Millennium Development Goals Report, New York.

32. Food donation: Leioa catering and hospitality college and Lagun Artean, an example of cooperation

C. Ascorbe Landa

Department Cooking and Baking, Leioa Catering and Hospitality College, Campus Universitario, Leioa, Spain; cristinaascorbe@hostelerialeioa.net

Abstract

Located on the Leioa University Campus, the state secondary school 'IES Escuela de Hostelería de Leioa' (Leioa Catering and Hospitality College) offers courses on hospitality and tourism, and the food industry. Practical experience is an essential part of its education programme: students and teachers prepare around 1,200 menus every day, which are served to the educational community on campus. Just like many other restaurants, the Catering and Hospitality College used to throw away any food that could not be reused in later sittings, either because the food was perishable or for organoleptic reasons. 2010 saw the advent of the idea to seek out a contributing partner with whom the surplus portions could be shared. In March of that same year, the college contacted Lagun Artean, a centre promoting social reintegration for homeless people that needed ready-to-eat food for people staying overnight on its premises. Given the mutual benefit of the initiative, a cooperation agreement was drawn up with clearly specified food safety conditions for both parties. Since signing the agreement, Lagun Artean has made daily collections, within the school calendar, of any portions that Leioa Catering and Hospitality College cannot reuse, thanks to which both organisations add value to their services: the Leioa Catering and Hospitality College meets the social challenge of not wasting food, whereas Lagun Artean has set up a service providing assorted high quality, nutritious dinners for the 32 homeless people who use its shelter on a daily basis. This initiative has had a positive impact in the city of Bilbao. Since they became aware of this good practice, Bilbao City Council and the Hontza Cáritas Centre have incorporated dinner services into their projects for people experiencing homelessness.

Keywords: food waste, food donation, collaboration agreement, good practice

1. Introduction

Food management can be improved by handing over surplus food wasted by the industry to organisations that distribute it to people in need. Following this good practice, the Leioa Catering and Hospitality College delivers 32 complete menus every day to Lagun Artean, a non-profit-making association that distributes them among homeless people or anyone at great risk of social exclusion.

2. Leioa Catering and Hospitality College

The Catering and Hospitality College is a public Vocational Training centre, depending on the Basque Government. This secondary school mainly gives courses for middle grade (sixth form) vocational training cycles: cooking and gastronomy; catering services; and bread-making, baking and confectionery; plus higher grade (further education) training cycles: kitchen management; and

Leire Escajedo San-Epifanio and Mertxe De Renobales Scheifler (eds.) *Envisioning a future without food waste and food poverty*
DOI 10.3920/978-90-8686-820-9_32, © Wageningen Academic Publishers 2015

service management. We also provide 'à la carte' training for hospitality staff from different public organisations such as the Bizkaia Local Government and Cruces Hospital and work in partnership with Lanbide (Basque Employment Service). More than 600 students are attending the college in the 2014-2015 academic year.

Brief history and location

This Hospitality School first opened in Plentzia during the 81-82 school year, providing grade II vocational training in the hospitality and tourism branch, specialising in hospitality. The philosophy behind vocational training recommends work experience as the only way of teaching an activity such as cookery so the University Campus was considered for the school's definitive location in 1983.

We believe that moving to the University of the Basque Country (UPV/EHU) campus was important and decisive in terms of defining the centre's character as it offered a series of advantages that have played a decisive role in how the school evolved. These advantages are as follows:
- improved teaching quality;
- better resources: facilities and primary materials;
- the customers provide a real reference point for students;
- joint work with different faculties on areas of common interest;
- our students form part of university life;
- high student capacity. It is the largest catering college in the Basque Country.

After long negotiations with the UPV/EHU and the Department of Education, the 2003/2004 school year began in the current building. This represented a great qualitative change in terms of infrastructures and equipment for the classrooms, workshops and dining rooms, putting us at the cutting edge not only within hospitality teaching but also in terms of catering companies.

These new facilities have helped us earn the ISO 9001, 14001 and the Q for Tourism Quality as well as setting up a self-monitoring hygiene system.

Centre's educational project: work experience

The students from this school undertake practical work experience as they make and serve over 1,200 menus every day, supervised by the teaching staff. We do not simply teach students how to do something at the School; they are learning by actually doing it.

This methodology is applied by fitting our students' training (practical subjects such as preparation, culinary techniques, bar-café services, bread-making, etc.) into the real organisation system of the food industry (sections for preparation, bread and cake baking workshops, industrial cookery, à la carte cooking, canteens, etc.)

The food made by the work experience students using these facilities would have to be thrown away if we were not trusted on a daily basis by the students and employees of the Leioa University Campus.

Social challenges

The Leioa Catering and Hospitality College practices good management of its surplus food and teaches students to 'reuse' leftovers. However, some dishes have to be thrown away either because of their expiry date (meeting the self-monitoring hygiene system) or due to organoleptic issues. In 2010, we got to work and tackled the social challenge of finding a collaborating partner who needed food. There is where Lagun Artean came in.

3. Lagun Artean

Lagun Artean is a non-profit-making association, connected to the Bilbao Diocese Caritas branch, that supports social incorporation processes for homeless people or anyone at serious risk of social exclusion.

Brief history and location

Lagun Artean began working in 1983 in the Bilbao neighbourhood of Deusto, promoted by a group of people from La Pasión Parish.

Lagun Artean's goals are as follows:
- Offering homeless people, or anyone at risk of social exclusion, a place of shelter, a reference point and somewhere to belong to, working for full social incorporation.
- Working for change and speaking out against the structural causes that lead to social exclusion.
- Promoting active participation and empowerment of people living in exclusion.

Lagun Artean currently has 20 employees and 70 volunteers and it offers the following resources and services:
- A hostel for 32 homeless people to spend the night. It offers the following services: accommodation, laundry, bag drop, showers, breakfast, dinners and social support. The dinners are provided by the Catering College. This over-night centre is connected to the Social Emergency Social Service (SMUS), depending on the city council, that centralises and manages all requests for accommodation, food and clothing from homeless people in the city of Bilbao.
- Open services: from Monday to Sunday, Lagun Artean opens its doors for anyone on the street who wishes to come for breakfast. In addition, from 2 to 4 pm anyone can drop in for coffee, laundry, showers and professional care.
- A day centre that offers personalised support for social incorporation through support in the fields of: training, health, finance, legal, documents, social work-employment, housing, etc.
- 13 sheltered flats with 70 beds caring for different collectives: women on their own or with accompanied minors (5 flats), drug addicts, people with chronic illnesses, social exclusion in general, immigration, etc.

Type of person attending the centre

According to the latest study, a nocturnal survey on 'The situation of persons suffering severe residential exclusion' carried out by Bilbao City Council on the night of 29th to 30th October 2014,

132 persons slept on our city's streets that night. Lagun Artean offers its service to these people. The occupation rate for our hostel is 100% as in the Uribitarte, Elejabarri and Ozanam hostels. This demonstrates that there are insufficient accommodation resources for people living on the streets. There are very few exceptional circumstances where anyone really chooses to spend the night on the street.

The profile of people experiencing homelessness continues to be mainly male, immigrant and young, although there is a significant increase among the native population and the number of women.

Food needs

Before receiving the dinners donated by the Catering College, people staying at the Lagun Artean over-night centre ate in social canteens. At the time, due to lack of space, the SMUS gave out canteen cards for lunch or for dinner but not for both. Consequently, many people arrived without having eaten dinner. Anyone who did have dinner had to walk across Bilbao from the Irala or Indautxu neighbourhoods (where the canteens are located) to Deusto. In parallel, providing these 32 dinners freed up 32 cards for other people in need.

The chance to have dinner where you are going to sleep is not only a better organisational model and more welcoming but above all, it is logical and standardising.

4. Collaboration agreement

In March 2010, the management of the IES Leoia Hospitality School got in touch with the manager of Lagun-Artean. Over the phone, they explained the social challenge they wished to address and invited them to visit the facilities. After analysing technical/organisational aspects, we prepared a collaboration agreement that would satisfy both parties and that, in turn, would avoid hygiene-public health issues.

Main points in the agreement and what actually happens in daily management:
- Food handling certificate. Everyone, both from the Leioa Catering and Hospitality College and from Lagun-Artean who was going to take part in food production, chilling and reheating, transport and/or serving the prepared food had food hygiene knowledge that could be demonstrated by a food handling certificate.
- Temperature control. A daily temperature record was kept monitoring the temperatures for food delivery by the College ($\leq 4\,°C$), the temperature of the food when the temperature-controlled van is opened at the Lagun Artean centre ($\leq 10\,°C$) and the food temperature when served ($\geq 70\,°C$ for hot food and $\leq 10\,°C$ for cold food).
- Delivery days and hours. Surplus meals are picked up on working days at 4pm (once the lunch service has finished) from the Leioa Catering and Hospitality College.
- Transport conditions. Students and teachers put the food in gastronorm trays. Lagun-Artean transports it in its own temperature-controlled container.
- Transport. The trip to the Catering College is made in a van owned by Lagun-Artean. Although a worker from the association initially did this job, it is now usually people on community service programmes who cover this daily transport, normally accompanied by a centre user.

- Cleaning and disinfection conditions. Lagun-Artean promises to keep the van, the temperature-controlled container and the gastronorm trays clean and disinfected to rule out any possible food contamination issues.
- Maintenance. Once the temperature-controlled containers have arrived at the hostel, the food temperature is checked and noted down in the temperature record and the trays are put into the cold room until dinner time.
- Consumption. Lagun Artean promises to throw away food if its temperature is >10 °C on opening the temperature-controlled container. Lagun Artean promises to serve the food on the same day it is delivered for the tea-dinner service. Under no circumstances will it be frozen or reheated for a later service. The dinner is served progressively by the actual users and volunteers from 7.30 pm onwards.

5. Social benefits

The benefits of this best practice have been felt ever since we began working together. Some of the most outstanding advantages are:

Benefits for the Leioa Catering and Hospitality College:
- No longer wasting food and helping to eradicate food poverty.

Benefits for Lagun-Artean:
- 32 homeless people have dinner every night, the main meal of the day for many of them. When the School is on holiday, the actual centre organises making dinners, obviously not as elaborate as dinners from the Catering College but a meal is served every night and the users help to make it.
- It can put money and energy into other activities that are more necessary for its organisation.
- It makes it possible to do work benefiting the community.

Benefits for the city of Bilbao:
- The most interesting aspect of this experience is that after approximately one year the rest of the municipal hostels in Bilbao set up the same model and began to serve dinner, using two employment enterprises: Peñaskal Foundation and Lapiko.

6. Conclusions

Small scale, local initiatives help to avoid waste and food poverty. Food donation, as carried out on a daily basis by the Leioa Catering and Hospitality College, is shown as a best practice that works and can be copied. As a consequence, the school's educational community obtains social benefits, along with Lagun-Artean users and the other municipal hostels in Bilbao.

33. Reducing food waste through charity: exploring the giving and receiving of redistributed food

E.G. Vlaholias*, K. Thompson, D. Every and D. Dawson
Central Queensland University, Appleton Institute, 44 Greenhill Road, Wayville SA 5034 Australia;
e.vlaholias@cqu.edu.au

Abstract

Food waste is a problem with serious environmental, social, financial, and moral implications. Reducing the amount of food sent to landfill is a key challenge to increase environmental sustainability. Over the past decade the food redistribution sector has grown with the rise of several new food bank and food rescue organisations. These organisations collect excess food and distribute it to welfare agencies that feed people in need. However, it is possible that by diverting food from landfill, a social problem is created or furthered for the people that receive this wasted food. Without first knowing if food redistribution organisations are causing unintentional harm to the recipients, we cannot develop and enhance them as a commercial food waste reduction strategy. There is a scarcity of critical research that explores the donors and recipients' experiences of giving and receiving redistributed food. Using ethnographic research methods of participant observation and interviews, this paper examines the perspectives of the multiple parties involved at each level of food redistribution. The findings of this study will recommend suitable methods to target and motivate people from the food industry to become donors. Furthermore, it will suggest ways to improve the experiences of recipients accessing the redistributed food. As a result, this paper will provide insight to optimise food redistribution organisations, with implications for the future policy and practice of food waste and food poverty interventions.

Keywords: food waste, charity, food security, gift theory, food redistribution

1. Introduction

The vast quantity of food waste produced by the global food system is an issue receiving increased consideration because of its economic, environmental, and social costs (Mason *et al.*, 2011; Papargyropoulou *et al.*, 2014; Reynolds *et al.*, 2014). The social challenges of food waste are entangled with its environmental impacts, and so academic and political discourse surrounding food waste is frequently discussed in relation to food poverty. The United Nations mid-range estimates of population growth suggest that there will be a further three billion mouths to feed by the end of this century (Fox and Fimeche, 2013; Mavrakis, 2014). These predications raise key questions about how to produce enough food to feed a rising population on a planet with finite resources. These questions are then furthered and confounded when the wastage of excess food and surplus resources are observed. Thus, government reports and academic studies recognise that 'the challenge of feeding a growing population includes addressing the issue of food waste' (Mavrakis, 2014).

Food redistribution, also known as food recovery or food rescue, has been recognised as one of the main ways for the food industry to tackle food waste as part of this broader food supply challenge. It is the practice of collecting surplus edible food, which would otherwise be discarded to landfill, and delivering it to charity agencies that feed people in need – the food insecure (Reynolds *et al.*, 2015). The growth of food redistribution over the last decade has been significant. There are currently food redistribution organisations in over 25 countries on six continents (Lipinski *et al.*, 2013; The Global FoodBanking Network Annual Report, 2013). The dominant ideology governing food redistribution in most Western countries, including Australia, is that food redistribution is a win-win situation for all involved. The concept of rescuing food from going to waste and gifting it to people in need has attracted broad approval from political representatives, the food industry, and the general public (Hawkes and Webster, 2000; Midgley, 2013). Gift terminology is used by the media, and articulated in food redistribution organisational brochures, information leaflets, and internet websites (Wells *et al.*, 2014). In these texts phrases such as 'our food donors generosity give', 'nourishing our country', and 'give a little love' are common. They are used to promote donations of surplus food, time, and money beyond marketplace exchange relations, and to preserve and enhance pro-social and pro-environmental behaviours, and to advance community values (Shaw and Webb, 2014).

At first glance food redistribution appears to address both the environmental issue of food surplus, and the social issue of food poverty, however, it may in fact perpetuate or even exacerbate inequality. The efficacy of food redistribution has been debated in other countries, but there still remains a scarcity of Australian research on the positive and negative impacts of food redistribution. Some scholars have criticised the fast development of food redistribution, suggesting that it could be a revealing symptom of our society where growing income inequality is causing some demographics to lack the income required to feed themselves (Booth *et al.*, 2014; Engler-Stringer and Berenbaum, 2007; Mulquin *et al.*, 2000; Riches, 2011; Tarasuk and Maclean, 1990; Michalski, 2003). Other scholars argue that food redistribution highlights the inefficiency of our food system – geared towards over production, with food redistribution assumed to be a 'moral' form of disposal (Poppendieck, 1999, 1994; Reynolds *et al.*, 2012; Tarasuk and Beaton, 1999). It is therefore possible that although donating surplus food benefits the food industry and the environment, it disadvantages the recipients, which it was envisioned to assist. Although these critiques hold merit, they often uncritically adopt a political activist viewpoint without considering the self-representations and self-understanding of the multiple parties involved in each stage of the food redistribution process (Lorenz, 2012). Only by understanding the donors' and recipients' emic perspectives of food redistribution can we explore its impacts, and determine whether providing food surplus to people in need is a socially desirable solution. Thus, there is an urgent need for critical research that explores the donors' and recipients' experiences of giving and receiving redistributed food.

2. Gift theory

Over the last few decades, the increase of theoretical interest in the phenomenon of the gift has intersected discussion around the meanings and implications of gift terminology for various donations, from financial to organ donations (Harrell-Bond *et al.*, 1992; Healy, 2010; Pitt *et al.*, 2002; Tutton, 2002). This paper proposes that gift theory may assist in critically evaluating the impacts of Australian food redistribution organisations, as it enables exploration into the multiple, mutually constituting, and political relationships between the 'givers' and 'receivers'. In particular, Marcel Mauss's seminal work, *The gift* (1990) and Jacques Derrida's *Given time. I. Counterfeit money* (1992) may be useful to examine food redistribution practices.

In Mauss's (1990 [1950]) theory of gift exchange in archaic societies, he questions the notion of 'the free gift', and he argues that the whole concept of the free gift is based on a misunderstanding. In other words, there is no free lunch. Through an extensive study of gift giving across various cultures and history, Mauss argues that all gifts are enchained in a circular system of giving-receiving-reciprocating. According to Mauss, it is integral that every gift is returned or countered to ensure the proper functioning of gift-exchange economies, although the gift transactions do not need to be immediate or equal. For Mauss, gift exchange creates relationships, in which the honour of the giver and receiver are engaged. Thus, failure to return a gift is to refuse to participate in moral and social life. His work is novel in relation to gift giving and charity donations. Mauss highlights that, 'to give is to show one's superiority, to be more, to be higher in rank ... [but] ... to accept without giving in return, or without giving more back, is to become a client and servant, to become small, to fall lower' (1990). Gift exchange therefore operates to promote social solidarity and offers a way to gain prestige or to rise up the social hierarchy (Shaw and Webb, 2014). The implications of a Maussian understanding of food redistribution as a form of gift exchange may mean that food donors receive a sense of pride and satisfaction for doing a good deed (what psychologists refer to as a 'warm glow'). At the same time, the food recipients may express gratitude for the food they receive, but they may also feel a loss of status, a sense of shame, and indebtedness for their gift.

In Derrida's (1992) interpretation of the gift, a gift is given without any need for reciprocation or obligation. In his conception of the gift, altruistic acts are undertaken without any expectancy of an external reward or reciprocation (Shaw and Webb, 2014). Derrida considers the gift as problematic because he finds it impossible to consider gifts as external to economic logic. He argues that even the reception involved in the concept of a gift appears to produce some residue of exchange, such as a value, a symbol, or an intention (Malo, 2012). According to Derrida, however, if there were to be a gift, it must not include any return. He explains, 'if the figure of the circle is essential to economics, the gift must remain *aneconomic*. Not that it remains foreign to the circle, but it must keep a relation of foreignness to the circle, a relation of familiar foreignness. It is perhaps in this sense that the gift is the impossible' (Derrida, 1992). Derrida argues that the gift must remain unrecognised by the donor and recipient in order to be a gift, but this results in a paradox. There appears to be a conflation of altruism with the gift in Derrida's thought. However, aspects of the gift as disinterested and unconditional are present in food redistribution rhetoric, which emphasise the ideal of altruism, as a voluntary act where food donors and volunteers may be able to present their unselfish concern and care for the needs of others.

Mauss and Derrida offer two theories of gift giving that may be relevant to understanding food redistribution practices. However, these theorisations of the gift may not perfectly fit the experiences of the multiple groups of people involved in food redistribution. Where theory and practice do intersect, people's experiences will not be identical or consistent (Shaw and Webb, 2014). The lived experiences of food recipients are likely to be qualitatively different to the experiences of food industry donors, and from the staff and volunteers engaged in various roles surrounding food donation, redistribution, and reception. In this paper, we present the views of the four different groups involved in food redistribution: (1) the food industry donors; (2) the staff and volunteers at the food redistribution organisations; (3) the staff and volunteers at the charity agencies; and (4) the food recipients. The respective groups were asked what food redistribution means to them, what their experience of food redistribution has been like, and how food redistribution has impacted their situation. From these questions and others, our aim is to evaluate the impact of food redistribution from the perspectives of those directly involved, and to consider the extent to which gift language sufficiently captures the multiple experiences of food redistribution. This knowledge is essential because food redistribution may have unanticipated negative consequences that should be countered. At the same time, there may be unidentified benefits that could be developed or enhanced, to support the overall aims of food redistribution organisations.

3. Methods

In this paper we use an ethnographic approach to qualitatively explore the experiences of the multiple parties involved in food redistribution. Participant observation was conducted at two food redistribution organisations in South Australia – a food rescue organisation and a food bank. Participant observation was also conducted at several charity agencies that received from the two food redistribution organisations. In-depth qualitative interviews were conducted with food businesses that donate their surplus food to the food redistribution organisations, with the staff and volunteers of the food rescue, food bank, and charity agencies, and with the food recipients that attended the charity agencies. Full ethics approval was granted for this project. The study commenced in October 2014 and is scheduled for completion in July 2015.

The managers of South Australian food redistribution organisations were contacted to determine whether their organisation was willing to be part of the study. The principal researcher discussed with the managers the most suitable way to recruit participants from the multiple parties involved in food redistribution. It was decided that the researcher would actively participate in various volunteer roles. For example, at the food rescue organisation, the researcher adopted the role of driver's assistant. This role involved being a passenger on the food rescue vehicle and helping to collect, measure, and deliver the redistributed food. At the charity agencies the researcher was involved in cooking and serving meals to the recipients. The food industry donors, staff and volunteers, and recipients were approached and recruited during these periods of observation. The principal researcher approached the food donors and recipients by handing them a flyer and asking for their cooperation. Staff and volunteers were also approached this way, and additionally through the food rescue organisation's monthly newsletter. Interviews were conducted at the charity agencies, the food businesses, or at a location convenient for the recipient (such as a coffee shop). All participants were self-selected. Written consent was obtained for all interviews and pseudonyms are used to protect confidentiality. The interviews lasted between 30 to 120 minutes, and were digitally recorded and developed into transcripts in order to conduct thematic analysis (Braun and Clarke, 2006).

The interview transcripts are currently being read and re-read, and the data is in the process of being coded using NVivo qualitative analysis software. The coding will be checked and discussed with all authors before being analysed thematically. At the conclusion of the study, thematic analysis will be used to explore themes surrounding gift giving, reciprocity, value, altruism, and others. Through exploration of the transcripts meaningful codes specific to all research questions will be produced from the data extracts, and then explored to identify and represent a patterned response of meaning throughout the data set (Braun and Clarke, 2006).

4. The giving and receiving of redistributed food

Previous research on gift giving and donations is biased towards the circulation of inorganic material objects. However, this paper argues that food can be understood as the universal gift. Not every culture gives clothes, shelter, money, and/or other inorganic materials, but throughout history, and across all cultures, there are ancient customs of sharing with strangers (Mauss, 1990 [1950]). Food is richly symbolic of our shared humanity. It is a source of identity, health, and wellbeing, and it can hold deep-rooted connections to culture and communities (Douglas, 1984; Marshall, 1995). When we eat, not only do our bodies respond to the nutrients of the ingredients, but the food also conveys meaning to others and ourselves about who we are. In this way, what we eat, how we eat, when, and why can reflect our collective identity, our material circumstances, and it can also reveal the complex cultural arrangements that exist in the structure of our society and its food system (Kittler *et al.*, 2011). When socio-cultural context of food is understood in this way, it becomes clear that the giving and receiving redistributed food needs to be understood differently to the donation of other inorganic materials.

Furthermore, food is in a constant state of decomposition. As a result, the temporal dimension of the gift giving of food is much more pressing and intense than with other more researched charitable materials, such as financial donation. The majority of research in this field is on money, but there is a particular need to understand how those theories apply to food – an organic, ephemeral material in a perpetual state of decomposition. Alexander and Smaje's (Alexander and Smaje, 2008) study of retail and food redistribution arrangements concerning surplus food and food waste in the UK, indicate the precarious task of donating surplus food. They explain that although food redistribution – where excess is directed to useful ends – appears to be an arrangement that suits all included parties, it can in fact be spoiled by conflicting aims. At each stage, where the food items are transferred from one party to the next, the ownership of either assets (edible food) or liabilities (waste) are similarly transferred. Alexander and Smaje highlight that 'in the case of perishable food, the temporal element separating assets from liabilities is particularly acute' (Alexander and Smaje, 2008). Thus, in order to understand the experiences of giving and receiving redistributed food, it is necessary to consider the material and cultural transformation of this food from edible to inedible, and how the temporal and surplus aspects of food impact upon donations.

5. Conclusions

Food redistribution highlights the societal challenges we face in envisioning a future without food waste and food poverty. The food redistribution sector requires more attention form social researchers and policy makers to examine the strengths and limitations of this sector, so that it does not develop unchecked or avert other upstream preventative actions. Although diverting food from landfill through food redistribution may have benefits for the food industry and the environment, it is possible that a social problem is created for the recipients of this surplus food. Without first knowing if food redistribution organisations are causing unintentional harm to the recipients, we cannot develop and enhance them as a sustainable response to food waste and food poverty. Thus, this paper argued that there is an urgent need for critical research that explores the donors' and recipients' experiences of giving and receiving redistributed food. This paper presented two theories on gift giving, and it then discussed the socio-cultural and temporal aspects of food, which may be relevant to understanding food redistribution exchange practices. Through this theoretically informed understanding of the donors' and recipients' emic perspectives of food redistribution we can explore its impact, and provide insight to determine whether providing food surplus to people in need is a socially desirable solution. As a result, the findings of this study/project will provide insight to optimise food redistribution organisations, with implications for future policy and practice of food waste and food poverty interventions.

References

Alexander, C., Smaje, C. (2008). Surplus retail food redistribution: An analysis of a third sector model. Resources, Conservation and Recycling 52 (11):1290-1298.

Booth, S., Whelan, C., Griffith, C.J., Caraher, M. (2014). Hungry for change: The food banking industry in Australia. British Food Journal 116: 1392-1404.

Braun, V.; Clarke, V. (2006). Using thematic analysis in psychology. Qualitative Research in Psychology 3 (2):77-101.

Derrida, J. (1992). Given time: I. Counterfeit money. University of Chicago Press.

Douglas, M. (1984). Food in the social order: Studies of food and festivities in three American communities. New York: Russell Sage Foundation.

Engler-Stringer, R., Berenbaum, S. (2007). Exploring food security with collective kitchens participants in three Canadian cities. Quality and Health Research 17 (1): 75-84.

Fox, T., Fimeche, C. (2013). Global food: waste not, want not. London, UK: Institution of Mechanical Engineers.

Harrell-Bond, B., Voutira, E., Leopold, M. (1992). Counting the refugees: gifts, givers, patrons and clients. Journal of Refugee Studies 5 (3-4): 205-225.

Hawkes, C., Webster, J. (2000). Too Much and Too Little?: Debates on Surplus Food Redistribution. Sustain London, UK.

Healy, K. (2010). Last best gifts: Altruism and the market for human blood and organs. University of Chicago Press.

Kittler, P., Sucher, G.K., Nelms, M. (2011). Food and culture. Cengage Learning.

Lipinski, B., Hanson, C., Lomax, J., Kitinoja, L.,Waite, R., Searchinger, T. (2013). Reducing food loss and waste. World Resources Institute.

Lorenz, S. (2012). Socio-ecological consequences of charitable food assistance in the affluent society: the German Tafel. International Journal of Sociology and Social Policy 32 (7/8): 386-400.

The Global FoodBanking Network (2013). Making a world of difference: the Global FoodBanking Network Annual Report 2013. Available at: http://tinyurl.com/pqf9ap3.

Malo, A. (2012). The limits of Marion's and Derrida's philosophy of the gift. International Philosophical Quarterly 52 (2):149-168.

Marshall, D.W. (1995). Food choice and the consumer. Springer Science & Business Media.

Mason, L., Boyle, T., Fyfe, J., Smith, T., Cordell. D. (2011). National food waste data assessment: final report. University of Technology, the Institute for Sustainable Futures. Sydney.

Mauss, M. (1990 [1950]). The gift: the form and reason for exchange in archaic societies. Trans. W. D. Halls. London: Routledge Classics.

Mavrakis, V. (2014). The generative mechanisms of 'food waste in South Australian household settings. Flinders University, South Australia.

Michalski, J.H. (2003). The economic status and coping strategies of food bank users in the greater toronto area. Canadian Journal of Urban Research 12 (2): 275-298.

Midgley, J.L. (2013). The logics of surplus food redistribution. Journal of Environmental Planning and Management 57: 1872-1892.

Mulquin, M., Siaens, C., Wodon, Q.T. (2000). Hungry for food or hungry for love? Learning from a Belgian Soup Kitchen. American Journal of Economics and Sociology 59, 2: 253-265.

Papargyropoulou, E., Lozano, R. Steinberger, J.K., Wright, N., Bin Ujang, Z. (2014). The food waste hierarchy as a framework for the management of food surplus and food waste. Journal of Cleaner Production 76: 106-115.

Pitt, L., Keating, S., Bruwer, L., Murgolo-Poore, M., De Bussy, N. (2002). Charitable donations as social exchange or agapic action on the internet: the case of Hungersite. com. Journal of Nonprofit & Public Sector Marketing 9 (4): 47-61.

Poppendieck, J. (1994). Dilemmas of emergency food: a guide for the perplexed. Agriculture and Human Values 11 (4): 69-76.

Poppendieck, J. (1999). Sweet charity? Emergency food and the end of entitlement. New York: Penguin Books.

Reynolds, C.J., Mavrakis, V., Davison, S., Høj, S.B., Vlaholias, E. Anne Sharp, Thompson, K., Ward, P., Coveney, J., Piantadosi, J. (2014). Estimating informal household food waste in developed countries: the case of Australia. Waste Management & Research 32 (12): 1254-1258.

Reynolds, C.J., Piantadosi, J., Boland. J. (2015). Rescuing food from the organics waste stream to feed the food insecure: an economic and environmental assessment of Australian food rescue operations using environmentally extended waste input-output analysis. Sustainability 7 (4): 4707-4726.

Reynolds, C., Thompson, K., Boland, J., Dawson, D. (2012). Climate change on the menu? A retrospective look at the development of South Australian municipal food waste policy. The International Journal of Climate Change: Impacts and Responses 3, 3: 101-112.

Riches, G. (2011). Why governments can safely ignore hunger: Corporate food charity keeps hunger off political agenda. CCPA Monitor 17 (8): 26-27.

Shaw, R.M., Webb, R. (2014). Multiple meanings of 'gift' and its value for organ donation. Qual Health Res:1049732314553853.

Tarasuk, V.S., Beaton, G.H. (1999). Household food insecurity and hunger among families using food banks. Canadian Journal of Public Health 90 (2): 109-113.

Tarasuk, V.S., MacLean, H. (1990). The institutionalization of food banks in Canada: a public health concern. Canadian Journal of Public Health 81 (4): 331-332.

Tutton, R. (2002). Gift relationships in genetics research. Science as Culture 11 (4): 523-542.

Wells, R., Caraher, M., Griffith, C.J., Cavicchi, A. (2014). UK print media coverage of the food bank phenomenon: from food welfare to food charity? British Food Journal 116, 9: 1426-1445.

34. Switching imperfect and ugly products to beautiful opportunities

M. Barba[1] and R. Díaz-Ruiz[2]*

[1]*Espigoladors Social Enterprise, C/Osi 35 3er 5a, 08034 Barcelona, Spain;* [2]*CREDA-UPC-IRTA, C/Esteve Terrades 8, 08860 Castelldefels, Spain; espigoladors@gmail.com*

Abstract

Espigoladors, based in Catalonia, is a social enterprise that fights food waste in a transformative and inclusive manner. We transform imperfect products into opportunities. The aim of the company is to have an impact on three social issues that are relevant nowadays: food waste; the lack of access to safe, healthy, and nutritious food in groups at risk of social exclusion; and the lack of opportunities for these groups. Espigoladors has a social enterprise model that directly affects the dignity of people who are in a critical situation while sensitizing the general public on a hot topic (food waste) in a transformative, participatory and inclusive manner. Our model involves the collection of fruits and vegetables from farms and wholesale markets that are rejected from the market for a variety of reasons, such as aesthetic standards or gluts. We donate part of this produce to organizations that provide food to people in need. The remaining fruit and vegetables are transformed into four types of product – jams, creams, juices and sauces – which are sold under our own brand, *'Es Im-perfect'*. *Es Im-perfect* works with groups at risk of social exclusion to create and to sell these products. We are creating a network of farmers, producers and food companies to work with Espigoladors in the fight against food waste, in which we recollect fruits and vegetables that would otherwise have been thrown out. All the enterprises that belong to the network are allowed to use Espigolador's certificate 'I don't waste: we are part of the solution'. This certificate is intended to give value to these enterprises' commitment and prove that they are aligned with their corporate social responsibilities.

Keywords: social enterprise, recovering fruits, new opportunities, imperfect fruits

1. Introduction

Espigoladors is a social enterprise that prevents food waste by empowering people in risk of social exclusion in a transformative, participatory, inclusive and sustainable manner.

Espigoladors, 'Gleaners' in English, is the name that refers to a practice-art that used to be practiced in the rural sector, a foretime. People with few resources used to come to collect what was left on the ground after harvesting with the permission of the landowner. These people, knelt into the ground, with dignity, to pick up food and transformed it into other products. Nowadays, exacerbated by the economic crisis, it is common to find a new generation of gleaners who glean dumpster and, contrary to prior, these are hidden and often have lost their dignity and self-esteem.

2. The main problem

We face the problem from three different standpoints: (1) food waste; (2) lack of access to nutrition; and (3) lack of opportunities to certain groups of people.

Food waste

Spain is the sixth country of the European Union with the higher amount of food wasted. As noted in the report of the BioIS (2011) 7.7 millions of tones food end up in the rubbish every year. This amount of food could had been consumed or used in alternatives ways. This food is discarded due to improper habits, high quality standards of the companies or poor businesses and citizens' planning. A fact that becomes even more irrational as the social damage caused by the crisis is increasing. In recent years, it has been detected an increasing demand of the soup kitchens' beneficiaries and the entities that work for the inclusion of people at risk of social exclusion and provide access to food are saturated with new requests.

Lack of access to safe, healthy and nutritious food of some vulnerable groups that are in vulnerable situations

The 27% of the Spanish population are at risk of poverty (INE, 2013), which represents approximately 13 million people. The situation affects more than one out of four citizens (specifically to 12,866,431; 27.3% of the population) according to the Idescat data in 2013 (Idescat, 2013). From the fourth study about the state of poverty in Spain, published by the European Anti-Poverty Network and Social Exclusion in Spanish State (Llano, 2015), analysing the evolution of the European Indicator Arope (income combined with the possibilities of consumption and employment) between 2009 and 2013, we highlight the following information on severe material deprivation: it is noted an increase of 38% of suffering people (from 4.5 to 6.2%) between 2009 and 2013, that is, there are nearly 3 million people, 800,000 more than before the crisis, struggling to feed themselves adequately, and who have to delay payments on their home, who cannot heat in winter or people who do not have capacity to cope with unexpected expenses. At the same time, it is noted that 18% of men and 16% of women over 18 suffer from obesity in Spain (INE, 2011-2012) in many cases due to a poor diet.

The lack of opportunities for these groups

Espigoladors focuses on the opportunities of young people who have left the educational system and women over 45 years. Spain is the country of the European Union with higher rate, 23.5%, of young people who do not continue their studies after the first cycle of secondary education, according to the data released by the agency Community statistics (2013). Spain recorded in 2013 the highest dropout rate in the EU. At the same time it is noteworthy that the youth unemployment rate is 51.6% and women over 45 years unemployment is 18% in Spain according to the National Employment Institute.

3. Espigoladors as a solution to tackle food waste: a social enterprise

Espigoladors develops a production model that impacts on and connects three social dimensions: food waste, the lack of access to safe, healthy and nutritious food of some vulnerable groups and the lack of opportunities for these groups. We aim to develop a social enterprise model that influences on the dignity of people who are in a critical situation; and, at the same time, to raise awareness and to promote a change in social awareness towards a culture of food utilization.

We provide a service to producers or companies that generate food surpluses and waste by means of a social and environmental alternative that tackles food waste. Nowadays, Espigoladors operates in the region of Catalonia and has the intention to replicate its model in other parts of Spain.

What if ugly and imperfect fruits and vegetables give opportunities to people at risk of social exclusion and vice versa? Espigoladors converts the hypothesis into reality by the creation of a brand: *Es im-perfect* which elaborates handmade and high quality canned products from food surpluses with the participation of groups at risk of social exclusion.

The social enterprise activity consists on the collection of fruits and vegetables that are discarded either by a decline in sales, for aesthetic reasons, for production surpluses, or because the fruit is ripe and the consumers no longer purchase them. Under no circumstances foods that are unfit for human consumption are collected. All the fruit and vegetables recollected are derive to two destinations: redistribution and transformation. Part of the 'gleaned' food is redistributed directly to local social entities that manage it for people with lack of access to food. The other part is aimed at the transformation to create *Es im-perfect* range of products.

Throughout the whole process, groups in risk of social exclusion are involved with a particular emphasis on women over 45 years old and young people who have left the educational system. The aim is to return them to the educational system and enhance their employment opportunities.

Es im-perfect is the first brand in Spain that sells high quality products elaborated from discarded fruit and vegetables due to their imperfections or market reasons. Groups at risk of social exclusion are involved throughout the whole process, from the gleaning to the processing of the products.

The brand *Es im-perfect* wants to place a value on products that are ugly and imperfect but are equally good, and with the same nutritional value than others. A range of jams, creams, sauces and canned vegetables are produced with close, environmentally friendly raw materials with a high social content. All the products are recommended by a renowned Catalan chef named Ada Parellada.

Es im-perfect brand aims to appraise those imperfect and ugly fruits and vegetables that are equally tasty and with the same nutritional value as 'standard' ones. Our range of products is displayed at the sales point with an exhibitor made from recycled fruit boxes. It broadcasts our roots and our differentiation: proximity, field, quality, social and environmental impact, invitation to be part of social change and a zero waste philosophy.

Gleaning: harvesting discarded fruit and vegetables

The collection of fruits and vegetables is done directly in the food sector companies or by gleaning the fields of producers. All the companies who are part of the Espigoladors' network are recognized by a certification (Figure 1). They could share the label on their websites or in their stores, placing value on their action and commitment in the fight against food waste and aligned with the corporate social responsibility of these organizations and producers. We want that the label to become a reference point in the fight against food waste while inviting and integrating all companies and producers to be part of the solution through a transformative way.

We are developing gleaning programs in two regions of Barcelona (Baix Llobregat, Maresme and Vallés Occidental) to create networks of producers engaged in the fight against food waste.

As they are consolidated, we will open new programs throughout Catalonia in order to replicate them in different Spanish regions. The gleaning program promotes the creation of gleaning communities formed by the beneficiaries who are the receptors of the redistributed food in the local charity entities. Therefore, people at risk of social exclusion are empowered and become active members of the process from the gleaning to receiving the fruits and vegetables and, besides, a safe and healthy diet is encouraged.

According to networking dynamics, Espigoladors is a member of the European network 'Gleaning Network EU' which drives farm gleaning. It is leaded by the Feedback Global NGO and belongs to FUSIONS FP7 Project. The main objective of the network is to promote the gleaning activity and share experiences in different countries of Europe: Belgium, Greece, France and Spain.

Social, environmental and economic impact

The dimensions of our impact are evaluated based on the following metrics:
- Decreasing of fruit and vegetable waste indicator; as well as the decreasing of CO_2 emissions.
- Providing opportunities to people at risk of social exclusion.
- Awareness and population consciousness change about food waste and to a culture of food utilization.
- Redistribution of healthy, fresh fruit and vegetables to social institutions giving access to food for people at risk of social exclusion.

During 2014 a four-month pilot study was launched where the business model was tested. Nowadays (May 2015) we have gleaned 46,805 kg of fruit and vegetables from wholesale sector and at the farm

Figure 1. Espigoladors label.

level. The gleaning program is comprised of 12 local producers. From June on is when larger quantities of waste occur in the fields due to the weather conditions. Our goal is to launch from three to six gleaning communities in the Catalan community these year (2015). In Table 1 we show the evolution of the indicators planned from 2014 to 2016.

Awareness campaign and dissemination

As an organization that promotes a change in society we carry out different seminars and activities to raise awareness and explain the project. One example is the *The Extraordinary Potato* campaign. It is an awareness campaign aimed at workers in enterprises and public administration. With *The Extraordinary Potato* action Espigoladors wants to invite participants to be part of the solution through a direct action with high social and environmental impact.

The aim of the campaign is to bring one kilo of extraordinary potatoes to workers with the history of the potato and the slogan: 'I am also im-perfect'. A challenge is proposed to the workers: change the potato destiny by proposing different recipes to do with the potatoes. The organizations promote an act of consciousness in which spread a culture of utilization of food, healthy eating and healthy and ethical values among workers.

Table 1. Indicators' evolution.

Indicator	2014	2015 (expected)	2016 (expected)
Fruit and vegetable recovered	22,000 kg	60,000 kg	130,000 kg
CO_2 savings[1]	83,600 kg	228,000 kg	456,000 kg
Economic value recovered	€ 12,250	€ 33,400	€ 66,800
Donated food	18,200 kg	44,000 kg	105,000 kg
Processed food	3,800 kg	16,000 kg	25,000 kg
Amount of 'Es-imperfect' products elaborated	3,000 jars	15,000 jars	23,000 jars
Groups participating	5 people	15 people	25 people

[1] CO_2 equivalent is based on UK figures and data are taken from the study by WRAP/WWF: http://tinyurl.com/qxlmst3.

4. Conclusions and future

Espigoladors presents an innovative and transformative model for several reasons:
1. An alternative model of donating food is proposed. This new way of offering an alternative that dignifies and promotes social change where participants retrieve the value of food as a resource and not as a residue. And also it generates opportunities for people who are at risk of social exclusion.
2. It is an action that aims to promote and provide access to safe and healthy food to all kind of people involved with positive connotations.

3. The high quality brand creation that uses surplus food and that is, currently, the first experience in Spain. Espigoladors not only proposes an innovative solution for the use of this food but also communicates the project in an original and attractive packaging that transmits the philosophy: ugly and imperfect fruits and vegetables transformed into high-quality products breaking the taboo that 'the ugly is not good'. *'Es im-perfect'* products are a declaration of intent which aims to bring changes to social, environmental and economic level. It has emerged as an innovation laboratory to give new opportunities to ugly fruit and vegetables to be placed on the market.

4. Espigoladors is a scalable and replicable model. There is the possibility to create a social franchise that use the same brand and methodology.

References

Bio Intelligence Service (2010). Preparatory study on food waste Across EU 27. Technical Report 2010-254.

Idescat Catalan Statistics Institute. Available at: http://www.idescat.cat/es.

INE Spanish National Institute. Available at: http://www.ine.es/en/welcome.shtml.

Llano, J.C.O. (2015). El estado de la pobreza 4° Informe. Seguimiento del indicador de riesgo de pobreza y exclusión social en España 2009-2013. EAPN-ESPAÑA.

35. Social supermarkets: a dignifying tool against food insecurity for people at socio-economic risk

M. De Renobales[1]*, L. Escajedo San-Epifanio[1] and F. Molina[2]

[1]University of the Basque Country – EHU, Urban Elika Research Group, Barrio Sarriena s/n, 48940 Leioa, Spain; [2]Berakah, Pastoral Unit of the Historic District of Vitoria-Gasteiz, C/ LasEscuelas 2 bajo, 01001 Vitoria-Gasteiz, Spain; mertxe.derenobales@ehu.eus

Abstract

The recent economic crisis has dramatically increased the number of people who live close to or below the poverty level in our cities. Thus, in recent years, a number of groups of citizens have promoted diverse initiatives to provide food to people at socio-economic risk. Among the best known and more widely distributed initiatives are food banks and soup kitchens. Although each of these organizations will function in different ways, both of them offer food free of charge to persons in need (sent by charities or social services), whether packaged to be taken home (food banks) or hot to be consumed at the premises (soup kitchens). In most cases beneficiaries cannot choose the types of food they are given because these organizations give out what they have. Receiving free food may be humiliating for many people, in addition to preventing them from having a balanced diet which may result in malnutrition. Social supermarkets are an interesting alternative to these initiatives. Basic food items are offered at a substantially reduced price to selected persons and families at socio-economic risk who pay for the food items they choose. In Spain, the number of these solidarity food markets has increased substantially in the last few years in an attempt to offer a more dignifying access to basic foods. In October 2014 the first National Congress of Spanish Social Supermarkets was held. In addition to food items, most of them offered a variety of services, such as nutrition education for better health through balanced diets, and otherwise help become integrated in society. An overview of these food markets will be presented.

Keywords: poverty, discount food store

1. Introduction

In the opening sentence of the Rome Declaration on World Food Security 'the right of everyone to have access to safe and nutritious food, consistent with the right to adequate food and the fundamental right of everyone to be free from hunger' is affirmed. Article 1 of the ensuing World Food Summit Plan of Action states that 'food security exists when all people, at all times, have physical and economic access to sufficient, safe and nutritious food to meet their dietary needs and food preferences for an active and healthy life' (Rome Declaration on Food Security, 1996).

Hunger, or food insecurity in politically correct language, is a consequence of poverty, of not being able to purchase the necessary food, both in quantity and quality, to live a healthy, active, productive and fully human life. Freedom from hunger is a fundamental right of everyone, consistent with the International Covenant on Economic, Social and Cultural Rights and other relevant United Nations instruments (FAO-WHO, 2014). The economic crisis that started in 2008 and the ensuing politics of austerity have caused a very large increase in the number of people who live at, or below, the poverty line in Europe, particularly in Cyprus, Greece, Italy, Ireland, Portugal, Romania and Spain (Caritas Europe Report, 2014). The 2015 Global Food Security Index (which takes into account 19 measures across the issues of availability, affordability and quality across 109 countries) is 78.9% for Spain (no. 19 together with Israel), down 1.6% points from 2014, indicating an increase in food insecurity (The Global Food Security Index, 2015). According to the Caritas Report, 28.2% of the population in Spain is at risk of poverty or social exclusion, affecting 12.9 million people, of which 9.6 million live below the poverty line. In addition, Spain is one of Europe's most unequal countries having an S80/S20 coefficient of 6.8. A recent new development is that 12.3% of the population has jobs but is poor, nevertheless (EAPN, 2014).

Poverty together with the low socio-economic status are the primary reasons for food insecurity in urban areas. Among the items considered in the 'indicator of material deprivation', 10.6% of the EU population reported that they could not *afford a meal with meat, chicken, fish (or a vegetarian equivalent) every second day* (Eurostat, 2015). In contrast, between 2005 and 2010, the percentage of people who could not meet that need decreased about 0.5% points each year (Loopstra *et al.*, 2015). Food insecurity is an urgent health problem. Not having access on a regular basis to sufficient food, both in quantity and quality, results in malnutrition for lack of enough daily calories and/or lack of certain nutrients, such as quality protein, vitamins and minerals. In addition, malnutrition is strongly associated with increased deterioration of mental health, difficulties to manage chronic diseases, and worsening of child health.

2. Typology of food providing organizations

Food banks and soup kitchens are perhaps among the most widely distributed initiatives promoted by groups of private citizens to provide food for those in need. Food Banks have been well established in North America for the last few decades, whereas in the European Union their numbers increased substantially since the 2006 crisis of food prices. Both of them, food banks and soup kitchens, are non-profit organizations contributing to redistribute surplus food and reduce food poverty. They are mostly managed by volunteers and beneficiaries are given available food items without being able to choose them. Food banks generally collect donated unsellable but still edible surplus food from the food industry and retail, but also from private individuals or organizations. Food is packaged and those receiving it must have kitchen facilities. Food banks distribute the food they receive to various charitable organizations and social services which, in turn, donate the food packages to the people in tight economic situations with whom they are in touch. For most people, turning to a food bank to receive free food is an emotionally difficult experience and thus it is the last thing to do after other strategies have been exhausted (reducing expenses, changing eating habits, asking family and friends for help, etc.) (Lambie-Mumford *et al.*, 2014). Yet, access to donated food may generate dependency in beneficiaries which, in turn, may discourage them from actively pursuing their integration in society. Soup Kitchens offer prepared meals to those in need who do not have kitchen facilities.

Social supermarkets (SSM) appeared in France in the 1980s and are now present in several countries in Europe, including Spain (Social Supermarkets Project). In 2011 the Institute of Retailing and Marketing at Vienna University of Economics and Business published the first sector overview of SSM focused on Austria (Holwerg and Lienbacher, 2011). Subsequent presentations at Efficient

Consumer Response Conferences revealed a strong interest in this concept by leading international retailers, such as Tesco, Carrefour, or Rewe as well as manufacturers, such as Unilever or Nestlé (Holweg and Lienbacher, 2011; Holweg and Steiner, 2012). A study published in 2013 (Lienbacher *et al.*, 2013) estimated that over 1000 SSM were operating in Europe, but determining their increase in the last 3 years is difficult. Many of those currently operating in Spain are less than 5 years old.

All SSM share several common characteristics: they sell food items at a substantial discount as compared to conventional supermarkets to selected beneficiaries with very limited economic resources. Their objective is to provide access to food in a dignifying manner because beneficiaries can choose the food items they need and pay for them, although prices can be as low as 20% of the price at a conventional store.

By choosing the necessary food items and paying for them in a just manner according to his/her economic resources, a person exercises his/her freedom and autonomy, and paying for them removes the stigma experienced by many when receiving food for free. When an older man who was supporting his daughter's family on his disability paycheck was asked how he felt about buying food at a social supermarket he said he was very thankful for being able to do so: 'I cannot go to a regular store because prices are too high for my income, and it is very hard and humiliating going to the food bank; I feel thankful for this store in which I can buy basic food items and pay for them, as anybody else pays for what (s)he buys'. Many SSM are set up in such a way that when families with children enter their premises children do not perceive any difference with conventional supermarkets and thus do not know that their families are buying at a discount.

3. The situation in Spain – First National Meeting of Social Supermarkets

Although it is not easy to have an exact count, in Spain there may be around 120-150 SSM spread over the entire country. Most of them are managed by church groups connected, primarily, with the Catholic Church through Caritas, an organization of the Catholic Church that advocates integral human development and justice to improve the quality of life of the poor and vulnerable (Caritas Europa). Religious orders, private groups of citizens, and private foundations are also involved in several SSM.

In an attempt to establish contact with one another, and to share what we have been learning, the first National Meeting of Social Supermarkets of Spain was held in Vitoria-Gasteiz, in October 2014. Twelve supermarkets attended, mostly from geographically close locations. An overview of the diversity of these social food stores is given.

The primary objective of the SSM attending the meeting was to dignify those in a precarious economic situation by facilitating their access to basic food and hygiene products at a price they can afford. The right to food, which is basically the right to life, is a fundamental human right, compromised by poverty, as stated above. In spite of government programs to provide assistance to those in need, an increasingly larger number of people do not fulfil all the requirements to obtain help from institutional assistance programs.

Although each social supermarket responds to the actual circumstances of the place it serves and to the characteristics of the organization behind it, they all share some common aspects:

- Management teams are almost always comprised of volunteers with a minimum number of hired persons. These teams coordinate: (1) relationships with food providers and periodic orders of food products; (2) in-store direct customer assistance; and (3) economic aspects and financing.
- Most of the supermarkets will buy the majority of food products at a discount through agreements with providers (food distribution chains, or small food industries) and sell them at an even higher discount ranging between 50 and 70% with respect to the prices in a conventional supermarket. Some accept donations, but for many social supermarkets donated food is, in turn, donated, not sold.
- Beneficiaries are always selected according to strict economic criteria. Often the selection process is shared with local institutions and/or charities and churches. In some independent supermarkets selection is carried by in-house volunteers. Beneficiaries receive an identification card which entitles them to purchase food in the supermarket. All personal data are securely kept according to strict confidentiality criteria, so that no sensitive information is accessible through the everyday use of the SSM.
- In all cases, beneficiaries pay for the items they select. Most of the times they pay cash. Other payment systems include points (of a total assigned each month), or food stamps received at the beginning of the month.
- Access to the supermarket is restricted to those with identification cards. In one notable exception, regular customers paying the prices of a conventional store (or slightly higher) are also welcome. This is a way to help finance the enterprise as described below.

The national background of beneficiaries depends on where the supermarket is located. Significant differences can be seen across Spain, from large to small cities or even in rural areas, or even according to gender, because poverty does have many faces. Immigrants are an important group in most places. Yet, in the last years a substantial increase in local people has been observed. Families with small children and older women living alone predominate. The situation of beneficiaries is periodically reviewed. Rotations may be established after a certain time so that others may also benefit from the discounts.

Actual facilities of SSM and their capacity to offer a variety of products differ substantially according to their economic support. Some social supermarkets may offer only basic food and personal hygienic products, whereas others offer a wider selection of products as in conventional stores. A substantial difference in criteria to offer discounted food products was apparent among the supermarkets attending the meeting in Vitoria. Some supermarkets offered the same discount for all products, whereas others did so only for those products considered basic from a nutritional point of view, maintaining full prices for the rest: the supermarket offered a variety of products but those which were nutritionally questionable (i.e. carbonated refreshment drinks) were available but not subsidized. The prevalent idea was to encourage consumption of nutritious foods on the one hand, and to learn to manage the family budget on the other, as part of wider educational programs.

The assistance with daily food availability provided by these supermarkets is, in the majority of the cases, part of wider program to empower the beneficiary by promoting his/her autonomy to access the necessary food, and to encourage participation in training and/or educational programs, such as debt management, local cooking, family budget management, language classes, clothes repair, or a variety of other programs. Several of them, in collaboration with government institutions, offer officially recognized training programs for a variety of jobs such as hotel cleaning person, caregiver for older persons, stock control, or supermarket cashier. In a few cases, when given access to the SSM, beneficiaries sign a contract by which they commit themselves to attend educational programs or to collaborate in various capacities with everyday tasks. Should the beneficiary not fulfil that commitment, his/her access to the SSM may be withdrawn. The ultimate goal is to facilitate his/her food, as well as, social inclusion. The supermarket facilitates regular contact with the individual families to encourage their assistance to educational activities which will be instrumental in finding a job and, ultimately, becoming integrated in society.

Nutrition education is a goal for most SSM, often carried out in collaboration with local universities which may offer a degree in human nutrition and dietetics. Students may help design nutritionally balanced menus according to beneficiaries' food preferences, and conduct workshops to explain the nutritional value of frequently consumed foods.

Because quite often the resources of a supermarket are limited, in all cases, the number of each food item beneficiaries can purchase in a given month is limited. Several reasons were given: (1) to promote management of daily menus and family budgets; (2) to curtail the possibility of negotiating with discounted items; and (3) to prevent buying at a discount for others who are not eligible for this service.

Financing a social supermarket is not easy without the support of a strong organization. Financing formulae are varied and mixed, and almost not one supermarket covers expenses completely. We could say that these are 'for-loss' businesses. In Spain most of them are backed by Caritas or a religious congregation, with a few cases depending on private foundations or a group of citizens. In many cases, the cost of a beneficiary's food basket is paid by the organization which sent him/her to the supermarket. All of them accept private monetary donations from individuals or enterprises, and organize fund-raising events, such as 'Buy a food basket for a family'-day.

In a nutshell, in addition to dignifying access to basic food and hygienic products for people at risk of socio-economic exclusion, SSM provide the organizations behind them with an excellent opportunity to actively promote their social inclusion.

References

Caritas Europe (2015). Policy and Advocacy. Available at: http://tinyurl.com/nkhb424.
Caritas Europe (2014). The European crisis and its human cost: a call for fair alternatives and solutions. Available at: http://tinyurl.com/puz3r55.
EAPN (2014). Dossier Pobreza de EAPN España 2014. Available at: http://tinyurl.com/ppgpd45.
Escajedo San-Epifanio, L., De Renobales, M., Dapena Montes, P., and Glz-Bascaran, I. (2015). Urban Elika, towards a better use of food resources in EU urban environments: focusing on decision making processes. In: D.E. Sumitras, I.M. Jitea and S. Aerts (eds.), Know your food: food ethics and innovation. Wageningen Academic Publishers, Wageningen, the Netherlands.
Eurostat (2015). Material deprivation and low work intensity statistics. Available at: http://tinyurl.com/omnkx6z.

FAO (1996). Rome Declaration on World Food Security, 13-17 November, 1996. Available at: http://tinyurl.com/o8tjbwm.

FAO-WHO (2014). Second International Conference on Nutrition. Rome, 19-21 November 2014.

Holwerg, C., Lienbacher, E. (2011). Social marketing innovation: the new thinking in retailing. Journal of Nonprofit and Public Sector Marketing 23, 307-326.

Holweg, C., Steiner, G. (2012). Profit by working together to prevent supply chain waste. Presentation at the 2012 ECR Europe Conference, Brussels.

Lambie-Mumford, H., Crossley, D., Jensen, E., Verbeke, M., Dowler, E. (2014). Household food security in the UK: a review of food aid. Available at: http://tinyurl.com/qamk9rq.

Lienbacher, E., Holweg, C., Schnedlitz, P. (2013). CSR in food retailing: what's really on customers' minds? In: A. Rindfleisch, J. Burroughs (eds.) Proceedings of the AMA WinterMarketing Educator's Conference. USA, American Marketing Association, pp. 235-243.

Loopstra, R., Reeves, A., Stuckler, D. (2015). Rising food insecurity in Europe. The Lancet 385, 2041.

Social Supermarkets Project (2015). Available at: http://socialsupermarkets.org.

The Global Food Security Index (2015). An annual measure of the state of global food security. The Economist, Intelligence Unit Ltd. Available at: http://foodsecurityindex.eiu.com/Resources.

36. New challenges against hunger and poverty: a food bank case study

O. Forcada[1], S. Sert[2]* and V. Soldevila[3]

[1]Fundació Empresa i Ciència, Universitat Autònoma de Barcelona. Edifici A Bellaterra, Spain; [2]Department of Management, Economics and Industrial Engineering; Via Lambruschini 4/b, 20156 Milano, Italy; [3]Departament d'Economia, Universitat Rovira i Virgili, Avda. Universitat 1, Reus, Spain; sedef.sert@polimi.it

Abstract

After being first implemented in the USA in the 1960s and later adapted to Europe in France in the 1980s, the first Food Bank in Spain was founded in Barcelona in 1987. The *Fundació Benèfica Banc dels Aliments* or, more common, Barcelona Food Bank (BFB) first mission was to cover very specific emergencies, providing assistance to non-profit organizations that supplied food as a sort of charity. However, since then, Spanish society and its challenges have changed. The economic crisis in Spain has been long and sustained. This situation makes necessary an evaluation of the current structure and management of the BFB, can it successfully adapt to the current economic situation and address new challenges? In order to answer this question, this paper explains the main changes the BFB performance has experienced and how its role in society has varied from the beginning of the economic crisis. Based on in-depth interviews with the food bank volunteers and workers and the reviewing of BFB annual reports, we present the BFB's current situation and determine the main challenges it is facing now and will likely face in the near future.

Keywords: food assistance, economic crisis, food waste, Spain

1. Introduction and literature review

In recent years, there has been an increasing concern about food insecurity in High Income Countries (HICs). In the EU, 10.5% of its total population is food insecure, that is, it is unable to afford a meal with meat, chicken, fish (or vegetarian equivalent) every second day (EUROSTAT, 2015). On the other hand, food waste is a huge problem for economic, environmental and social reasons. According to the estimations given by Gustavsson *et al.* (2011), food waste is around 280-300 kilograms per capita per year in Europe and North America.

Due to the existence of food waste and food insecurity in HICs, food banks have emerged all over the world. They recover food from the food industry, public aid programs or from private donors, and redistribute it to non-profit organizations (NPOs), which support the most deprived. In this way, they face both food waste and food insecurity.

The first food bank was created in the USA in 1967 and later adapted to Europe in the 1980s. Food banks have taken attention from the researchers since the 1990s, especially in the USA and Canada, where at that time they were already fully implemented (Poppendieck, 1994; Tarasuk and Davis, 1996). In Europe, academic studies about food banks were mainly carried out in the UK and pointed

Leire Escajedo San-Epifanio and Mertxe De Renobales Scheifler (eds.) *Envisioning a future without food waste and food poverty*

DOI 10.3920/978-90-8686-820-9_36, © Wageningen Academic Publishers 2015

291

out the relation between the reduction on welfare state and the increasing relevance of food banks in order to alleviate food poverty (Lambie-Munford, 2013; Lambie-Munford and Dowler, 2014; Perry *et al.* 2014). Although most of the European food banks were created in the 1980s, they took little attention from researchers because they had a marginal role in food assistance. However, the situation has dramatically changed in recent years. According to Gentillini's estimations, 19 million people in the EU are food bank users and, as he pointed out, probably these data are an understatement (Gentillini, 2013). The increasing role of food banks has risen research interest about food banks in Italy (Santini and Cavicchi, 2014), Spain (González-Torres and Coque, 2015) as well as in Australia (Booth *et al.*, 2014; Butcher *et al.*, 2014) and Korea (Kim, 2015), in recent years.

The current literature on food banks comes from different approaches such as analyzing how to manage food surplus in order to prevent food waste (Garrone *et al.*, 2014; Schneider, 2013); focusing in their role to alleviate poverty (Lindberg *et al.*, 2014; Perry *et al.*, 2014); or studying the beneficiaries, both according to their socioeconomic situation (Perry *et al.*, 2014) and to their health conditions (Garthwaite *et al.*, 2015; Tarasuk *et al.*, 2014).

Food bank 'is a sensitive and emotional topic' (Gentilini, 2013) and a 'dilemma' to anti-hunger activists (Poppendiek, 1994). It could be seen as a neoliberal instrument to address food poverty or as a community-based organization that takes responsibility of people in need. Although a favorable vision of food banks remains, some criticisms have been posed (Poppendriek, 1994). Among them, two main criticisms are in the center of the debate: food banks limitations in achieving food security and 'replacement' of welfare state functions by food banks. Achieving food insecurity means to provide enough and/or adequate food for a healthy diet and to satisfy the beneficiaries dietetic requirements (Butcher *et al.*, 2014; Garthwaite *et al.*, 2015; Tarasuk *et al.*, 2014), while replacement of welfare state functions refers to doubts about government role on the issue of poverty (Riches, 2002, 2011).

The paper examines how the ongoing economic crisis and increasing concerns about food poverty are pushing Barcelona Food Bank to adapt to new challenges. To this aim, in the following sections research design, results of the study, discussion and suggestions for future research are presented.

2. Research design

Case study method is suitable when contemporary phenomenon is investigated and the researcher has little control over events within real-life context (Eisenhardt, 1989; Yin, 2003). In fact, case studies are frequently used as an adequate methodology to increase the knowledge about how food banks work (Lindberg *et al.*, 2014; Santini and Cavicchi, 2014, among others) and it has allowed seeing different typologies of food banks and different strategies adopted by them to face food poverty. As Gentilini remarks, food bank models, activities and operations require further review, especially in Europe (Gentilini, 2013).

Being a particular case, Barcelona Food Bank is selected as a unit of analysis. Although single cases have limits on the generalizability of the obtained conclusions, they provide a greater depth (Voss *et al.*, 2002), coherent to the aim of this study.

In total 8 interviews have been conducted covering all the main departments of the food bank (supply, relationships with NPOs, volunteers, relationships with supermarkets, studies and logistics) as well as the president and vice-president of the organization. Each interview lasted around 1 hour and was conducted by three researchers. For confidentiality reasons it was decided not to record the interviews, thus the role of interviewers was critical. One researcher was responsible for asking the questions while the other two were taking notes. Archival records are used, in conjunction with the interviews, as secondary sources for data triangulation.

3. Results

The *Fundació Benèfica Banc dels Aliments* (Barcelona Food Bank; BFB) is the first food bank in Spain, operating since 1987. Its main role is to collect and redistribute surplus food to non-profit organizations that supply food to people at risk of social exclusion (frontline NPOs). The collection of food was done by the food industry, EU aid programs, individual donations and campaigns or it may also be directly purchased. The surplus food collected by industry is certified to allow donors to receive tax benefits. On the other side, the collaboration with frontline NPOs ensures providing food to final beneficiaries.

Interviewers pointed out the economic crisis as a turning point in the food bank's operation. According to the interviews analysis, five main adaptation areas have been identified.

Aim

In its first twenty years of existence (1987-2007), BFB fought mainly to prevent surplus food to become waste, but never worrying if the amount of food collected was enough or if the food was nutritionally adequate. In fact, in the 1980s and 1990s, food insecurity was not an issue in Spain. But with the economic crisis starting at 2008, the number of people in need increased dramatically and the risk of poverty in Catalonia has reached to 20% of its total population (Eurostat, 2015). The crisis brought more potential beneficiaries to food aids and at the same time it made companies more aware of their surplus food generation which led to more prevention procedures being put in place and less available surplus food for the food bank. This has pushed the food bank to become an active food fund raiser with a main aim on increasing the amount of collected and distributed food year after year together with the increasing concern about the nutritional adequacy of the distributed food.

Sources

The sources of food bank have evolved through the years, both for surplus food collected and for financial funding. Figure 1 shows the evolution between 2009 and 2013 of the food collected by the food bank (Fundació Banc dels Aliments, yearly reports 2009-2013). In 2009 the main contributor was the EU, whose help reached its peak in 2011 and diminished later in 2012 and 2013. The second biggest source of collection came directly from the food industry, which donated surplus generated in manufacturing and distribution. As mentioned before, manufacturing surplus has slowly decreased year after year to 12.45% due to prevention strategies, however the food bank has established collaborations with retailer's since 2009-10 and in 2013 it has achieved to obtain 8.14% of its total food collected. Finally, private donors have increased their importance in food collection

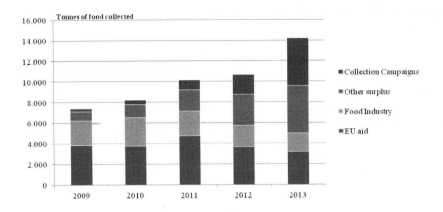

Figure 1. Food managed by the food bank, amount and origin (Fundació Banc dels Aliments, 2009-2013).

through different channels as campaigns (6.46%), direct donations (5.54%) and starting in 2009 'The Great Collection' (19.94%), which has grown from merely obtaining over 200 tons of food on the first year to almost reaching three thousand tons in 2013. The BFB works with other food banks in Spain (and in Europe), exchanging information and food (minimal 0.65%) to enhance their supply spectrum.

Increasing concerns about diet variety and nutritional contents have forced the food bank to look for sources of fresh products. These products are hard to manage so they are to be promptly distributed to the final beneficiary or transformed into longer expiration products (juices, sauces, etc.). Agricultural products obtained by the food bank came from the Barcelona central market (Mercabarna), where the food bank has a collecting and distributing warehouse (4.38%), or from the market regulator (SERMA), who subtracts production from the market to maintain stable prices (11.63%).

Financial sources have also changed. The public financial aid received by the food bank in 2013 was 32.68% of the total income, a much smaller proportion of the total budget if compared to 2009, in which the total public amount received (€ 636,058.42 in 2013 for € 656,732.22 in 2009) represented 80.30% of the total budget. Food bank efforts move towards increasing private donations as it has been difficult to claim for public support in the current situation. During these five years the total budget of the food bank grew 138% as private donations were multiplied by 7.63.

Size

These increases in collected food and funds, have allowed the food bank to increase the attended beneficiaries from 57,381 in 2008 to 152,489 in 2014.

Volunteers and warehouses have grown during these years. Volunteers are the main workforce of the food bank. In 2014, the food bank had 210 volunteers; while in 2007 only 63. The activities of BFB have been spread among all the provincial territory and, nowadays, it has a delegate in each

Barcelona province county. In certain key positions, not susceptible to the goodwill of volunteers, BFB is obligated to employ workers. It is remarkable how BFB runs with only 7 workers (the same number as in 2007). Warehouses cover more than 4,700 m² and the food bank has two trucks and five vans at its disposal.

Internal structure and management

The use of volunteers increases the degree of personal involvement with the food bank's global objectives although it adds complexity to its internal management. According to the interviews, smaller departments with more concrete responsibilities are better coordinated when using volunteers. So, there was a spread of small departments: from 7 departments in 2007 to 21 departments in 2014. Even at this moment the BFB has 21 different departments, the internal structure can be grouped in four main areas:

- *Supply*. Since the beginning of the economic crisis the food supply has been diversified as the food bank had to find alternative suppliers to cover the increasing social food demand. Each new source of food donation is in charge of a new department (campaigns, losses – in charge of the supermarkets program – Mercabarna, SERMA, company events and company collections).
- *Logistics*. This is the core department to ensure the proper collection and distribution of donated food directly managed by the food bank in its warehouse. Most employed workers are located in this area because of their mandatory technical expertise, necessary to plan logistics and manage machinery. Fresh products have increased complexity in logistics, making it harder to guarantee food safety. Specialized departments are addressing these problems – cold chain and food safety.
- *Distribution*. Once supply ensured food availability, distribution seeks its most efficient social reach by contacting non-profit organizations with enough capacity to handle and dispense food to people in need. After reaching a collaboration agreement, this area provides managerial assistance and maintains periodic visits to ensure these organizations follow the food bank's guidelines.
- *Support and management*. While the other three areas are centered on food, support and management are focused on assisting and guiding the other three areas. Support contributes to the proper performance of the rest of areas through administration, accounting and communication. To increase social awareness and spreading the food bank's social role and activities, new departments completed this communications functions: schools, advising, surveys and waste.

Public awareness

The BFB has gone from being a charity non-profit organization only known by a few volunteers and collaborators, to an organization that is in the centre of social awareness. Social sensibility about food poverty has grown and has made the food bank's mission part of the daily culture in society. Increases in food collected, funds obtained and the number of volunteers would have not been possible without the increasing efforts in communication.

4. Discussion and further research

The food bank has changed its internal functioning to adapt to the new social needs and to cover food security for thousands of people. By these means, the food bank has grown and now is much more popular than ever before. While the importance of the food bank on ensuring food security for people at risk of social exclusion has been capital, this adaptation has enlarged its role to a similar one present in other countries in which the coverage given by food banks might have started by need but are now a core part of the welfare system (Canada, Australia, UK).

The BFB could be considered as an example of success, as it has been able to increase the amount of food donations and find new ways of financial support. On the other hand, it has been sensitive to the necessity to increase the variety of offered products to its beneficiaries, as showed by the increasing efforts in the collection of fresh products or in the transformation of fruits coming from SERMA into high-quality juices.

Yet, at the current time, the main issue remains at the scope of questioning whether public office, independently of the food banks existence, want or would be able to fulfil the needs of people living in poverty or with risk of social exclusion. Reality has shown this is not the case. Public office is highly dependent on a third sector more and more financed through private good will (monetary or volunteering).

References

Booth, S., Whelan, J. (2014). Hungry for change: the food banking industry in Australia. British Food Journal, 116 (9): 1392-1404.

Butcher, L.M, Chester, M.R, Aberle, L.M., Bobongie, V.J., Davies, C., Godrich, S.L, Milligan, R.A, Tartaglia, J., Thorne, L.M., Begley, A. (2014). Food bank of Western Australia's healthy food for all. British Food Journal,116 (9): 1490-1505.

Eisenhardt, K.M. (1989). Building Theories from Case Study Research. The Academy of Management Review, 532-550.

Eurostat (2015). Statistics Database. Available at: http://ec.europa.eu/eurostat/data/database.

Fundació Banc dels Aliments (2007, 2009, 2010, 2011, 2012, 2013). Annual Reports. Available at: https://www.bancdelsaliments.org/ca/biblioteca.

Garrone, P., Melacini, M., Perego, A. (2014). Surplus food recovery and donation in Italy: the upstream process. Bristish Food Journal, 116 (9): 1460-1477.

Garthwaite, K.A, Collins, P.J., Bambra, C. (2015). Food for thought: An ethnographic study of negotiating ill health and food insecurity in a UK food bank. Social Science and Medicine, 132: 38-44.

Gentillini, U. (2013). Banking on food: the state of food banks in high-income countries. IDS Working Paper, 415. CSP Working Paper, 008.

González-Torres, P., Coque, J. (2015). How is a food bank managed? Different profiles in Spain. Agriculture and Human Values. DOI: 10.1007/s10460-015-9595-x.

Gustavsson, J., Cederberg, C., Sonesson, U., Van Otterdijk, R., Meybeck, A. (2011). Global food losses and food waste: extent, causes and prevention. Rome, FAO. Available at: http://tinyurl.com/lyrsdrf.

Kim, S. (2015). Exploring the endogenous governance model for alleviating food insecurity: Comparative analysis of food bank systems in Korea and the USA. International Journal of Social Welfare, 24: 145-158.

Lambie-Munford, H. (2013). 'Every town should have one': emergency food banking in the UK. Journal of Social Policy, 42 (1): 73-89.

Lambie-Munford, H., Dowler, E, (2014). Rising use of 'food aid' in the United Kingdom. British Food Journal, 116 (9): 1418-1425.

Lindberg, R., Lawrence, M., Gold, L., Friel, S. (2014). Food rescue – an Australian example. British Food Journal,116: 1478-1489.

Perry, J., Williams, M., Sefton, T., Haddad, M. (2014). Emergency use only. Understanding and reducing the use of food banks in the UK. Available at: http://tinyurl.com/okdzl4m.

Poppendieck, J. (1994). Dilemmas of emergency food: a guide for the perplexed. Agriculture and Human Values, 11(4): 69-76.

Riches, G. (2002). Food banks and food security: welfare reform, human rights and social policy: Lessons from Canada? Social Policy and Administration, 36 (6): 648-663.

Riches, G. (2011). Thinking and acting outside the charitable food box: hunger and the right to food in rich societies. Development in Practice, 21 (4-5): 768-775.

Santini, C., Cavicchi, A. (2014). The adaptive change of the Italian Food Bank foundation: a case study. British Food Journal, 116 (9): 1446-1459.

Schneider, F. (2013). The evolution of food donation with respect to waste prevention. Waste Management, 33:755-763.

Tarasuk, V., Davis, B.(1996). Responses to food insecurity in the changing Canadian welfare state. Journal of Nutrition Education, 28(2): 71-75.

Tarasuk, V., Dachner, N., Loopstra, R. (2014). Food banks, welfare, and food insecurity in Canada. British Food Journal, 116(9): 1405-1417.

Voss, C., Tsikriktsis, N. and Frohlich, M. (2002). Case research in operations management. International Journal of Operations and Production Management, 22 (2): 195-219.

Yin, R. (2003). Case study research design and methods. London, UK: Sage Publications

Part 8. Food sovereignty and food security

37. The growing role of contract farming for food security

A. Rossi

Doctoral School on the Agro-food System, Università Cattolica del Sacro Cuore, Via Emilia Parmense n. 84, Piacenza, Italia, arianna.rossi@unicatt.it

Abstract

The author analyses the increasing role of contract farming, which is gaining ground in developing countries because of its significant opportunities for economic and social development, in particular in order to sustain local producers with access to markets and support them with technology transfer and credit facilities; this is because contract farming may improve the quality and the quantity of production. For that reason, this type of contract is further seen as an alternative tool for reducing poverty, providing development in rural areas and increasing food security. In fact this tool contributes to: (1) guaranteeing the selling of yield; (2) establishing the retail price; and (3) selling on the market food with certain agreed quality requirements, reducing also administrative costs. According to this theory, contract farming can promote not only a fair market and a certain quality of food (e.g. food safety), but can also give the opportunity to increase the access to food, to provide better nutrition and to face food poverty (e.g. food security). Despite this positive frame, the author will show that this contract has its problem from a legal point of view.

Keywords: development, farming arrangements.

1. Introduction

Agriculture is one of the major livelihoods in most developing countries, where the smallholder agriculture is the main source of the food and business sector. Because of the increasing management among production, processing and distribution activities along the food chain, contract farming needs to be used by developed countries more and more as well as by developing countries, because the early management phase of the agro-food sector can develop in different countries. Nowadays most agribusinesses are organised in value chain and always more frequently farmers produce only if they have a contract when the counterpart is a multinational company. In fact through this vertical contract, the activity of farmers and buyers becomes highly aligned and this creates advantages for both the parties under the contract, particularly if they are developing countries.

The role that this type of contract plays in the agro-food sector is generally recognised. Recent years are bearing witness to an increasing attraction of considerable academic and policy maker attention (Brons-Stikkelbroeck, 2011; Da Silva and Rankin, 2013). Contracts, mostly used by food processing companies, are signed for a period of time and specify the quantity and the quality of the sold product. On the one hand, thanks to this contract, it is possible for the farmer to be certain of selling a certain amount of his product and at a certain price on the market. On the other hand, for the company buyers it reduces the uncertainty of buying crop in an open market and it gives the possibility, as we will see, to have some control over the production process. In an ideal world, this situation is the perfect arrangement, which provides a *win-win* scenario for the farmers, as well as for the buyers

Leire Escajedo San-Epifanio and Mertxe De Renobales Scheifler (eds.) *Envisioning a future without food waste and food poverty*   *301*

DOI 10.3920/978-90-8686-820-9_37, © Wageningen Academic Publishers 2015

(World Farmers' Organisation, 2013). In a time of the long food chain, contracts represent a flexible means to tackle many issues, as we will see.

The first section of this work provides the definition and description of this contract; it will describe also its development all around the world. The second section describes the advantages and disadvantages of it from a critical point of view. The third section presents the international organisations interest in it.

2. Contract farming

The market is a net of exchanges and contract relationships, through which, buyers (demand) and sellers (supply) interact and meet; therefore the discipline of the contract has a strict impact on the market, because the former influences the functioning of the latter. For that reason countries, by setting their contract rules, intervene also in the market exchange. Because of the importance given to this contract inside the market, it becomes necessary to classify that instrument.

The substance of contract farming rises for the need of firms (generally the one at the end of the food supply chain) to come into contact with farmers, in order to have a greater control over the production process and to invest directly in certain production, as well as for the need of farmers, who want to sell and have a guarantee of the sale of their product.

The definition of contract farming

Literature has given many definitions of what contract farming is; the first were Mighell and Jones (1963), but then there were definitions by authors such as Minot (2007) and Rehber (2007). Authors give more relevance to the definition of Rehber:

> a contractual arrangement between a farmer and a firm, whether oral or written, which provides resources and/or specifies one or more conditions of production, in addition to one or more marketing conditions, for an agricultural product, which is non-transferable.

That definition gives the possibility to distinguish between at least three types of contract: (1) *market specification contracts*, in which farmers maintain a full control over production; (2) *resource providing contracts*, which are drafted between the firm and the farmer and requires quality standard; and (3) *production-management contract*, in which the firm has the highest control over the farmer, because it binds the farmer to produce certain high quality product (Institute of Development Policy and Management, 2012; Jia And Bijiman, 2013).

With these contracts, farmers, food processors and distributors may be ensured of quantity, quality, retail price and time of the selling product. In particular, by using the latter form of contract, the distributor can ask the farmer to also use certain type of technical methods, which means a higher control over the production. Naturally, as we will see, the major transnational corporations may abuse their strong economic position.

Sometimes governments state the requirements of this contract and set specific rules for it: for example they may impose a certain form or a minimum content, such as in Italy, where article 62, co. 1, L. 27/2012 states that the contract must be written and contain the duration of the contract, the quality, the quantity, the retail price and the payment condition and method, under penalty of being declared null. In all of this it is, of course, necessary to ensure that, by producing and selling under a contract, producers and buyers are well-informed of their future obligation. So, contract farming is a long-term form of vertical integration within agricultural commodity chains (Galazzi and Venturini, 1999) that may build stable and fair relationships, based on clear and sure compliance, when properly drafted.

The development of contract farming

The expansion of contract farming has taken place all over the world. One of the first countries to use contract farming was the USA, where *market agreements* started to be used around the 1950s; these agreements are long-term contracts that bind the farmer to sell all the production to the counterparty, with a certain quality standard, at a certain price and for a certain period of time. Due to the competitive injury and the unfair practises, the USA government set the *Farm Bill* of 2008. But in 2010 the *Antitrust Division* started to investigate whether or not the market agreement leads to a price manipulation and to an abuse of economic dependence because of a contractual imbalance. That gave the opportunity to the government to intervene on the previous act (Sirsi, 2013).

But contract farming has also been gaining ground in developing countries for many years (Glover and Kusterer, 1990). For example, Latin America has seen a rapid increase in the use of this contract, as well as South Asia, that has seen this growth in recent decades. In India the Agricultural Produce Market Committees Act was created by the government; it is a model act for regulating the marketing of agricultural products, which sets binding standards that the parties must respect, leaving to the parties the freedom of setting terms and conditions (Institute of Development Policy and Management, 2012; Pultrone, 2012).

Contract farming is also taking off in Sub-Saharan Africa. For example in Kenya there is the Dairy Industry Act, which sets rules also for the agri-business sector and has published of a standardized contract. In particular, the growth has been seen in the cotton sector, that it is one of the most important agricultural export commodities in the African continent, besides coffee and cocoa (for a deeper analysis see FAO, 2013; Kirsten *et al.*, 2009).

3. Positive and negative effect of contract farming

For farmers and firms the positive effects of this contract are simple to understand. On the one hand farmers may: (1) be allowed access to a reliable market, increasing the welfare of the area in which the farmer works; (2) be guaranteed the allocation of the production at a certain price; (3) be given stable incomes, because the farmer knows previously the amount and the price of the product they will sell; (4) have greater access to bank credit (World Bank, 2007); and (5) benefit from technology transfer even in remote areas, because of the need of the buyer, as well as of the drafted contract. On the other hand, firms may: (1) reduce administrative costs; (2) buy at a lower price; (3) have more control over the production; (4) invest directly in the production (such as providing technology transfer); (5) increase reliability of food quality; and (6) allocate the risk onto farmers, in many cases (Institute of Development Policy and Management, 2012; Pultrone, 2012).

In my opinion this wide-range of benefits can be split in three different groups, according to their effect on different issues. The first group affects food safety; that's because the quality of food sold may increase and it gives major support to food control activities and makes products more traceable, because the contract already states the quality requirements of the sold product. Through this relationship, it may also support farmers in meeting sanitary and phyto-sanitary standards (FAO, 2013). The second group influences food security, because it strengthens the relationship between industrialised and developing countries, increasing the general welfare, e.g. access to credit, better nutrition and reduction in poverty. The last group, highly aligned to the second group, concerns the market, because thanks to these positive effects, the whole value chain can benefit. Contract may, in fact, increase profit, technology transfer and the trade market of remote rural areas, giving them access to a reliable market (Glover, 1994).

This should be a perfect solution, a *win-win* contract (World Farmers' Organisation, 2013). In reality, however, this contract may lead to the following negative side effects: (1) the difficulties to convert specific investment in case of a breach of contract; and (2) the abuse of bargaining power, in particular in case of *production-management contract*.

In fact, in case of long-term relationships, such as the one in contract farming, circumstances may change during the contract performance and the contract may end before it expires because of market conditions or business interests of the parties (in general not because of the farmer interests). This long-term relationship usually requires the farmer or the firm to make specific investments and, for that reason, many authors speak about '*vertically-integrated investments*' (Kirsten *et al.*, 2009). This investment could be done with the sole scope of satisfying the counterparty of the contract, both by the firm and by the farmer, and, in case of a premature end of the contract, it means upset returns on investments and it may cause strong financial difficulties to the party which has invested in the contract. Obviously this system of production exposes farms to the risk of abuse because of the position of economic dependence in which they find themselves when, in order to adapt their production to the demand of the counterparty, they make specific investments, which are hard to convert. These investments are difficult to translate and parties, linked to the previous contract because of the investment, may not find alternatives on the market. Typically the distributor and multinational firms can spread their business risk among many subjects, but small farmers, engaged in an exclusive relationship with the firm, cannot spread the risk and don't have the opportunity to contract with other subjects either. For that reason, this contract can be a disadvantage and it becomes important to be careful while drafting it and it is essential to insert specific clauses to protect both parties. This situation raises the need to develop legal instruments (Pultrone, 2012) which, in respect of the market mechanism, may prevent and remedy imbalances caused by abuse of bargaining power.

The other pathologic effect of this long-term relationship is the bargaining power and the risk of abuse of economic dependence. This is the main aspect of this contract. It may happen that one of the parties has a greater influence on the other in the phase of negotiation of the contract.

This may lead to unfair trading practices in situations where an imbalance of bargaining power exists between the parties. This imbalance may be seen in the whole food supply market, which is a different situation, called abuse of dominant position. In both cases it is important to understand what happens to the contract and to its relative's clauses (term, price, quality) if one of the parties drafted the contract only because of the influence of the other; it becomes necessary to know if the weak party didn't have the ability to find satisfactory alternatives to the drafted contract.

Therefore, this raises the need to develop, even in terms of interpretation, legal instruments, also judicial, which, in respect of the market mechanism, could prevent and remedy imbalances caused by the abuse of bargaining power. Also the authorities, mainly the antitrust authorities, have a potential role (Sepe, 2013), in order to face this issue because of potential anticompetitive effects that some of these practices may have in the long term, as well as on consumer welfare.

Furthermore, this system gives a higher protection to the freedom of contract and requires the governments' intervention only in case of abuse or in order to face cases of anti-competitive behaviour. So, it is important that countries coordinate the parties to draft contracts in order to achieve good result, but without governments taking the place of the contract parties. That intervention, also judicial, can be legitimized when one of the parties of the contract, although they are both professional subjects (called B2b contract), cannot obtain beneficial effects because of a contractual imbalance.

4. The international frame work

As aforementioned, many contract-farming schemes in less developed countries are set by international organisations. In fact, this contract became useful only when properly drafted and when they are fair and based on mutual compliance.

The FAO advises that '*when properly structured, contract farming may offer farmers the opportunity of a secure revenue through guaranteed market access, including to specialized high-price segments*', as it was said by José Angelo Estrella Faria, the Secretary-General, International Institute for the Unidroit (World Farmers' Organisation, 2013).

The International reference of this type of contract is the United Nations Convention on Contracts for the International Sale of Goods, the background legal framework for many sales contracts. Furthermore the International environment has been intervening in that area with *soft law* settlements. The Unidroit and the FAO are formulating international legal guidance concerning contract farming, which will be submitted by the Governing Council of Unidroit by mid-2015, which can have its impact on agricultural contracts by influencing domestic laws, by applying it as general principles of International law or by introducing them inside the contract as clauses. On this ground, the purpose of the Unidroit legal guide is to assist and advice governments and parties on how to draft contracts, in order to have more stable and balanced relationships. The main idea is that it is necessary to improve commercial relationships in the food supply chain with principles that must guarantee business sustainability, e.g. continuity and competitiveness (Unidroit, 2015).

In this contest, where the aim is to put products arriving from developing countries on the market, a good practice reference may provide more information for legislators and policy makers, but also to farmers and firms using this type of contract.

5. Conclusions

As we have seen, the system of contractual relationship, that, through the meeting of demand and supply allocates the profits of the food chain fairly, is indeed a necessary prerequisite not only to increase the quantity and the quality of goods, but also to ensure the lasting sustainability of agriculture production and its capability to fulfil the needs of people. In fact this contact can only be beneficial when properly drafted by technical experts and when the farmers are fully informed about the nature of the agreement they are entering into. The international organisations are setting new contract models, but what is more difficult is to implement these agreements (World Farmers' Organisation, 2013).

Although there is a strong interest of many authorities, what is most important, in order to achieve food security, is the fairness, transparency and the mutual cooperation among the different parties of the contract, which may end in a *win-win* contract. The interest of national policy makers and international organization must also be the promotion of sustainable contract farming models in order to increase agricultural production, to improve the livelihood of poor areas and to achieve food security worldwide. All this intervention can promote a fair market and give the opportunity to increase access to food, to provide better nutrition and to reduce food poverty, by increasing market access in developing countries.

Knowing that contracts affect the market, only a fair contract may allow the market to be effective and fair.

References

Brons-Stikkelbroeck, E. (2011). Franchising strengthens the use of private food standards, Private food law. In: Van der Meulen, B.M.J. (ed.), Governing food chains through contract law, self-regulation, private standards, audits and certification schemes. Wageningen Academic Publishers, pp. 255-264.

Da Silva, C., Rankin, M. (2013). Contract farming for inclusive market access: synthesis and findings from selected International experiences. FAO, Rome. Available at: http://www.fao.org/3/a-i3526e.pdf.

FAO (2013). Contract farming for inclusive market access. Rome, Italy. Available at: http://www.fao.org/3/a-i3526e.pdf.

Galazzi, G., Venturini, L. (1999). Vertical relationship and coordination in the food system. Physa-Verlag, Heidelberg, Germany.

Glover, D.J., Kusterer, K.C. (1990). Small farmers, big business: contract farming and rural development. Macmillan, London, UK.

Glover, D.J. (1994). Contract farming and commercialisation of agriculture in developing countries. In: Von Braun, J., Kennedy, E. (eds.) Agriculture, commercialisation, economic development and nutrition. John Hopkins Press, Baltimore, Maryland, pp. 166-175.

Jia, X., Bijman, J. (2013). Contract farming: synthesis themes for linking farmers to demanding market. Available at: http://www.fao.org/3/a-i3526e.pdf

Kirsten, J.F., Dorward, A.R., Poultron, C., Vink, N. (2009). Institutional Economics Perspective on African agricultural development. International Food Policy Research Institute, Washington, DC.

Institute of Development Policy and Management, University of Antwerp (2012). Contract farming in developing countries – A Review. Available at: http://tinyurl.com/pcabyvn.

Mighell, R.L., Jones, L.A. (1963). Vertical coordination in agriculture. Department of Agriculture, Economic Research Service, Washington DC.

Minot, N. (2007). Contract farming in developing countries: patterns, impact, and policy implications. Case Study #6-3 of the Program: 'Food Policy for Developing Countries: The Role of Government in the Global Food System'. Ithaca: Cornell University, New York. Available at: http://tinyurl.com/ofefd2u.

Pultrone, C. (2012). An overview of contract farming: legal issue and challenges. Uniform Law Review, pp. 263-289. Available at: http://tinyurl.com/prdq8dy.

Rehber, E. (2007). Contract farming theory and practice. ICFAI Press, Hyderabad.

Sepe, G. (2013). Il controllo del potere di mercato nella filiera agro-alimentare: profili concorrenziali e ruolo dell'AGCM. Rivista di diritto alimentare, 1, 35-39.

Sirsi, E. (2013). I contratti del mercato agro-alimentare: l'esperienza USA. Rivista di diritto alimentare, 1, 40-47.

Unidroit, FAO and IFAD (February 2015). Legal guide on contract farming. Available at: http://tinyurl.com/pw4s6lt.

World Bank (2007). World Development Report 2008. Agriculture for Development, Washington.

World Farmers' Organisation (2013). Farmers' integration in the value chain: fair terms need fair contracts. Issue n. 19 September 2013. Available at: http://tinyurl.com/npntjqw.

38. Organic farming: food security of small holding farmers

S.Z. Ali[*], G.P. Reddy and V. Sandhya

Agri Biotech Foundation, Rajendranagar, Hyderabad – 500 030, Telangana State, India; skzali28@gmail.com

Abstract

Agriculture is the backbone of Indian Economy as it contributes about 13.9% to the total GDP and provides employment to over 54.6% of the population. Over the last few decades India has successfully transformed itself from a food deficient country to one which is essentially self-sufficient in availability of food grains. This success resulted from the 'Green Revolution' (GR), technological interventions in agriculture. Expansion of irrigation, hybrid crops and use of chemical fertilizers and pesticides were the major technological interventions of GR which boosted Indian agriculture. Although GR has played a leading role in making the country self-sufficient in food grains, it has created some adverse effects, which are of serious concern. The negative effects includes, soil degradation, increased salinity, desertification, destruction of soil fertility, micronutrient deficiency, soil toxicity, insect resistance to pesticides and contamination of water bodies which are challenging the sustainability of conventional agriculture. In this regard, organic farming, a holistic production management system for promoting and enhancing the health of the agro-ecosystem, has gained wide recognition as an alternative to conventional agriculture and ensures safe food for human consumption. Organic farming avoids/largely excludes the use of chemical fertilizers, pesticides, growth regulators and relies on legumes, green manure, crop rotation, crop residues, animal manures, biofertilisers, biopesticides, bioherbicides, etc. Agri Biotech Foundation (ABF) in India is a leading nonprofit Organisation in organic farming. ABF earlier carried out the Andhra Pradesh Netherland Biotechnology Programme for about 12 years and delivered 24 new technologies including technologies that have implications for promoting organic farming, biofertilizers, biopesticides, etc. ABF trained men and women farmers, students, government officials, extension agents etc. on organic farming. This paper explains some of the innovative technologies popularised by ABF to promote organic farming in urban and peri-urban areas through community efforts.

Keywords: biofertilizers, biopesticides, vermicompost, vegetables

1. Introduction

Agriculture is the backbone of Indian economy as it contributes about 13.9% to the total GDP and provides employment to over 54.6% of the population. It was highly gratifying that India has successfully transformed itself from a food deficient country to one which is essentially self-sufficient in availability of food grains. This success resulted from the 'Green Revolution' (GR), a technological interventions in agriculture widely adopted by Indian farmers. Expansion of irrigation to cover rain-fed areas, popularization of hybrid varieties of crops, and use of synthetic fertilizers and pesticides were the major technological interventions of GR which enabled India to become self-sufficient in food grain production (Sharma, 2005).

Leire Escajedo San-Epifanio and Mertxe De Renobales Scheifler (eds.) *Envisioning a future without food waste and food poverty*
DOI 10.3920/978-90-8686-820-9_38, © Wageningen Academic Publishers 2015

The introduction of synthetic chemical fertilizers and pesticides has boosted productivity per hectare and helped increase food production. However, these increases in production have slowed down in recent years, and in some cases there are indications that production is going down (Rundgren, 2006). The main reasons for this negative side effects are secondary salinity, decreases in soil fertility, damage to bio-diversity and environment, growing insect resistance to pesticides, degradation or destruction of water resources, pesticides residues in food products, groundwater contamination and increased costs of production, all of which are challenging the sustainability of conventional agricultural production at high levels (Balak Ram, 2003; NASS, 2005; Rundgren, 2006). There is an urgent need to take a holistic view of this problem to curb its negative impact. Organic agriculture (OA) is a major pillar for sustainable agriculture and an answer to our problem of environment degradation, unsafe food, polluted water, degraded land and wide range of illness due to conventional agriculture practiced in the recent past.

According to the International Federation of Organic Agri-culture Movements 'organic agriculture is a production system that sustains the health of soils, ecosystems and people. It relies on ecological processes, biodiversity and cycles adapted to local conditions, rather than the use of inputs with adverse effects. Organic agriculture combines tradition, innovation and science to benefit the shared environment and promote fair relationships and a good quality of life for all involved' (Shetty *et al.*, 2013). The recent UNEP-UNCTAD (2008) report says that 'organic agriculture is a good option for food security in Africa'. The Food and Agriculture Organization of the United Nations conference on OA and food security, May 2007, observed that OA has the potential to contribute to sustainable food security through improved household nutrient intake, resilience to food emergency situation and contribution to healthy diets (Scialabba, 2007).

However, there is very little information available in the public domain on OA in relation to food security and livelihood improvement of smallholder farms in India. This paper explains some of the innovative technologies popularised by Agri Biotech Foundation (ABF) to promote OA in urban and peri-urban areas through community efforts.

2. Materials and methods

The research was designed as a study covering three districts in two states of India – Telangana State and Andhra Pradesh. The district selected for the present study are Ranga Reddy and Nalgonda district of Telangana State and Srikakulam district of Andhra Pradesh. The selected villages in these districts are located in the vegetable growing belt with higher application of synthetic chemical fertilizers and pesticides. More than 40% of the population belongs to scheduled castes and schedule tribes who are either marginal farmers or landless agricultural labourers.

ABF wanted to change the above poverty scenario and to improve the life of the farmers and also to convert the farmers from conventional agriculture to organic agriculture. The activities were performed in a cluster of three villages in Nalgonda district, two villages in Ranga Reddy district and two villages of Srikakulam district. Farmers volunteered after a series of introductory meetings in the concerned villages. In these meetings the concept of OA was explained. Farmers feared loss of production due to complete elimination of chemical fertilizers. The elimination of pesticides was considered less of a problem by the farmers, partially due to the fact that they had been exposed to integrated pest management technologies to a certain extent. Project staff provided an alternative cropping system and a comprehensive outline of crop management, which showed that the same levels of nutrient application could be achieved through organic means and that various tested organic options for disease and pest management existed. This convinced farmers that they would not end up in loss.

For skill development at farm level ABF conducted training programmes at farmer fields in the selected villages. They were given brochures and handouts in local language (Telugu) containing the information on technologies like pro-tray technologies for raising organic vegetable seedlings and vermi composting. Farmers were trained on the use of low external inputs like biofertilizers (*Azospirillum, Azotobacter,* phosphate solubilizing bacteria and *Rhizobium*) and biopesticides (*Trichoderma* and *Pseudomonas*) followed by their application methods (seed, seedling and soil application as well as foliar spray) and mass multiplication at farmers level (Figure 1). The training

A

B

Figure 1. Women farmers practicing (A) seed application of biofertilizers and biopesticides, (B) mass multiplication of biopesticides, Trichoderma.

Figure 2. Distribution of biofertilizers and biopesticides to farmers

programmes were followed by distribution of biofertilizers, biopesticides and vermi compost (Figure 2). ABF staff continuously monitored the farmers' fields who were utilizing the bio-products and their impact on yield and income.

3. Results and discussions

A total of 542 farmers were given awareness and training and 20 farmers were selected from seven villages based on their interest for OA. Among 20, 16 were interested in cultivation of organic vegetable and four were interested in raising organic vegetable seedlings. Field surveys were conducted in seven villages by interacting with the farmers and visiting their fields (Figure 3) in order to check the improvement if any with respect to organic vegetable cultivation and raising organic vegetable seedlings.

All the 20 farmers used biofertilizers, biopesticides and vermi compost. The mode of application of

Figure 3. Agri Biotech Foundation staff monitoring a farmers' field.

these bio-inputs was by three different methods, *viz.*, seed and seedling or soil application. But most of the farmers applied at seedling stage though a few followed seed application too, and some even applied these bio-inputs to the soil (Figure 4). For organic vegetable seedling production farmers were trained in preparation of seedling mixture, filling the mixture, seed treatment, sowing of seeds in pro trays and management practices in different stages (Figure 5).

Organic vegetable cultivation

In terms of production, farmers could find better results using these bio-inputs. They observed good growth, increase in number of crop cycles and yield. In case of pest and disease a few could notice crops free from disease and pest. Further, they could find variation in their fields compared to the fields without biofertilizers and biopesticides.

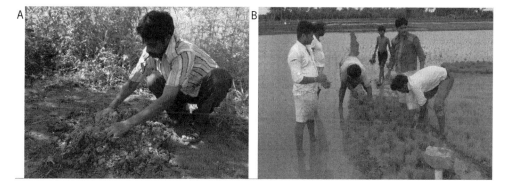

Figure 4. Farmer preparing biofertilizers and biopesticides for (A) soil application and (B) seedling application.

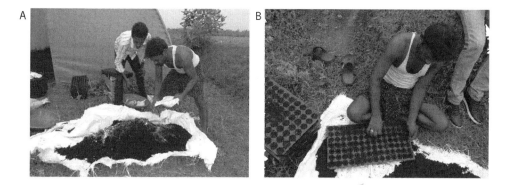

Figure 5. Farmer (A) preparing organic seedling mixture for rising vegetable seedlings and (B) sowing biofertilizers and biopesticides treated seeds in pro-trays.

In the cropping season prior to this programme, chemical fertilizers such as urea, potash, and ammonium phosphate were used by the farmers. By this programme the use of biofertilizers and biopesticides in crop production was introduced.

Case study 1

Farmer 1, a 37 year old male farmer from Ramdaspalli village (Ranga Reddy district of Telangana state), applied biofertilizers supplied by ABF in his vegetable field add 10 kg/acre via seed and soil application. During the first year the yield was equal to conventional agriculture but, from the second year onwards he got an enhanced yield in tomato, ridge guard and radish which increased (Figure 6A) by 20%. Additional income of Rs. 22,500/year (€ 323) was gained due to production and reduction of chemical fertilizers.

Figure 6. Organically grown tomatoes at (A) Ranga Reddy and (B) Nalgonda district of Telangana.

Case study 2

Farmer 2, a 28 year old female farmer from Veldanda village (Nalgonda district of Telangana State) applied biofertilizers and biopesticides in her vegetable field adding 8.5 kg/acre via seed, seedlings and soil application. During the first and second year the yield was low, compared to third year (Figure 6B), which increased by 25%. Due to which an additional income of Rs. 14,000/year (€ 200) was gained with reduced use of chemical fertilizers.

Protray technology for raising organic vegetable seedling nurseries

The traditional system of raising vegetable seedling using soil beds may lead to a decrease in germination percentage, sick seedlings and loss of seedlings during transplantation. Further farmers also depend on the unscrupulous nursery men and purchase the seedlings from them with little or no concern about the quality and the healthiness of the plants. The sick seedlings lead to heavy gap filling and low crop yield. Therefore, there is a need to empower the vegetable growing farmers technologically for achieving higher economic growth. Vegetable seedling production using advanced scientific technology can make a significant difference to the livelihoods of the farmers. It empowers them both socially and economically giving better returns.

Case study 3

A total of four farmers from four villages of Srikakulam district of Andhra Pradesh state came forward to adopt the pro-tray technology for organic vegetable seedling production. As guided and trained by the ABF staff about protray technology, farmers started raising organic vegetable seedlings. They adopted the technology and showed enthusiasm in growing the organic vegetable seedlings in an efficient way. At all the project sites i.e. Boddam, Kothavalasa, Mandawakuriti and Chintalapeta villages of Srikakulam district fully developed organic seedlings were (Figure 7) sold to the surrounding farmers who were growing vegetables. The income generated by selling the seedlings by the four farmers is shown in Table 1

Figure 7. Organically grown vegetable seedlings at (A) Kothavalasa and (B) Mandawakuriti villages of Srikakulam district of Andhra Pradesh.

Table 1. Additional income generated from vegetable seedlings.

Beneficiary	Place	Additional income generated/year in Indian Rupees
Farmer A	Boddam	6,200 (€ 89)
Farmer B	Kothavalasa	9,110 (€ 130)
Farmer C	Mandawakuriti	6,635 (€ 95)
Farmer D	Chintalapeta	7,400 (€ 106)

4. Conclusions

The farmers in the present study expressed that the conversion from conventional to organic agriculture had improved their livelihoods in a range of ways. Enhancement in natural assets such as improved soil structure, improved water holding capacity and increased abundance of beneficial organisms, as a positive effect of the conversion to organic agriculture. Enhanced natural assets were said to allow production with less amounts of external inputs. They pointed out that over the long term the conversion had improved their net-farm incomes, reduced the risk of pesticide poisonings, lead to more self-sufficiency and improved food safety and security.

Acknowledgements

The authors are thankful to the Department of Biotechnology, Government of India for the financial assistant in the form of projects 'Popularization of Softer Biotechnologies for Agricultural Improvement' and 'Socio-Economic upliftment of the Rural and Peri Urban SC/ST Population of Srikakulam District through Agri Biotechnologies'.

References

National Academy of Agricultural Sciences (2005). Organic Farming: Approaches and Possibilities in the Context of Indian Agriculture. Policy Paper 30.

Ram, B. (2003). Impact of human activities on land use changes in arid Rajasthan: Retrospect and Prospects. In: Narain, P., Kathju, S., Kar, A., Singh, M., Praveen-Kumar, P. (eds.) Human impact on desert environment. Scientific Publishers, Jodhpur, India, pp. 44-59.

Rundgren, G. (2006). Down side of conventional agriculture. IFOAM Dossier Organic Agriculture & Food Security, Germany, pp. 19-22.

Scialabba, N. (2007). Organic agriculture and food security. In: International conference on organic agriculture and food security. May 3-5, 2007. Food and Agriculture Organization of the United Nations, Italy. (OFS/2007/5). Available at: www.fao.org/organicag.

Sharma, A.K. (2005). The potential for organic farming in the drylands of India. Soil management for drylands. Arid Lands News Letters, Issue 58.

Shetty, P.K., Hiremath, M.B and Murugan, M. (2013). Status of organic farming in agro ecosystems in India. Indian Journal of Science and Technology 6: 5083-5088.

United Nations Environment Programme-United Nations Conference on Trade and Development (UNEP-UNCTAD) (2008). Organic agriculture and food security in Africa (UNCTAD/DITC/TED/2007/15). Available at: http://www.unep-unctad. org/cbtf.

39. Human right to food: some reflections

A. Corini
Doctoral School on the Agro-Food System, Università Cattolica del Sacro Cuore, Piacenza campus, Via Emilia Parmense 84, 29122 Piacenza, Italy; antonia.corini@unicatt.it

Abstract

The eradication of extreme poverty and hunger is one of the Millenium Development Goals to be achieved within 2015. The human right to food is recognized in International sources: the Universal Declaration on Human Rights (Article 25), the International Covenant on Economic, Social and Cultural Rights (Article 11, General Comment 12). Human right to food means right to a food secure situation, where both the access to food and the adequacy of food are ensured. The latter 'meaning' of food security recalls another important 'dimension' related to the food consumption: food safety. The EU is involved in ruling and implementing policies on food safety in order to guarantee to consumers a high level of protection: from farm to fork. However, a question is raised: is the right of access to food a goal certainly achieved in the EU? On the one hand, the recent economic crisis, the many labour insecurities and the connected increase of new situations of poverty have made the right to food security less and less secure even in the EU. On the other hand, issues related to food waste and food loss require measures of intervention in order to face the present problems and to be prepared for future challenges (i.e. to feed an ever growing population). The human right to food is not expressly recognized in the EU Charter of Fundamental Rights, however it can be derived from the fundamental right to dignity (Title I Charter) that is one of the values on which the EU is founded (Article 2 TEU). The hope for the future, also considering the possible accession of the EU to the European Court of Human Rights, is that human right to food will be the core of EU food policies both on food safety and consumer protection issues and on food waste prevention techniques.

Keywords: food security, EU, unemployment, poverty, future challenges

The year 2015 is a crucial moment in which to reflect on the global enjoyment of the Human right to food. Now, more than ever, an analysis of the history of the recognition of this right and of all the legal instruments and programmes adopted for a real achievement of the human right to food is extremely important. The latter is recognized in International sources, mainly in the Universal Declaration on Human Rights (Article 25) and in the International Covenant on Economic, Social and Cultural Rights (Article 11, General Comment 12).

In particular, Article 25 of the Universal Declaration on Human Rights (of 1948) states that:

> (1) Everyone has the right to a standard of living adequate for the health and well-being of himself and of his family, including food, clothing, housing and medical care and necessary social services, and the right to security in the event of unemployment, sickness, disability, widowhood, old age or other lack of livelihood in circumstances beyond his control. (2) Motherhood and childhood are entitled to special care and assistance. All children, whether born in or out of wedlock, shall enjoy the same social protection.

From the provision itself it appears that the right to food is a 'single component' of the right of a standard of living adequate for the health and well-being of a person and of his/her family. The latter human right is protected if all the components of the right itself are ensured.

Differently, in Article 11 of the International Covenant on Economic, Social and Cultural Rights (of 1966), a step further is taken. This is evident by analysing Article 11 itself, which affirms that:

> 1. The State Parties to the present Covenant recognize the right of everyone to an adequate standard of living for himself and his family, including adequate food, clothing and housing, and to the continuous improvement of living conditions. The State Parties will take appropriate steps to ensure the realization of this right, recognizing to this effect the essential importance of international co-operation based on free consent. 2. The State Parties to the present Covenant, recognizing the fundamental right of everyone to be free from hunger, shall take, individually and through international co-operation, the measures, including specific programmes, which are needed: (a) To improve methods of production, conservation and distribution of food by making full use of technical and scientific knowledge, by disseminating knowledge of the principles of nutrition and by developing or reforming agrarian systems in such a way as to achieve the most efficient (b) Taking into account the problems of both food-importing and food-exporting countries, to ensure an equitable distribution of world food supplies in relation to need.

It is clear that Article 11 recognizes not only the right to an adequate standard of living, including adequate food, but also establishes concrete commitments on the part of the States Parties to the International Covenant. This in order to effectively ensure the enjoyment of the right to food and other important related rights such as the right to be free from hunger as well as the need to ensure an equitable distribution of food in the world.

The legal concept of human right to food has experienced an evolution towards the concept of food security (Gualtieri, 2013). In effect, in 1996, the World Food Summit took place and the Rome Declaration on World Food Security and the World Food Summit Plan Action were adopted. The concept of food security was defined by the World Food Summit as follows: food security is existing '*when all people at all times have access to sufficient, safe, nutritious food to maintain a healthy and active life*'.

During the Twentieth session of 1999 the Committee on Economic, Social and Cultural Rights issued the General Comment n. 12 to the International Covenant on Economic, Social and Cultural Rights, with the objective of '*a better definition of the rights relating to food*' and with the purpose of '*monitoring the implementation of the specific measures provided for in article 11 of the Covenant*', as stated in the General Comment 12 to the Right to adequate food (Article 11).

Since Article 11 sets forth concrete commitments on the part of State Parties, General Comment n. 12 highlights the objectives which these concrete actions should be aimed to, by stating that '*the right to adequate food is realized when every man, woman and child, alone or in community with others, has the physical and economic access at all times to adequate food or means for its procurement*'.

It has to be noted that in the Declaration on World Food Security of 1996 the objective to halve the number of undernourished people no later than the year 2015 was provided. In effect, the eradication of extreme poverty and hunger is included, as goal number one, in the Millenium Development Goals (established in 2000) to be achieved within 2015.

Moreover, 2015 has been designed, at EU level, as the year of development. Today, in 2015, an urgent question is raised, is the right to food a goal really achieved in the World?

The Report on the New Millenium Goals adopted in 2014 underlines both pros and cons concerning the achievement of the goal by stating that '*The proportion of undernourished people in developing regions has decreased from 24% in 1990-1992 to 14% in 2011-2013. However, progress has slowed down in the past decade. Meeting the target of halving the percentage of people suffering from hunger by 2015 will require immediate additional effort, especially in countries which have made little headway*'.

As can be seen in the report, the data on the proportion of undernourished people, indicated in percentage and in relation to the period 1990-1992 and 2011-2013, mainly refers to developing regions. Concerning the developed regions, it appears, from what is stated in the 2014 report, that the level of undernourishment has not changed, levelling out at <5%.

However, other factors should be taken into account when analysing the right to food in the EU: poverty, social exclusion and unemployment.

These are pressing problems in the EU, where the poverty and social exclusion, according to the data provided in relation to the matter of poverty and social exclusion in the EU, have reached nowadays shocking levels: 24% of all the EU population is at risk of poverty or social exclusion, 10% of Europeans live in households where no one has a job and 12 million more women than men are living in poverty in the EU. Furthermore, as estimated by Eurostat, even if the percentage of unemployment in the EU is being reduced, 23.748 million men and women are still unemployed (data of March 2015).

Since poor, socially excluded and unemployed people are experiencing more and more problems on the economic access to food, new challenges should be faced by the EU in the domain of food security. In effect, food security is not only a matter of food supply (Gualtieri, 2013) but involves other aspects such as the economic access to food (Gualtieri, 2013).

EU policies have been oriented, since 2002, mainly towards the objective of food safety. In that moment, after the crisis provoked by the spread of the bovine spongiform encephalopathy, commonly known as mad cow disease, it was really necessary to provide rules and instruments with the purpose of dealing with any food safety problems '*in order to ensure the proper functioning of the internal market and to protect human health*' (Recital No. 10, Regulation (EC) No. 178/2002). Several regulations and instruments exist in the EU in order to guarantee a high level of food safety and an important policy on food safety has been implemented both at EU level and at national level.

On the other hand food security has always been considered an objective already achieved at EU level. However, in the present situation, this certainty has disappeared.

Therefore, it is really important to take into consideration if the present EU sources may reach the objective to (immediately) overcome the existing serious situation of 'food insecurity', as has been defined (Jannarelli, 2010).

A question comes to the fore: are the existing tools sufficient in order to deal with the growing problem of economic food insecurity in the EU?

The human right to food is not expressly recognized in the EU Charter of Fundamental Rights, but it can be derived from the fundamental right to dignity (Title I Charter) that is one of the values on which the EU is founded (Article 2 Treaty of the EU).

This proclamation of the fundamental value and right of dignity is extremely significant. The Treaties represent the primary source of law in the EU and the rights recognized in the Charter guide the actions of the EU in the areas of its competence and of the Member States when they transpose EU Law.

Thus, in the EU the right of dignity, which has to include the right to food, should create the basis of all secondary legislation. Moreover, there are no specific provisions that establish the objective of food security in the EU yet. Probably, this has to be linked to the fact that solutions cannot be reached only at EU level. The situation of economic food insecurity may derive from problems connected to employment and to the promotion of social rights. As referred to in Article 4, par. b, of the Treaty on the functioning of the EU the competence of *'social policy, for the aspects defined in the Treaty'* is shared between the Union and the Member States. As to employment, the treaty on the functioning of the EU dedicates Title IX to it and it states, in Article 145, that *'Member States and the Union shall, in accordance with this Title, work towards developing a coordinated strategy for employment (...)'.* Regarding social policy, the Treaty on the functioning of the EU dedicates Title X to it and, in Article 151, it affirms that *'the Union and the Member States (...) shall have as their objectives the promotion of employment, improved living and working conditions'.* Therefore, a strong commitment is needed both on the part of the EU and the 28 Member States towards an improvement of working conditions and social policy in the EU.

Furthermore, the present food insecurity becomes increasingly alarming if we think about future challenges such as the growth of the world population and the reduction of food availability for people and/or the increasing level of disparities in the access to food.

To conclude, measures of intervention on the part of the EU Member States and of the EU itself are required. In effect, as Ms. Hilal Elver, the special rapporteur on the right to food, stated in the report concerning the *access to justice and the right to food: the way forward,* of 12 January, 2014, *'Wealthy States not only have moral obligations to address poverty and hunger beyond their borders, they are also legally obliged to do so under international law. International cooperation and development assistance must become the legal norm in an increasingly global world'.*

References

Committee on Economic, Social and Cultural Rights (26 April-14 May 1999, Twentieth session). Substantive issues arising in the implementation of the International Covenant on Economic, Social and Cultural Rights: General Comment 12, The right to adequate food (Article 11). Available at: http://tinyurl.com/csyydtr.

European Commission (2002). Regulation (EC) No 178/2002 of the European Parliament and of the Council of 28 January 2002. Laying down the general principles and requirements of food law, establishing the European Food Safety Authority and laying down procedures in matters of food safety. Official Journal L 31, 1 February 2002, 1-24.

European Commission (2010). EU Charter of Fundamental Rights (30 March 2010). Official Journal C 83, 30 March 2010, 389-402.

European Commission (2012). Consolidated version of the Treaty on the Functioning of the European Union. Official Journal C 326, 26 October 2012, 47-390.

European Commission (2015a). Poverty and social exclusion. Available at: http://ec.europa.eu/social/main.jsp?catId=751.

European Commission (2015b). Our world, our dignity, our future / 2015 European Year for Development. Available at: https://eudevdays.eu/theme/our-world-our-dignity-our-future.

Eurostat (2015). Euro area unemployment rate at 11.3%. Available at: http://tinyurl.com/oagred9.

Gualtieri, D. (2013). Right to food, food security and food aid under international law, or the limits of a right-based approach. Future of Food, 1, 2, 18-28.

Jannarelli, A. (2010). La nuova food insecurity: una prima lettura sistemica. Rivista di diritto agrario, I, 565-606.

UNHR (1966). International covenant on economic, social and cultural rights (16 December 1966), Available at: http://www.ohchr.org/EN/ProfessionalInterest/Pages/CESCR.aspx.

United Nations (1948). The Universal Declaration on Human Rights (1948). Available at: http://tinyurl.com/cgnkmq

United Nations (2014a). Report of the Special Rapporteur on the right to food, Ms. Hilal Elver (12 January 2014).

United Nations (2014b). Millenium Development Goals Report. Available at: http://tinyurl.com/pfz4kn9.

Van der Meulen, B., Rătescu, I. (2014). Food prints on human rights law paradigms. European Food and Feed Law Review, dossier n. 6, 2014: 372-381.

World Food Summit (13-17 November 1996). Rome Declaration on World Food Security and World Food Summit Plan of Action. Available at: http://tinyurl.com/o4z7kjl.

40. The contribution of community food gardens to food sovereignty in Johannesburg, South Africa: a look at localisation and democratisation

B. Kesselman

University of KwaZulu Natal, Development Studies, Memorial Tower Building, Howard College, Durban 4041, South Africa; brittany.kesselman@gmail.com

Abstract

This paper addresses the question of how urban community gardens contribute to food sovereignty. While the concept of food sovereignty was developed by peasant movements to address the injustices in the current globalized industrial food system, it has also come to be applied to urban contexts. This paper focuses on two key elements of food sovereignty, namely food system localization and democratization, while also considering the importance of scale in assessing impacts on food sovereignty. The research was conducted at two case study gardens in socioeconomically deprived areas of Johannesburg, South Africa, with additional research at other gardens around the city. The research found evidence of food system localization at the case study gardens, in the form of face-to-face relationships between customers and farmers; shortened food value chain; and a contribution to the local economy. There is limited evidence of food system democratization at the local level, in terms of increases in food choices and engagement with policy makers. However, this democratization has not been scaled up to regional or national levels. While community food gardens do contribute to food sovereignty in a number of ways, their transformative potential is not being fully realized. In order to transform the food system, additional education and improved networking opportunities will be required so that the impact of food gardens can be enhanced.

Keywords: food democracy, local food system, scale

1. Introduction

The paper addresses the question of how urban community gardens contribute to food sovereignty. The concept of food sovereignty grew out of a recognition of the limitations of the concept of food security, particularly its silence with regard to how food is produced, by whom, and who controls the food system (Schanbacher, 2010). While the concept of food sovereignty was developed by peasant movements to address the injustices in the current globalized industrial food system, it has also come to be applied to urban contexts. Food sovereignty emphasises rights, democratic processes, local control and transformed social relations free of oppression and inequality (Nyéléni Declaration on Food Sovereignty, 2007). While food sovereignty has many components, this paper focuses on food system localisation and democratisation, while also considering the importance of scale in assessing impacts on food sovereignty.

2. Methods

Research was conducted at two case study gardens in socio-economically deprived areas of Johannesburg, South Africa, with additional research at other gardens around the city. The research methods include participant observation, key informant interviews, food diaries and life history interviews. The primary research complemented an extensive review of relevant academic literature as well as policy documents.

3. Context

Johannesburg is a city of contrasts, with extreme wealth alongside severe poverty, unemployment and food insecurity. The overall rates of poverty (estimated at 21.6% of households in 2008) and unemployment (officially 26.7% in 2015) smooth over serious inequalities, with geographical 'deprivation clusters' suffering from much lower levels of income, employment, education, health and a poorer living environment (City of Johannesburg, 2013; Statistics South Africa, 2015).

There is sufficient food available in Johannesburg to meet the requirements of all residents. However, food security in the city is a matter of access to food, rather than availability, so high levels of poverty and unemployment translate into high levels of food insecurity (Rudolph *et al.*, 2012). One survey of food insecurity, conducted in three relatively poor areas of the city found that 56% of households were food insecure, with 27% severely food insecure, using the Household Food Insecurity Access Scale (Rudolph *et al.*, 2012). Another study focusing on the most deprived wards of the city found rates of food insecurity as high as 90% (De Wet *et al.*, 2008).

The two case study gardens are located in areas covered by the first food insecurity survey mentioned above. Suthani community garden (the name means 'make yourselves full' in isiZulu) is in East Bank, Alexandra township, in the north of Johannesburg. It has been operating since 2010, on an empty lot under high-tension electrical wires, owned by the national power utility, Eskom. It is registered as a cooperative, with seven members, of whom five regularly work in the garden. Those five members are all elderly pensioners. The other garden, Bambanani Food and Herb Cooperative (meaning 'hold each other' in isiZulu), is located in Bertrams, in the inner city. It started in 2006, on land owned by the City of Johannesburg, and there are six members of the cooperative, of whom only two regularly work in the garden.

4. Food system localisation

Amongst the six principles of food sovereignty developed by food producers, consumers and activists of the food sovereignty movement at Nyéléni, is that it localizes food systems. This is defined as follows:

> Food sovereignty brings food providers and consumers closer together; puts providers and consumers at the centre of decision-making on food issues; ... protects consumers from poor quality and unhealthy food, inappropriate food aid and food tainted with genetically modified organisms; and resists governance structures, agreements

and practices that ... give power to remote and unaccountable corporations (Vía Campesina, 2007).

This definition from Nyéléni is more or less aligned with the generally-celebrated elements of localisation as: (a) favouring small-scale producers and alternative distribution networks with shorter food supply chains over corporate concentration and globalisation; (b) re-embedding the market in face-to-face social relations; and (c) giving more power over the food system to food providers and consumers, starting at (but not limited to) the local level (Feagan, 2007; Feenstra, 1997; Hendrickson and Heffernan, 2002). Other arguments in support of food system localisation cite the environmental benefits, as well as the development of the local economy (Kloppenburg *et al.*, 2008; Maretzki and Tuckermanty, 2007; Pothukuchi *et al.*, 2002).

The un-reflexive celebration of the local has been heavily criticized. Aside from challenges in defining the local, critics have pointed out its potential to obscure local power relations, exclude certain groups and to conflate spatial relations with social relations (Bellows and Hamm, 2001; DuPuis *et al.*, 2006; Feagan, 2007). As Born and Purcell (2006) argue, there is nothing inherently desirable about the local, as a scale. These critics, however, do not find anything inherently undesirable about the local, either, but rather call for a reflexive approach that focuses on social relations rather than spatial ones (Hinrichs, 2000).

In terms of the desired outcome of food sovereignty, the issue is whether local food initiatives 'can effectively introduce any measure of democratic control over economic systems that are essentially nondemocratic or whether meaningful agrifood system change can only be accomplished by first transforming the larger society as a whole' (Ostrom and Jussaume, 2007). This question applies not only to democratic control, but to other desired outcomes such as environmental sustainability as well.

The research found evidence of food system localisation at the case study gardens, in the form of face-to-face relationships between customers and farmers, shortened food value chains, and a contribution to the local economy. Many customers who come to the case study gardens to buy food have gotten to know the members of the cooperatives working there, and find them 'friendly' and 'helpful' (customer interviews). They indicate that the produce is fresher and of better quality than in the supermarket, and that it 'still has vitamins' (customer interviews).

In poor South African townships, supermarket penetration is still relatively uneven, with many South Africans still buying a significant portion of their food from street traders and small informal shops (called 'spazas') (Battersby and Peyton, 2014; Pereira, 2014). There is no supermarket in the immediate vicinity of Suthani community garden, though there are some informal vendors in the area. While some of these vendors stock some fresh produce, the selection is limited. In addition, most of them get their fresh produce from the main Johannesburg Fresh Produce Market in the south of the city, a massive market trading in produce from commercial farms all over the country (key informant interviews). This raises questions about what makes a 'local' food system: If local vendors sell food produced by distant commercial farms, sourced through a massive 'middleman' located outside of the community, that is hardly a shortened food value chain. Thus Suthani fills an important need for fresh produce in the area, providing a greater selection and saving customers the expense, in time and money, of a trip to the supermarket. It also represents the shortest possible food value chain, and all of the money spent by customers on food produced there remains in the local community (at least until the gardeners go to the supermarket).

Bambanani, located in central Johannesburg, is near a number of other food retail outlets, including a supermarket. One of the main draws for customers at Bambanani is the variety of produce sold, including some traditional indigenous vegetables that are not commonly sold in supermarkets. Local residents have even brought seeds to Suthani in the past, so that the garden could grow particular vegetables that they missed (participant observation and key informant interviews). In addition to direct sales to passers-by, Suthani also supplies the local supermarket with spinach – this relationship contributes further to the localisation of the food system, although the majority of the supermarket's products are sourced through a national distribution system (participant observation and key informant interviews).

5. Food system democratisation

The concept of food system democratisation is included in two of the principles of food sovereignty developed at Nyéléni; the one discussed above as well as the next, which states that food sovereignty 'puts control locally' (Vía Campesina, 2007). As Hassanein (2003) defines it, food democracy 'is about citizens having the power to determine agro-food policies and practices locally, regionally, nationally, and globally'. Thus democratisation of the food system should occur at multiple scales, and should involve active participation of community members in decisions about their food system. This once again raises the question asked above, as to whether local food initiatives can contribute to democratising an undemocratic global system (Ostrom and Jussaume, 2007).

South Africa is a democracy, with a constitution that recognizes the right to food (Republic of South Africa, 1996). Despite this, individuals have little say in the development of policies and legislation that affect the food system. The recent adoption of the new Integrated Food Security and Nutrition Policy (gazetted in 2014) provoked anger from many civil society groups due to the secrecy and lack of consultation involved in the process (South African Food Sovereignty Campaign, 2015). For the average citizen, the complex maze of policies, laws and regulations that govern the food system – from agricultural production to food processing to retail distribution – are unknown and inaccessible. Even for this researcher, some policies have been incredibly difficult to acquire, despite rigorous searches and requests to those tasked with implementing them. Thus the majority of the South African population is not actively involved in shaping food system policy.

Both gardens receive some support from two levels of government: the municipal level through the Johannesburg Department of Social Development (DSD) and City Parks, and the provincial level, through the Gauteng Department of Agriculture and Rural Development (key informant interviews). Through interaction with government officials, the gardeners have opportunities, at least theoretically, to shape government policies and programmes at local or provincial level. In the case of Bambanani, cooperative members have regularly engaged with the head of DSD and have even met the mayor (participant observation and key informant interviews). It is unclear how much impact, if any, this interaction may have had on policy development, and it certainly has not been scaled up to higher levels.

At the level of food practices, however, we can see the impact of the community food gardens. One of the two case study gardens is committed to organic, permaculture methods of production, while the other uses limited chemical inputs (key informant interviews and participant observation). Gardeners are able to determine exactly how their food is produced, as well as what foods they produce. By growing indigenous vegetables not commonly found in shops, the gardens expand the food choices available in their communities. Community members are also able to influence production, either directly by requesting certain vegetables or indirectly through purchasing practices.

6. Discussion

The issue of scale is critical in assessing the contribution of community food gardens to food sovereignty. Are they shifting the balance of power with regard to food in their immediate communities? Can they have an impact at higher scales (e.g. regionally or nationally)? The question of whether food system changes can be made at a local level, or whether the local food system is largely determined at higher levels of governance (e.g. national laws and regulations, international trade arrangements and practices of transnational food corporations) impinges directly on the potential impacts of community food gardens.

From the literature and the field research in gardens in Johannesburg, it appears that some impact at the local level is possible. The two case study gardens do change the nature of food purchasing for their customers, through face-to-face interaction between producers and consumers. They share their knowledge with their customers, and also accept suggestions from customers on what to produce. The quality and availability of produce in the community is better as a result of the gardens, resulting in improved and expanded food choices. They also contribute to the local economy, by keeping at least some food spending in the community rather than having it drain out through large corporate retailers.

In terms of democratisation, the impact appears fairly limited. It seems unlikely that the gardeners, even though their interaction with local and provincial government representatives, have much influence on the development of government policies and programmes. In only one of the case study gardens is the contact with government frequent and consistent. In the other case study garden, and most of the other gardens surveyed, visits by city or provincial government representatives are sporadic and brief (garden interviews). While the gardens do appear to have at least some impact on food practices, there is no indication that this influence extends beyond the immediate community of gardeners and their customers.

Hinrichs and Barham (2007) argue that scaling up local food initiatives can be challenging. However, they point to the development of federations and other networks as a promising development, that enable individual local initiatives to form groups that can influence policy at higher levels. In Johannesburg, there have been some efforts to form regional fora of gardeners, though these have largely been the result of city initiatives aimed at streamlining its engagement with gardens (key informant interviews). When one of the members of Suthani cooperative attempted to start a forum in the area, other gardeners treated the idea with suspicion (key informant interview). In addition, expensive and relatively poorly connected public transport system in Johannesburg is a real obstacle to gardener organisation.

For many gardeners, 'the local' is their immediate community, with other parts of the city largely out of reach. This is particularly apparent at Suthani, which is only about 7.5 km from Sandton, the wealthy commercial centre of Johannesburg. The contrast between the mansions, luxury hotels and expensive retail shops in Sandton and the government housing, shacks and illegal dumping grounds in Alexandra is stark. Despite being only 7.5 km away, it would take about 30-40 minutes to get there by public transport, at a not-insignificant cost. It is, for all practical purposes, a world away, and not accessible even as a market for the garden's produce. In light of these kinds of socio-spatial barriers to scaling up, the non-governmental organisations and government departments that support community gardens may need to provide them with greater networking opportunities, improved transport access and additional education and training, in order to empower gardeners to play a more active and informed role in decisions affecting them.

References

Battersby, J., Peyton, S. (2014). The geography of supermarkets in Cape Town: supermarket expansion and food access. Urban Forum 25: 153-164.

Bellows, A., Hamm, M. (2001). Local autonomy and sustainable development: Testing import substitution in localizing food systems. Agriculture and Human Values 18: 271-284.

Born, B., Purcell, M. (2006). Avoiding the local trap: scale and food systems in planning research. Journal of Planning Education and Research 26: 195-207.

City of Johannesburg. (undated a). Jozi: A City @ Work. 2012/16 Integrated Development Plan, 2014/15 Review.

City of Johannesburg. (undated b). Summarised Version of the 2013/16 Integrated Development Plan (IDP).

City of Johannesburg. (2013). City of Johannesburg: 2013/16 Integrated Development Plan (IDP).

De Wet, T., Patel, L., Korth, M., Forrester, C. (2008). Johannesburg Poverty and Livelihoods Study, Johannesburg, RSA.

DuPuis, E.M., Goodman, D., Harrison, J. (2006). Just values or just value? Remaking the local in agro-food studies. In: Marsden, T., Murdoch, J. (eds.), Between the Local and the Global: Confronting complexity in the contemporary agri-food sector, Vol. 12, Elsevier JAI Press, Oxford, UK, pp. 241-268.

Feagan, R. (2007). The place of food: mapping out the 'local' in local food systems. Progress in Human Geography 31: 23-42.

Feenstra, G. (1997). Local food systems and sustainable communities. American Journal of Alternative Agriculture 12: 28-36.

Hassanein, N. (2003). Practicing food democracy: a pragmatic politics of transformation. Journal of Rural Studies 19: 77-86.

Hendrickson, M., Heffernan, W. (2002). Opening Spaces through Relocalization: Locating potential resistance in the weaknesses of the global food system. Sociologica Ruralis 42: 347-369.

Hinrichs, C. (2000). Embeddedness and local food systems: notes on two types of direct agricultural market. Journal of Rural Studies 16: 295-303.

Hinrichs, C., Barham, E. (2007). Conclusion: a full plate. challenges and opportunities in remaking the food system. In: Hinrichs, C., Lyson, T. (eds.) Remaking the North American food system: strategies for sustainability. University of Nebraska Press, Lincoln, USA, pp. 345-356.

Kloppenburg, J., Hendrickson, J., Stevenson, G.W. (2008). Coming in to the foodshed. In: Pretty, J. (ed.) Sustainable agriculture and food, Volume 3. Earthscan, London, UK, pp. 363-374.

Maretzki, A.N., Tuckermanty, E. (2007). Community food projects and food system sustainability. In: Hinrichs, C., Lyson, T. (eds.) Remaking the North American food system: strategies for sustainability. University of Nebraska Press, Lincoln, USA, pp. 332-344.

Nyéléni (2007a). Nyéléni Declaration on Food Sovereignty. Nyéléni, Mali. Available at: http://tinyurl.com/obhm63u.

Nyéléni (2007b). Nyéléni Synthesis Report. Nyéléni, Mali. Available at: http://tinyurl.com/neaoq88.

Ostrom, M.R., Jussaume, R. (2007). Assessing the significance of direct farmer-consumer linkages as a change strategy in Washington State: civic or opportunistic? In: Hinrichs, C., Lyson, T. (eds.) Remaking the North American food system: strategies for sustainability. University of Nebraska Press, Lincoln, USA, pp. 235-259.

Pereira, L. (2014). The future of South Africa's food system: what is research telling us? South Africa.

Pothukuchi, K., Joseph, H., Burton, H., Fisher, A. (2002). What's cooking in your food system? A guide to community food assessment. Venice, CA.

Republic of South Africa (1996). Constitution of the Republic of South Africa.

Rudolph, M., Kroll, F., Ruysenaar, S., Dlamini, T. (2012). The state of food insecurity in Johannesburg. Kingston and Cape Town.

Schanbacher, W.D. (2010). The politics of food: the global conflict between food security and food sovereignty. ABC-CLIO, p. 148.

South African Food Sovereignty Campaign. (2015). Statement from 1st National Coordinating Committee Meeting. Johannesburg.

Statistics South Africa. (2015). Quarterly Labour Force Survey. Quarter 1, 2015. Pretoria.

41. Caught in the middle: mapping the actors negotiating food security

C. Godet

Université Saint-Louis Bruxelles, Boulevard du Jardin Botanique 43, 1000 Brussels, Belgium; claire.godet@usaintlouis.be

Abstract

Who negotiates food security? Scholars have brought different answers to this question depending on their main area of focus (trade, local development, agricultural policies, etc.). Most of the time, however, they segment the negotiation space by focusing on one or two types of actors (e.g. private actors, national governments or international organizations) or one specific issue. The regime complex theory, on the other hand, highlights the different regimes – and thus, systems of values and interests – in which an actor can be involved. However, it does not take a close look to each single actor and its behaviour. This contribution aims at mapping the actors involved in the negotiation process at the international level and at understanding the contradictions and compromises they have to deal with. Using the regime complex theory combined to a sociological approach of international relations, it focuses on the actors that negotiate policies against food price volatility. Based on official documents and scientific literature on the negotiations at the World Bank and at the Food and Agriculture Organization, it analyses how a same actor may be required to defend contradicting interests in negotiation processes. This contribution might help to understand why, even though all the actors seem to agree on the fact that food security is a fundamental issue, international cooperation fails to produce effective agreements to implement it.

Keywords: international organisations, regime complex, FAO, World Bank

1. Introduction

Who negotiates food security? Scholars have brought different answers to this question depending on their main area of focus (trade, local development, agricultural policies, etc.) and on the actor they examine. Recently, Margulis (2010, 2013) showed that international organisations (IOs) had a continuously evolving role to play in negotiating food security policies. His map of the regime complex of food security presents the main IOs involved in the global governance of food security. The aim of this contribution is to point a magnifying glass to this map to identify who are the actors within the IOs that negotiate food security, and in particular the actors who work to limit food price volatility. It focuses on two IOs: the UN Food and Agriculture Organisation (FAO) and the World Bank. Indeed, these two actors, despite their many differences, have both reaffirmed the necessity to fight against food price volatility and taken new measures to encourage discussions on the matter. However, they seldom work together and the links between them are not clearly established.

The first part of this contribution shows why it is important to study IOs as autonomous actors in international relations. The second one presents briefly the initiatives taken to limit food price volatility and the third part will identify the actors involved in the negotiation process.

2. International organisations as autonomous actors

IOs are conceptualised as actors since the 1990s. Beforehand, they were mainly seen as arenas in which states defend their interests (Morgenthau, 1948; Waltz, 1979) or as a means to limit transaction costs (Keohane and Nye, 1989). Constructivists developed the idea that states were not the only actors on the international scene and brought a new light on IOs. This contribution adopts the constructivist definition of IOs; it states that, as any other social object, IOs are human constructions resulting from social interactions. Griffith and O'Callaghan (2002, p. 51) define them 'as fundamentally cognitive entities that do not exist apart from the actors' ideas about how the world works'. They see IOs 'as social networks and patterned sets of interactions that take on a life of their own' (Oestreich, 2011, p. 168). According to this perspective, they are actors with their own values and norms; and the production of international norms is not as much the consequences of the states' interests than the results of shared values and beliefs transmitted through socialization. Rather than questioning why the IOs were created or how to measure their efficiency, this literature intends to understand whether the IOs do what they have been created to do. It adapts the principal-agency theory developed in industrial economics to the relationship between IOs' bureaucracies and member states (Haftel and Thompson, 2006; Hawkins, 2006). It emphasizes the characteristics and mechanisms that explain the IOs' power, autonomy or their propensity to 'pathological behaviours' (Barnett and Finnemore, 1999, p. 700). This contribution does not evaluate the autonomy of the IOs but it still considers IOs as social networks and it describes the interactions between these different networks.

The interactions between these networks have been broadly drawn by Margulis (2010) in his design of the regime complex of food security. A regime complex is 'a network of three or more international regimes that relate to a common subject matter; exhibit overlapping membership; and generate substantive, normative, or operative interactions recognized as potentially problematic whether or not they are managed effectively' (Orsini *et al.*, 2013, p. 29). In the regime complex of food security, conflicts arise notably to the diverging views on trade liberalization. Despite the common ground, divergent norms can engender problems of trust between institutions and undermine multilateral efforts (Margulis, 2013). This contribution takes a closer look to the FAO and the World Bank to understand who are the persons involved in the negotiation process and which norms they bring to the table. This sociological approach sheds a new light on these intertwined networks.

3. Limiting food price volatility

Since the food crises of 2008 and 2010, global food security has remerged as an urgent matter and IOs have increased their attention to the issue. The creation of the UN High Level Task Force on the Global Food Security Crisis, the decision by the World Bank to make food security a priority or the G8's new global food security financing mechanisms show that IOs have adapted their behaviour to the recent crises (Margulis, 2010, p. 33). Among the new topics brought to the surface by these events, the limitation of food price volatility has been one of the most prominent. Indeed, IOs seem to agree that the crises were not only due to particularly high prices but rather to dramatic variations in prices (UN, 2011, p. 62). Since then, several initiatives have been taken to better the limitation of price volatility. Olivier De Schutter, former UN special rapporteur on the right to food fought against speculation on food commodities; the World Bank has created Food Price Watch to follow the evolution of food prices; the G20 requested an international report on the matter; and the FAO has organised several high level meetings and ministerial conferences on the issue. However, the evolution is slow and food prices are still not under control. Event though these IOs work for a same objective, they seem to have difficulties working together. It is thus necessary to understand their interactions to identify the crux of the problem.

4. Mapping the actors involved in the negotiation process: expected results

The research focusses on the actors in two particular IOs: the FAO and the World Bank. Both of them pertain to the regime complex of food security (Margulis, 2013). Both of them also work on other development issues and have insisted on the necessity to cooperate to limit food price volatility (FAO, 2011; World Bank, 2012). However, the research results will show that they do not often work together and their direct interactions are limited. The main nexus of collaboration between these two actors is via their common members and partners. Even though the FAO and the World Bank do not often talk face to face, they have many common interlocutors. This poster intends to show who these interlocutors are and how they link two very different IOs.

References

Barnett, M., Finnemore, M. (1999). The politics, power and pathologies of international organizations. International Organization, 53, 4: 699-732.

Food and Agriculture Organisation (FAO) (2011). Price volatility and food security. A report by the High Level Panel of Experts on Food Security and Nutrition. Available at: http://tinyurl.com/3kjjbpz.

Griffiths, M., O'Callaghan, T.O. (2002). International relations: the key concepts. Routledge, London, 464 pp.

Haftel, Y.Z., Thompson, A. (2006). The independence of international organizations: concept and application. Journal of Conflict Resolution, 50, 2: 253-275.

Hawkins, D.G. (2006). How agents matter. In: Hawkins, D.G. (ed.). Delegation and agency in international organizations. Cambridge University Press, Cambridge: 199-228.

Keohane, R.O., Nye J.S. (1989). Power and interdependence. Scott, Foresman and Company, Glenview, 315 pp.

Margulis, M.E. (2010). International (re)organization: the emergence of the Global Food Security Regime. In: REPI Workshop on Issue-linkage and regime complex, Université Libre de Bruxelles, Brussels.

Margulis, M.E. (2013). The regime complex for food security: Implications for the global hunger challenge. Global Governance, 19: 53-67.

Morgenthau, H.J. (1948). Politics among nations: The struggle for power and peace. Alfred Knopf, New York, 688 pp.

Oestreich, J.E. (2011). Intergovernmental organizations in international relations and as actors in world politics. In: Reinalda, B. (ed.). The Ashgate research companion to non-state actors. Ashgate, Farnham and Burlington: 161-172.

Orsini A., Morin, J-F., Young, O. (2013). Regime complexes: a buzz, a boom or a boost for global governance? Global Governance, 19: 27-39.

United Nations (UN) (2011). The global social crisis. Report on the world social situation 2011. United Nations, New York, 110 pp.

Waltz, K.N. (1979). Theory of international politics. McGraw-Hill, New York, 251 pp.

World Bank (2012). Responding to higher and more volatile world food prices. Report n° 68420-GLB. Available at: http://tinyurl.com/o3jsvua.

42. Who are behind the food security initiatives in Nicaragua – a comparative network analysis across public policy cycle

M. Moncayo Miño[1] *and J.L. Yagüe Blanco[2]*

[1]*Technical University of Madrid, School of Agricultural Engineering, Complutense Avenue 3, 28040 Madrid, Spain;* [2]*Planning and Management of Sustainable Rural Development Research Group, Technical University of Madrid, School of Agricultural Engineering, Complutense Avenue 3, 28040 Madrid, Spain; marco.moncayo.mino@gmail.com*

Abstract

Latin America has made considerable advances in terms of developing formal agreements and the establishment of a legal framework that safeguards the right to food. Some countries like Nicaragua are promoting new local working bodies for fighting hunger. These aim to provide greater governance in food security and strengthening policies and programmes. The Nicaraguan food security law was adopted in 2009 and allows local governments to implement a Municipal Committee of Food Security to liaise with stakeholders and enhances an ongoing dialogue between the central government and civil society. The aim of this research is to recognise how this network is related, showing the actors who have more control over the development and implementation of public policies that promote the right to food. Employing policy network analysis as methodology, actors of these committees in seven municipalities were interviewed. Their interaction in the public policy cycle was evaluated, resulting that Municipal Committee represents an arena within which institutional and societal interests interact, establishes priorities and defines resource mobilisation from public and private sources. The central government institutions are active participants over the entire cycle, which is the principal component of the network and remains a central actor in these processes along with other actors. Local government interacts in the cycle but has no main responsibilities, while some institutions prefer to work with civil society organisations directly. The agenda-setting and implementation phases have more interaction including cooperation and academic actors. In formulation, approval and evaluation phases the State institutions are the key actors defining objectives and goals. The results show how actors interact at municipal level and allow future joint initiatives to take into account the particularities of this research.

Keywords: governance, COMUSSAN, policy network

1. Introduction

The Latin American country that has the greatest progress in the fight against hunger is Nicaragua. It has gone from having a prevalence of malnutrition of 54.4% of its total population in 1990 to 16.6% in 2014 (FAO, 2015). Various efforts have been made to achieve these results, including the implementation of the Law on Sovereignty and Food Security and Nutrition (LSSAN). This law promotes a model of governance that seeks to create conditions for greater involvement of the private sector and civil society.

After five years since the entry into force of the LSSAN, it has not yet managed to form all areas of the interagency interaction as proposed and there are several factors that have affected these processes. At the municipal level, unlike other territorial levels, it has managed to build methods of participation in 55 municipalities. These spaces known as Municipal Committees of Sovereignty and Food Security and Nutrition (COMUSSAN) gather civil society, private and public actors to coordinate joint actions. Among the main responsibilities that the law gives, the COMUSSAN perform the following: coordinating the development, implementation and evaluation of public policies, planning and evaluating food security operations, among others

Pierre and Peters (2005) mention that networks of actors that are better connected with government institutions responsible for policy have a greater public policy influence. In this respect the process of implementing the law has become a process of institutional learning, in which the actors have undertaken efforts to break through the traditional barriers between public, private and civil society sectors, introducing new thinking of the role of the state and its interaction with society in decision-making, changing established norms and generating new government processes.

These networks are formed from common interests and are presented as a mediation mechanism and influence of actors (Grimes, 2013; Kickert *et al.*, 1997), which involves generating interactions to seek consensus and cooperation in the implementation of public policy (Duit and Galaz, 2008; Giest and Howlett, 2014). The beginning of the dialogue and joint activities within the COMUSSAN is recognition by those involved that there is no single actor who has the knowledge and resources necessary to address the fight against hunger unilaterally and requires a coordinated effort with common goals.

This research aims to identify the behaviour of the network of actors that comprise COMUSSAN and to recognise which are the most representative stakeholders in the creation of public policies.

2. Methodology

Selection of the survey respondents

The selection of political, technical and academic stakeholders was based on the premise of who cooperated in the LSSAN elaboration and are participating in the law execution. A snowball sampling was applied and 40 actors from 6 COMUSSAN (Somotillo, El Jicaral, El Sauce, Santa Rosa del Peñón, Totogalpa and San Juan de Limay) were interviewed.

Public policy cycle

The respondents filled a standardised survey to measure their engagement with the network. The variable used was interaction in public policy cycle. Five phases of this cycle were identified (Jann and Wegrich, 2006): (1) agenda-setting; (2) formulation; (3) approval or decision-making; (4) implementation; and (5) evaluation. Each actor identified in which phase or phases and with which actor or actors of the network interacts. If there exists any interaction a value of 1 was assigned, otherwise the interaction was marked with a 0. From 40 surveys 3 were deleted due to lack of information. The actors were grouped in 9 categories according to the institution they represent:

MAGFOR (Ministry of Agriculture and Forestry); MEFCCA (Ministry of Family Economy); MINED (Ministry of Education); MINSA (Ministry of Health); INTA (National Institute of Agricultural Technology); CIUSSAN (Universities' Committee); local governments (municipalities); partners (NGOs and international agencies) and agro producers associations. These categories takes into account the conformation of COMUSSAN proposed by the LSSAN. Actors from the same category may have different kinds of interactions; therefore it is difficult to compare just one actor across the network. To obtain the most representative value of each category Statistical Mode was used. This research seeks to determine the behaviour of actors within the municipal general level, not in a particular municipality. Although the community cabinets (cabinets of the family, community and life) is an actor whom may be a guest in COMUSSAN according to the LSSAN. This research was included because all actors mention it.

Social network analysis

The data were analysed in UCINET (Borgatti *et al.*, 2002). We worked with a square matrix (10×10), which consists of 9 actor categories and community cabinets. From community cabinets, two tests were conducted to obtain their response: (1) matching results to those granted by the network; and (2) giving the value of 0 (this value was used for this research). In both cases the degree centrality and betweenness centrality had no changes. Network graphs were developed with NetDraw in each phase. Black colour represents governmental actors and grey colour non-governmental actors. The structural importance of actors was assessed through betweenness centrality to contrast their power in the network (Ingold, 2014). It calculates the number of times an actor is on the path between two non-interlinked actors, high betweenness means superior access to network-based resources, or social capital (Lin, 2001) and may serve as a broker or mediator between or among actors; and degree centrality, shows the whole number of interactions in which an actor participates. In policy networks, actors with high betweenness centrality and degree centrality are better integrated in the network and can have more influence into the planning process and the policy resource management (Lienert *et al.*, 2013; Papadopolou *et al.*, 2011).

For each phase the density network was obtained dividing the sum of the degree centrality of each actor per phase by the total amount of possible interactions of a square matrix of 10 actors (10×10 − 10 = 90). The actors are connected by undirectional arrows in the event that only one player recognises the interactions, and arrows are bidirectional when the two stakeholders recognise interaction.

3. Results

There are differences of interaction during the cycle of public policy. Although the law does not provide specific functions to each institution that forms the COMUSSAN, certain actors assume greater responsibility. The phases that have greater interaction are implementation (80% network density) and agenda-setting (51.1% network density).

Andrews (2013) mentions that more effective institutional network comes from the identification of the problem that needs to be solved and the development of institutional solutions, probably in an incremental way and growing out of the existing institutional context. As shown in Table 1, in agenda-setting the actors that have greater degree centrality are MAGFOR, INTA, municipal

Table 1. Degree centrality and betweenness centrality of COMUSSAN network.[1]

	Agenda setting		Formulation		Approval		Implementation		Evaluation	
	Degree	Betweenness	Degree	Betweenness	Degree	Betweenness	Degree	Betweenness	Degree	Betweenness
MAGFOR	6	4.78	6	15.25	6	15.50	8	2.00	0	0.00
MEFCCA	4	0.49	3	0.92	2	0.00	8	0.48	1	0.00
MINED	2	0.38	1	0.00	0	0.00	7	0.33	0	0.00
MINSA	5	1.92	2	0.00	1	0.00	6	0.14	0	0.00
INTA	6	2.92	3	0.33	5	9.50	9	2.64	1	0.00
CIUSSAN	2	0.31	3	1.08	1	0.00	3	0.00	1	0.00
Local government	6	2.21	2	0.00	3	7.00	7	0.14	1	0.00
Partners	5	3.48	5	6.33	1	0.00	9	2.64	1	0.00
Producers associations	6	7.01	6	12.08	1	0.00	8	0.48	1	0.00
Community cabinets	4	0.51	1	0.00	2	0.00	7	0.14	0	0.00
Total relations	46		32		22		72		6	
Network density (%)	51.11		35.56		24.44		80.00		6.67	

[1] COMUSSAN = Municipal Committees of Sovereignty and Food Security and Nutrition; MAGFOR = Ministry of Agriculture and Forestry; MEFCCA = Ministry of Family Economy; MINED = Ministry of Education; MINSA = Ministry of Health; INTA = National Institute of Agricultural Technology; CIUSSAN = Universities' Committee; local governments = municipalities; partners = NGOs and international agencies.

governments and associations of producers. These can influence the prioritisation of initiatives around the agro sector, leaving aside the participation in other areas and losing sight of the multiple causes of hunger. In this stage producer associations are a driving force behind introducing greater network of betweenness centrality. In the formulation phase MAGFOR and producer associations are of greater degree centrality who have and can give rise to sectorial proposals. Also at this stage these two actors have greater betweenness centrality, which shows that the connection points are the most important in the entire network. In the approval phase, the MAGFOR has more influence reaching the highest degree centrality and beetweenness centrality of the entire network. The latter institution that provides the most amount of budget is the one that takes the most important decisions when approving initiatives. The implementation phase has the most interaction with 72 relations. The actors with greater degree centrality and betweenness centrality are INTA and the partners. This shows that NGOs and international organisations play an important role in the functioning of COMUSSAN. Although MAGFOR is the institution that is granted more funding it is not the most interactive, suggesting initiatives executed individually. The assessment is carried out without generating specific interaction actors, which may generate individually targeted initiatives and are in no interaction. The actors performed individually thus hindering the transparency of the process affecting governance.

Scores of betweenness centrality and degree centrality show that the most important actor in the public policy cycle is the MAGFOR and the second player in relevance is the INTA. This proves that COMUSSAN activities have promote a greater specification in the agricultural field. Producer and cooperation associations are continuing in importance, showing that these actors have great influence on the decisions taken at the municipal level. In fifth place are the municipal governments, showing that local authorities have been relegated to promote food security within their territories, as exemplified in centralised and decentralised decision-making in the execution stage. It is important to understand that Institutions have formal and informal elements. The formal are visible, but they depend heavily on a substructure of informal relationships and practices, which may be invisible.

In Figure 1, we see that the MAGFOR is the most important actor during the public policy cycle, uniquely not participating in the evaluation stage. This indicates that greater involvement of other

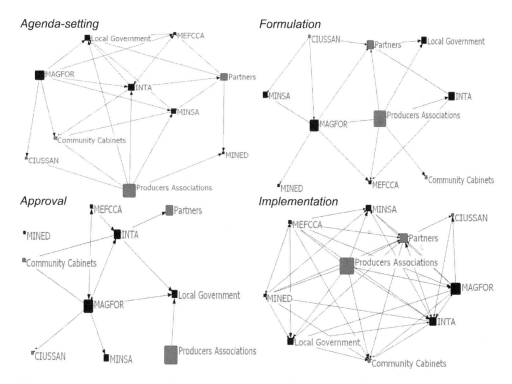

Figure 1. Interaction network of Municipal Committees of Sovereignty and Food Security and Nutrition actors in the agenda-setting, formulation, approval and implementation phases of the public policy cycle. Black nodes represent governmental actors or the public sector, the grey nodes represent social organisations, producer associations, universities and cooperation agencies. The size of the nodes indicates the betweenness centrality of each actor (Table 1). An arrow represents every interaction from the actor who recognises the input to another in their actions. MAGFOR = Ministry of Agriculture and Forestry; MEFCCA = Ministry of Family Economy; MINED = Ministry of Education; MINSA = Ministry of Health; INTA = National Institute of Agricultural Technology; CIUSSAN = Universities' Committee; local governments = municipalities; partners = NGOs and international agencies.

ministries is required to strengthen joint efforts to provide solutions to the multiple causes of hunger. It is observed that partners are important in municipal level, although at the national level they may not have much public policy incidence. These two mentioned actors are the main sources of financing actions of COMUSSAN. Producer associations recognise that they work with various actors in the political and development agenda, which means they do impact the proposals.

Local governments are recognised by political agenda, approval and implementation phases. However this actor does not have a high importance in the joint network, for its limited technical and financial resources allow for other actors to take a greater role. Their involvement is important because the most successful programs have to be implemented by local governments that are responsible and lead the implementation of their activities (Mansuri and Rao, 2013). Community cabinets have an important role in the agenda-setting and implementation phases, ensuring that future beneficiaries influence the prioritisation of initiatives that will be implemented so that civil society can exercise programme oversight.

4. Conclusions

COMUSSAN still remain limited and relatively isolated at the municipal level. It is important to study the relationship between institutions at this level for understanding how they can be scaled up. Initiatives at this level should have a balanced participation of stakeholders to prevent the implemented actions responding to a single sector and to maintain a multidisciplinary approach in combating hunger.

For the functioning of such bodies involved in systems of centralised decision-making, a high level of state decentralisation is required and also that there is adequate coordination between the public sector. The figure of COMUSSAN has become a group of joint planning, covering the technical weaknesses of local (municipal) governments. From the operational point of view, these spaces allow the complement of human and financial resources, which means greater impacts and increasing the coverage of the food security and nutrition programmes.

NGOs and international organisations play an important role in the implementation of initiatives in municipalities. It is important that these institutions articulate their actions to achieve goals established within the network of actors. With initiatives like the COMUSSAN, civil society in general have greater participation in the generation of public policy, its participation is important with special emphasis on the support they can take to ensure that Food Security and Nutrition actions are implemented properly and serve the population most in need. Undoubtedly linking policy network analysis with the public policy cycle, allows us to recognise which players have the most impact on food security, which steps to take to expand interaction with other actors, and how they should be linked with civil society to have greater effect their participation.

References

Andrews, M. (2013). The limits of institutional reform in development: changing rules for realistic solutions. Cambridge University Press, Cambridge.

Borgatti, S.P., Everett, M.C., Freeman, L.C. (2002). UCINET for Windows: software for social network analysis. Analytic Technologies, Harvard MA.

Duit, A., Galaz V. (2008). Governance and complexity – emerging issues for governance theory. Governance 21: 311-335.

FAO (2015). The state of food insecurity in the world. Food and Agricultural Organisation of the United Nations. FAO, Rome.

Giest, S., Howlett, M. (2014). Understanding the pre-conditions of commons governance: The role of network management. Environmental Science and Policy, 36: 37-47.

Grimes, M. (2013). The contingencies of societal accountability: examining the link between civil society and good government. Studies in Comparative International Development 48: 380-402.

Ingold, K. (2014). How involved are they really? A comparative network analysis of the institutional drivers of local actor inclusion. Land Use Policy, 39: 376-387.

Jann, W., Wegrich, K. (2006). Theories of the policy cycle. In: Fischer, F., Miller, G.J., Sidney, M.S. (eds.) Handbook of public policy analysis. CRC Press, Boca Raton, FL, USA, pp. 43-62.

Kickert, W.J., Klijn, E.H., Koppenjan, J.F.M. (eds.) (1997). Managing complex networks: strategies for the public sector. Sage. London.

Lienert, J., Schnetzer, F., Ingold, K. (2013). Stakeholder analysis combined with social network analysis provides fine-grained insights into water infrastructure planning processes. Journal of Environmental Management, 125: 134-148.

Lin, N., Cook, K.S., Burt, R.S. (eds.). (2001). Social capital: theory and research. Transaction Publishers. New Jersey.

Mansuri, G., Rao, V. (2013). Localizing development: does participation work? World Bank Policy Research Report. The World Bank, Washington DC, USA.

Papadopoulou, E., Hasanagas, N., Harvey, D. (2011). Analysis of rural development policy networks in Greece: Is LEADER really different? Land Use Policy 28: 663-667.

Pierre, J., Peters, B.G. (2005). Governing complex societies: Trajectories and scenarios. Palgrave Macmillan, Basingstoke.

43. Impact of energy crops on food sovereignty in Ecuador

L. Toledo Pazmiño and *A. Velasco Arranz*
Polytechnic University of Madrid (UPM), Department of Agricultural Economics and Social Sciences, 28040 Madrid, Spain; lenin_t18@hotmail.com

Abstract

Actually there is a continuous debate in Ecuador on renewable energy and the production of the basic food crops due to the Ecuadorian government's decision to promote the crop of African palm and to increase biodiesel production to favour alternative energy sources to fossil mineral oil. The influence of industrial companies on this debate is very important, as well as the conflicts of interest between these companies and agriculturists, regarding the use and cultivation of the African palm tree. Therefore it is also necessary to revise possible changes in Ecuadorian politics to promote investment in this kind of energy crops, in order to mitigate the negative impact on rural areas. The aim of this paper is to advance some results obtained in the doctoral thesis research where the economic and social problems regarding the production of biofuel have been explored. Here we analyse the financial data of the production of biofuel of African palm in Ecuador between years 2006 and 2012, in that period took place an important growth of African palm cultivation. The method that will be used in this empirical study is a quantitative analysis of financial and economic data and the hypothesis is that the production costs reaching to proportion of 40% in relation to Free on Board exports in millions of dollars. These data are based on our review of policy objectives of the Ecuadorian government, regarding the development of biofuel and the improvement of advantage economic. We conclude that economic and social problems regarding the production of biofuel will be explored because African palm oil is sensitive to – highly volatile – oil prices.

Keywords: African palm oil, financial data, debate, Ecuador

1. Introduction

The term 'food sovereignty' refers to a political framework designed from a rural-development perspective in order to tackle food deficit. It is based on the right for food production to serve as a mechanism for eradicating hunger and malnutrition, and for legitimising long-lasting and sustainable sources of food at community level. In order for food sovereignty to be achieved, a range of alternatives needs to be built into agricultural policies (Windfuhr and Jonsén, 2005), whereby states must guarantee there is enough food on the market, whether by way of import or domestic production. This challenge must be addressed in conjunction with political support to guarantee access to food: a concept known as food security.

Current energy policy incentivises the production of biofuels as an alternative to reduce dependency on fossil fuels such as oil. This focus on biofuels is justified in both economic and environmental terms, in light of the potentially positive impact this alternative could have on climate change by reducing greenhouse gases. These considerations have led many governments to subsidise the production of energy crops, including at times for electoral purposes, and to enter into big contracts with biofuel industries (Tangermann, 2007). This approach to crop production has led to significant inflation in these raw food materials, generating a debate on food security, since if it is theoretically possible to satisfy the global demand for food, competition for these products from the energy sector represents a worrying threat when compared to many countries' lack of purchasing power to buy these foods. In addition, we are also seeing a growing division among countries, as some are competing for energy products for automotive use while others need these same products for food, thus creating a paradigm in terms of who uses biofuels and generating a dependence on this type of fuel (Daño, 2009). Industrialised countries have enough productive capital to satisfy internal demand and to produce biofuels as well; however, the transformation of food products into energy crops would have a negative effect in developing countries and countries with less purchasing power.

Biofuels can be a source of income for developing countries such as Ecuador, Colombia, Brazil or Malaysia, yet we must ask ourselves if the debate on energy crops in these countries should focus on the way in which wealth is distributed and how priorities are selected in legislative and economic policies. On a European level, energy policies have included both tax exemptions and subsidies, and studies have also suggested potential competition for use of land (Banse *et al.*, 2008). Therefore, can investing in the cost of biofuel production over the cost of food crops be justified?

Although farmers could increase their income through subsidy-based incentives for producing energy crops, the production of these crops also creates expenses at a national level.

It is important to note that we are still in the early stages of researching the aspects that counterbalance incentives for producing energy crops against challenges regarding rural development, environmental sustainability or empowerment of local farmers in Ecuador. As a result, our contribution to this debate aims, on the one hand, to present biofuel policies in simple terms and explore how these are linked to Food Sovereignty and, on the other, to offer some findings from analysing the financial and economic efficiency of African palm oil production in Ecuador between years 2006 and 2012.

2. Food sovereignty as a political priority in Ecuador

In its 2008 Constitution, Ecuador included 'food sovereignty' as a fundamental pillar of development policy; Article 13 on rural development, entitled 'El Buen Vivir' (the good way of living), states that guaranteeing food sovereignty for the entire population is an obligation of the State. In addition, Articles 281 and 282 of this constitution establish the legal framework that upholds this priority: the use of and access to land (farming land) is one of the main focal points. Large estate farming and the concentration of land is forbidden, along with the monopolising or privatisation of water, in order to guarantee communities and villages can achieve food self-sufficiency. These articles also outline the creation of agricultural policies aimed at small- to medium-size farmers, in order to ensure the production of food for self-sufficiency purposes. Furthermore, Article 281 emphasises the need to encourage agricultural modernisation through the use of technology and creation of sources of funding, such as access to low-interest loans, in order to boost production, in addition to the use of fiscal tools and protectionist laws.

Therefore, the Ecuadorian legislative framework establishes national sovereignty as a central priority that brings together: policies to redistribute land-based and water-based resources; the diversification of production; promotion of small- and medium-size production units; prevention of the speculative practices of monopolies; and the creation of sale and distribution networks that foster equality between all Ecuadorians (Hidalgo *et al.*, 2014).

3. Biofuel policies on African palm oil in Ecuador

In 2004, the Ecuadorian Ministry of Agriculture declared the production, sale and use of biofuels as a national interest (Executive Decree No. 2332: Official Record No. 482 of 15 December 2004), in an attempt to reduce dependency on oil and in response to the need to create new fuels for export. This marked a shift in production practices in favour of increasing the export of raw materials and returns on production through the use of research.

In 2012, President Rafael Correa, in Executive Decree No. 1303 (Official Record 799, 2012), expressed his support of the systematic production and sale of biofuels, describing it as an inclusive strategy for rural development that guarantees Food Sovereignty and environmental sustainability. The production of biofuel became subject to State-controlled commercialisation. The goal of this move was to obtain 10% of biodiesel from vegetable sources in Ecuador, in a mixture of diesel fuels for use in the automotive sector and, at the same time, to open up the production and sale of vegetable biofuel to the open market. However, this decision was not immune to potential threats to food sovereignty, such as the promotion of single-crop farming of African palm oil resulting in the substitution of basic food stuffs, soil degradation, deforestation, displacement of indigenous communities in the Amazon to other internal regions, etc. These are precisely the development aspects that must be taken into account in the respective analyses.

4. Methodology

Here, we will focus on financial analysis of palm oil production in Ecuador during the period 2006-2012. This analysis is based on the following hypothesis: the production costs reaching a proportion of 40% in relation to Free on Board (FOB) exports in millions of dollars. The results are complemented by another methodological stage regarding a legislation-based study of rural and energy policies in Ecuador, elements of which have been touched on in the foregoing paragraphs. Our research will include interviews to be carried out in the coming months with farmers and sector representatives in order to understand the social response to the rise of biofuel crops. These interviews will form part of the body of work for the doctoral thesis[5].

5. Economic/financial analysis

In Ecuador, both large and small exporters of African palm oil are grouped together in the National Federation of Palm Oil Farmers. According to the figures from the Central Bank of Ecuador, FOB

---

[5] The first author is a doctorate student in the Polytechnic University of Madrid (UPM), E.T.S. Ingenieros Agrónomos, Agricultural Economics and Social Sciences Department. The title of the thesis is 'The impact of the energy cultivations in the alimentary sovereignty of Ecuador'. The project is in the second year and supervised by the last author.

exports of African palm oil in the study period rose by 285,539,580 million dollars, or 362% compared to 2006. Comparing 2012 (official estimate) to 2006, FOB palm oil exports increased their share of GDP by 0.19%. This rise from 0.13 to 0.32% (Figure 1) was due to factors such as the positive effect on the sector of a hike in oil barrel prices during this period.

However, return in terms of total production (million tonnes) compared to harvested hectares shows a downward trend in African palm oil production (Figure 2). In order to achieve a profitable return in current terms, producers and exports of this crop need to be more efficient and increase return through prices or by expanding the production area.

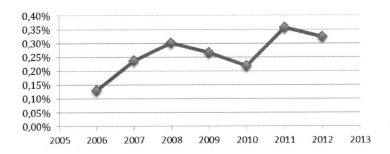

Figure 1. Percentage exports free on board/gross domestic product (FOB/GDP) (BCE, 2015).

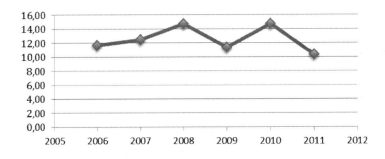

Figure 2. Total African palm oil production harvested hectares (BCE, 2015).

6. Sensitivity analysis regarding prices

Two scenarios have been used to perform this sensitivity analysis regarding the price variable in metric tons of exported African palm oil, since this variable has proven to be highly dependent on oil prices. Ecuador is a country that accepts the international prices of large-scale producers.

As indicated in Table 1, the cash flow analysis for 2006-2012 results in a net present value (NPV) of $ 73,462,204 and an internal rate of return (IRR) of 24%, which is higher than the 10% discount rate applied to Fabril shares (the leading palm oil exporter in Ecuador). Therefore, we found that a

Table 1. Data financial analysis.

Weighted average cost of capital	10%	
Internal rate of return (IRR)	24%	
Net present value (NPV in USD ×1,000,000)	73,462,204	
Sensitivity analysis regarding prices	IRR	NPV (×1,000,000 USD)
Positive variation 5% (scenario 1)	29%	101,216,428
Negative variation -14% (scenario 2)	9%	-4,249,621
No variation	24%	73,462,204

positive price variation of 5% would result in IRR rising by 5% and NPV increasing by $ 27,754,224; this result is based on an optimistic scenario (what would have happened during this period if prices had risen by 5%).

On the contrary, as indicated in the sensitivity analysis, if prices had dropped by 14%, in a negative scenario and as actually happened in 2008, 2009 and 2012, palm oil production would not have been so profitable. With a 14% downturn in oil prices, IRR falls to 9%, which is no longer higher than the discount rate, and NPV takes on a negative value of $ 4,249,621. This negative scenario is not that far removed from reality, considering the global food, financial and environmental crisis that began in 2008 and from which no sector emerged unscathed.

If we then add the environmental and social costs of palm oil production, this second scenario becomes even more negative and would certainly not be of benefit to the country, as the land could be used for other products in order to ensure Food Security.

7. Conclusions

A process of change is occurring in production practices in Ecuador; moving away from the current importer model to an exporter model, in which energy or biofuel crops are being integrated into agricultural activity. This type of production has been declared of significant national interest in the 2008 Constitution and in the agricultural policies on biofuels to emerge from this constitution. Political resolutions and legislation in this area all present incentives for palm oil production that are linked to rural development policy or the right to a good way of living and to sustainable environmental development.

On the other hand, this has sparked a debate about the extent to which palm oil production is actually sustainable from an environmental and food-security perspective. The interconnected problems of deforestation and land use – discriminating against small-scale producers – also need to be addressed, especially with regard to the large industries, which stand to benefit the most from investing in African palm oil. However, this debate has its fair share of contradictions and is hindered by a breach between palm oil industries and indigenous farmers. This breach will be analysed at a later stage in our research when the findings have been compiled from our surveys or interviews with farmers.

Finally, it is clear from the abovementioned financial analysis on the price of this crop that production of African palm oil is not only a highly profitable investment but also a very sensitive product that is greatly dependent on oil prices and the corresponding demand for biodiesel. Therefore, since this product is sensitive to – highly volatile – oil prices, we can expect a profitable return in a positive scenario for both products. However, in a negative scenario, it would be interesting to see an increase in the production of biofuels in Ecuador when the land could be used to produce basic food stuffs.

References

Banse, M. (2008). Consequences of EU biofuel policies on agricultural production and land use. Journal Agricultural and Applied Economics Association 23(3): 22-27.

Banco central del Ecuador (BCE) (2015). Servicio de Estadisticas. Available at: http://www.bce.fin.ec.

Daño, E. (2009). Biofuels: false solution, disastrous consequences. Women in Action 1: 112-119.

Hidalgo, F.F., Houtart, F., Lizárraga, A.P. (eds.) (2014). Agriculturas campesinas en Latinoamérica: propuestas y desafíos. Quito: Editorial IAEN.

Registro oficial (R.O) suplemento No. 799 (2012). Decreto Ejecutivo No 1303 declarase de interés nacional el desarrollo de biocombustibles en el País como medio para el impulse de fomento Agrícola. Emitido 17 de Septiembre de 2012.

Tangermann, S. (2007). Biocarburants et sécurité alimentaire. Économie rurale: 100-104. Available at: http://economierurale.revues.org/2260.

Windfuhr, M., Jonsén, J. (2005). Soberanía Alimentaria. Hacia la democracia en sistemas alimentarios locales. FIAN-Internacional: 1-59. Available at: http://tinyurl.com/qf33nzp.

Authors index

Printed in the United States
by Baker & Taylor Publisher Services